AF740448

NEWTON AND RELIGION

ARCHIVES INTERNATIONALES D'HISTOIRE DES IDÉES

INTERNATIONAL ARCHIVES OF THE HISTORY OF IDEAS

161

NEWTON AND RELIGION

Context, Nature, and Influence

edited by
JAMES E. FORCE
and
RICHARD H. POPKIN

Founding Directors:
P. Dibon† (Paris) and R.H. Popkin (Washington University, St. Louis & UCLA)
Director:
Sarah Hutton (The University of Hertfordshire, United Kingdom)
Assistant-Directors: J.E. Force (Lexington); J.C. Laursen (Riverside)
Editorial Board: J.F. Battail (Paris); F. Duchesneau (Montreal); A. Gabbey (New York); T. Gregory (Rome); J.D. North (Groningen); M.J. Petry (Rotterdam); J. Popkin (Lexington); G.A.J. Rogers (Keele); Th. Verbeek (Utrecht)
Advisory Editorial Board: J. Aubin (Paris); B. Copenhaver (Los Angeles); A. Crombie (Oxford); H. Gadamer (Heidelberg); H. Gouhier (Paris); K. Hanada (Hokkaido University); W. Kirsop (Melbourne); P.O. Kristeller (Columbia University); E. Labrousse (Paris); A. Lossky (Los Angeles); J. Malarczyk (Lublin); J. Orcibal (Paris); W. Röd (München); G. Rousseau (Los Angeles); H. Rowen (Rutgers University, N.J.); J.P. Schobinger (Zürich); J. Tans (Groningen)

NEWTON AND RELIGION

Context, Nature, and Influence

edited by

JAMES E. FORCE

University of Kentucky,
Lexington, U.S.A

and

RICHARD H. POPKIN

University of California,
Los Angeles, U.S.A

KLUWER ACADEMIC PUBLISHERS

DORDRECHT / BOSTON / LONDON

A C.I.P. Catalogue record for this book is available from the Library of Congress.

ISBN 0-7923-5744-2

Published by Kluwer Academic Publishers,
P.O. Box 17, 3300 AA Dordrecht, The Netherlands.

Sold and distributed in North, Central and South America
by Kluwer Academic Publishers,
101 Philip Drive, Norwell, MA 02061, U.S.A.

In all other countries, sold and distributed
by Kluwer Academic Publishers,
P.O. Box 322, 3300 AH Dordrecht, The Netherlands.

Printed on acid-free paper

All Rights Reserved
© 1999 Kluwer Academic Publishers
No part of the material protected by this copyright notice may be reproduced or
utilized in any form or by any means, electronic or mechanical,
including photocopying, recording or by any information storage and
retrieval system, without written permission from the copyright owner

CONTENTS

FOREWORD —*J. E. Force* vii

INTRODUCTION —*R. H. Popkin* ix

ESSAY 1 / Alchemy and Eschatology: Exploring the Connections between John Dee and Isaac Newton —*D. E. Harkness* 1

ESSAY 2 / Newton and the Rosicrucian Enlightenment —*A. P. Coudert* 17

ESSAY 3 / From Paracelsus to Newton: The Word of God, the Book of Nature, and the Eclipse of the "Emblematic World View" —*J. J. Bono* ... 45

ESSAY 4 / "Acceptable to inquisitive men": Some Simonian Contexts for Newton's Biblical Criticism, 1680-1692 —*J. A. I. Champion* 77

ESSAY 5 / Those "Whose Business It Is To Cavill": Newton's Anti-Catholicism —*R. Iliffe* 97

ESSAY 6 / Newton, Corruption, and the Tradition of Universal History —*R. Markley* 121

ESSAY 7 / Newton's *Of the Church*: Its Contents and Implications —*M. Goldish* 145

ESSAY 8 / The Seven Trumpets and the Seven Vials: Apocalypticism and Christology in Newton's Theological Writings —*S. Hutton* 165

ESSAY 9 / Interpretive Strategies in Newton's *Theologiae gentilis origines philosophiae* —*K. J. Knoespel* 179

ESSAY 10 / Newton's Apocalypse —*M. Murrin* 203

ESSAY 11 / Newton and the Guaranteeing God —*G. A. J. Rogers* 221

ESSAY 12 / Newton, the "Ancients,"and the "Moderns" —*J. E. Force.* 237

ESSAY 13 / The Logic of Millennial Thought: Sir Isaac Newton Among His Contemporaries —*R. Smolinski* 259

INDEX 291

FOREWORD

JAMES E. FORCE
University of Kentucky

In 1990, when Richard H. Popkin and I published our *Essays on the Context, Nature, and Influence of Isaac Newton's Theology* (Dordrecht: Kluwer Academic Publishers), my friend and mentor, the "grand sceptic of the West Coast," Richard H. Popkin, wrote in the *Introduction* that "perhaps, some years hence, we will have greater distance from the material, and from the exciting chases involved in uncovering it, can go back and write a consecutive narrative." The time for such a comprehensive overview has not yet arrived. The indefatigable Professor Popkin and I still find ourselves gripped by the excitement of the chase. Too many new and fruitful perspectives beckon before any such synthetic, comprehensive overview might become possible. In this interim volume, we take pride in presenting essays which point to some of the most enticing new directions in Newton scholarship. The authors of these essays attempt to trace aspects of *both* the wider intellectual context in which Newton wrote—including predecessors as diverse as John Dee, Paracelsus, and contemporary archaeologists (or, "antiquarians")—*and* the nature and, occasionally, the influence of Newton's religious thought upon thinkers and schools as disparate as the Anglican Church, English anti-Catholics, the Alchemists, and Cotton Mather.

All of these papers have, as their original point of genesis, a conference on the topic of "Newton and Religion" held on February 9th and 10th, 1996, at the William Andrews Clark Memorial Library under the sponsorship of the University of California, Los Angeles, Center for 17th- & 18th-Century Studies which is directed by Professor Peter H. Reill of the University of California, Los Angeles. For scholars discouraged by the bland, business-as-usual, pretend-professionalism of "junk bond" conferences put on by "professional" organizations which are really more akin to meat markets than to any genuine marketplace of ideas, we heartily recommend the conferences put on at the William Andrews Clark Memorial Library as a sovereign anti-dote. An audience and a group of speakers composed of people who have respect and affection for, broad knowledge of, and critical talent in the history of ideas make for a genuinely unique and exciting intellectual venue. We do not mean to claim that one had to be there to appreciate or understand the following essays; finally, by their essays, you shall know them. But we cannot deny the very special ambiance of the Clark

Library and the critical contributions, within that noble venue, of such commentators as Peter H. Reill of UCLA and Hartmut Lehmann of the Max-Planck-Institut für Geschichte. Finally, we would like to extend our thanks to the extremely thorough and hard-working experts chosen by Kluwer as the outside referees for this volume. Their critical diligence has done much to improve this volume and to make it as good as it has, with their help, become.

INTRODUCTION

RICHARD H. POPKIN
Washington University and the University of California, Los Angeles

This collection of essays grew out of a conference held at the William Andrews Clark Memorial Library of UCLA in 1996. Over the years several of the people involved in this conference had discussed research developments concerning Newton's alchemical and theological papers. During the last twenty-five years—since the very large collection in Jerusalem became publicly available and supplemented the Keynes collection at Cambridge and some other, smaller collections—the beginnings of a new picture of what Newton might have been up to has emerged.

The 1936 Sotheby auction of the Newton papers, held by the Earl of Portsmouth and described in the auction catalogue, revealed the breadth and depth of Newton's interests in alchemy and theology. A very large number of the items listed disappeared from public view into private collections. The largest number of these manuscripts, purchased by the Palestinian Jewish Arabist, A. S. Yahuda, was carted to the United States in 1940 where Yahuda became a refugee. He and his friend Albert Einstein tried to get Harvard, Princeton, or Yale to house the collection but to no avail. Yahuda and Einstein were told that nobody would care about Newton's opinions on these subjects, that these papers were not scientific, and also that there was a war on. (The letters from these institutions to Einstein are themselves in the Yahuda collection.) So the Newton papers remained in Yahuda's house in New Haven along with his very very rich collection of Middle Eastern documents.

Yahuda had been the first Jewish Palestinian Zionist but had broken with the movement at the time of the Balfour Declaration. Over the years, he became more and more anti-Zionist. He privately printed a polemical work against Chaim Weitzman and inveighed against other Israeli leaders and scholars. Yahuda debated about to whom he ought to leave his manuscript collection. Upon Yahuda's death-bed, one of his students apparently convinced him to leave it to the National Library of Israel in spite of his grave misgivings about the new country. When Yahuda died in 1951, his family tried to block sending the bequest to Jerusalem. A long court case ensued in New Haven which ended in 1969. Crates and crates of papers and books were then shipped to the library in Jerusalem. Fortunately for us all, the Newton scholar, David Castillejo was there when the boxes were opened. He compiled an inventory which has made it possible for later scholars to find their way through the many, many files.

Professor Richard S. Westfall, the late Newton biographer, read about the Yahuda collection and wrote to Jerusalem to obtain a microfilm of the holdings. Westfall, with his microfilm which he received in 1972, inaugurated a new era of Newton studies in which scholars attempted, for the first time, to incorporate Newton-the-alchemist and Newton-the-theologian into the commonplace view of Newton as the revolutionary progenitor of modern science. Westfall, I. B. Cohen, Frank Manuel, and others looked at the manuscripts and gave their views on Newton's religion and its relation to his science. The late Betty Jo Teeters Dobbs studied the alchemical materials in her books on the subject and revolutionized our understanding of this side of Newton's work.

Since the eighteenth century, the problem had existed of understanding how one and the same person, the greatest genius of his age, could have written such brilliant mathematics and physics and still have dabbled in arcane and occult nonsense such as alchemy, astrology, polytheistic mythologies, early Christian Church history, radical religious heresies such as Arianism, and the interpretation of Biblical prophecies.

Voltaire and others in the eighteenth century had offered theories that the great man became senile and stopped being a scientist. Perhaps, it was suggested, an accident, when he tripped over his cat, led to this premature senility. Another suggestion was that the great Newton was poisoned by mercury when doing some silly alchemical experiment and that this addled his brain and led to his writing on nonsensical topics. This kind of simplistic "two-Newton" interpretation—one a young, brilliant scientist, the second a senile religious nut case—went by the boards when the Yahuda materials, and others, which were auctioned at Sotheby's in 1936 showed that Newton was working on these "unscientific" topics throughout his adult life and was, in fact, working on them at the very times that he did his greatest scientific writings. A large part of the scholarly world up to this very day continues to insist that no matter why Newton worked on alchemy, astrology, religious topics, or Bible prophecies, it had nothing whatever to do with his scientific achievements and, therefore, did not deserve serious study by historians of science. Wesrfall himself developed a version of the "two-Newton" position in which he attempted to take into account all the diverse topics which interested Newton during his long life but concluded that, towards the end, Newton was, blessedly, moving to a rational, deistic view and was discarding his earlier bizarre interests. Westfall offered as his proof texts lines from an unfinished collections of notes and writings in the Yahuda collection entitled *Theologiae gentiles origines philosophiae*. Others have contended that this work is really a jumble of notes from various commentators on the varieties of religions at the time and that

Newton was not really breaking any new ground in this vast compendium of notes. In the last decade or so there have been many studies arguing for and against Westfall's final view. In our collection of essays in this volume, there are some new readings of the Yahuda manuscripts which offer new hypotheses about how they may fit with Newton's overall views.

In recent years, quite a few younger scholars have emerged who have been examining the vast corpus of unpublished Newton papers, a process made more easily possible by the generally available Chadwick-Healey microfilm project, a large number of microfilm reels containing most of the unpublished papers including the Bodmer Library manuscript in Geneva (which had not been available to Westfall.) More and more people have examined the great Yahuda manuscripts in the Jewish National and University Library in Israel as well as on the microfilm copy. Finally, the Clark Library's ninety-two page Newton manuscript attacking St. Athanasius and Trinitarianism was studied seriously, especially by James E. Force and Rob Iliffe.

The interest and concern in the context and background of what interested Newton in these areas seemed to provide a much richer background for interpreting what Newton had written and how it might relate to his mathematical and scientific concerns. Finally, James E. Force, Kenneth Knoespel, and I decided to plan what we thought would be an exciting conference which would bring together scholars in the history of alchemy, religion, astrology, science, philosophy, and theology. The Center for 17th- and 18th-Century Studies at the University of California, Los Angeles, through its director, Peter H. Reill, agreed to fund the conference and to hold it at the lovely facilities of the Clark Library.

In preparing for such a conference, we felt the need to place what Newton was doing in his most puzzling papers on alchemy, astrology, theology, and various religious subjects into the context of the mix of disciplines that existed and co-existed all during the 17th century. In the period often interpreted as the scientific "revolution" when, according to long hallowed tradition, rational science "rose" and replaced medieval superstitions, we must remember some facts. This was also the period when Johannes Kepler could be one of the makers of modern astronomy as well as a leading astrologer; when the prophetic interpreter, John Dury, could discuss how best to discover a basis for certainty with René Descartes, whether in mathematics or in the discovery of the proper method for interpreting Scriptural prophecies; when Isaac Newton and Henry More could have had a monumental falling out over how to interpret the vials and the trumpets in the Book of Revelation; when Newton could undertake a trip to visit John Locke just to present him a chart for interpreting the symbols in Revelation and the chronological consequences thereof; and when Newton's chosen successor as Lucasian Professor of Mathematics at Cambridge,

William Whiston, could deliver the Boyle Lectures on *The Accomplishment of Scripture Prophecy*, using the best modern mathematical techniques to estimate the probabilities as to when the remaining unfulfilled prophecies would be fulfilled. All of these events probably seemed normal to the 17th-century participants though we today, now brainwashed by Whiggish interpretations of the Enlightenment, may find it difficult to fathom how our intellectual heroes of the past could have been so misguided.

We sought, in this conference, to look into the background of so-called esoteric and prophetic religious thought from the late-17th century until the time of Newton. We decided to bring together younger, as well as established, scholars in the history of science, the history of religion, and the history of theology so that they could present and discuss their views about the "new" Newton who was emerging after twenty-five years of studying his manuscripts: the Newton who wrote the *Principia* and the *Opticks* and who also did alchemical experiments and, for about sixty years as a scriptural exegete, tried to figure out the prophetic messages in the books of Daniel and Revelation. The Clark Library provides a delightful forum for such exchanges, with its formal drawing room, elegantly appointed in art treasures, providing a setting for lectures and discussions, and the garden providing an informal place for further discussion over lunch and refreshments.

The problem of whether there is just one Newton, or two, or more still needs to be worked out. We hope that the range of topics dealt with in these papers will contribute to this task by showing some of the context, concerns, and influences of Newton's contributions in his time. These essays represent the authors' refined version of their presentations based on the lively discussions at the Clark conference as well as editorial suggestions from some very helpful referees.

In the first essay, Deborah Harkness, a student of the late Betty Jo Teeter Dobbs, presents the achievements of the great British astrologer, alchemist, and natural philosopher, John Dee, who was active a century before Newton. Dee tried, through conversations with angels over many years, to understand Nature and to learn how to perfect it and to prepare for the conclusion of the divine cosmic drama. Dee saw comets and special astronomical configurations as important portents of God's messages to mankind. His accounts of his conversations with angels might constitute a merely isolated case of a late-Renaissance magus run amok were it not for the fact that much of his corpus of conversations was published in 1659 and was studied and discussed by several of the natural philosophers in the circle of the Royal Society in the latter decades of the seventeenth century. Newton, with his great interests in matters occult and supernatural, might have felt less alone if he was aware of Dee as a partial predecessor.

In the second essay, Allison Coudert, who studied at the Warburg Institute in London with Frances Yates and D. P. Walker, undertakes to place Newton's work into the context of a revised version of the Yates thesis according to which modern science "rose" out of occultism. Yates had claimed that it was revived Hermeticism which launched the modern scientific world. In the last decades there has been much criticism of the Yates thesis from all sides. Coudert offers a different interpretation. She argues that it was not Hermeticism but rather a revived Gnosticism in combination with ancient science which launched the outlook of Bacon, Comenius, Hartlib, and, later, Newton. According to this revived synthesis of Gnosticism and science, man was able to comprehend the universe by returning to a *prisca theologia* or *prisca sapientia* uncorrupted by the centuries of pagan metaphysics, kabbalah, etc.

In the third essay, James Bono examines the primary way of conceiving God's "Book of Nature" during the Renaissance, the Emblematic World View, in which natural things—in terms of their descriptions, associations, habits, names, representations—are seen as conveying something about the prelapsarian knowledge of things and their essences. Paracelsus and many others studied natural things in terms of their emblematic significance. This way of looking at the natural world—with its embedded, emblematic representations of the original divine truths—gradually gave way to seeing the world in mathematical terms. Bono argues that Newton, though one of the great practitioners of mathematical science, was also firmly looking back to the emblematic way of seeing nature in his search for origins of human religious behavior, science, etc., in ancient records and in mythological expressions. Newton was, as Betty Jo Dobbs said, "Janus-faced", looking forward and backward at the same time.

In the fourth essay, Justin Champion examines the role of Father Richard Simon's biblical criticism on both Newton and John Locke. Newton had sent Locke some letters about the fact that there was no scriptural authority for the Doctrine of the Trinity. Locke forwarded these letters to Jean LeClerc in The Netherlands for publication without telling LeClerc the name of their author. Through Locke, Newton withdrew them from publication probably because he feared that he would be recognized as the author and punished for his anti-Trinitarianism. Two of these letters became public only in the mid-18th century and the other in the mid-20th. Leclerc told Locke that the "anonymous" author would profit from the study of Simon's *Critical History of the New Testament*. Champion follows Leclerc's advice to the unknown author, i.e., to Newton. Champion surveys Simon's critical examination of the Scriptures and shows how both Newton and Locke, strong Protestants, could find aid and comfort for their views in Father Simon's evidence that there was no original or authentic text of the Bible and that private interpretation must be permitted. Champion shows that both

Newton and Locke had the major works of Simon in their libraries and used them extensively while also following Simon's advice to study many Bibles in many languages.

In the fifth essay, working mainly from unpublished manuscript texts, Rob Iliffe examines the nature and profound depth of Newton's anti-Catholicism. Newton's basic rejection of idolatry in its ancient pagan forms led him to see Catholicism as a new and even worse form of idolatry with its worship of saints, its adoption of various pagan elements, its corruption of Scripture (to justify Trinitarianism), and the many more corruptions introduced by Athanasius when he crushed the Arians. Newton's life-long study of church history provided him with an enormous amount of ammunition about the wickedness of Popery from ancient times to the present. Iliffe shows that Newton's deep-seated anti-Catholicism led him to refuse to take seriously Jesuit criticisms of some of Newton's scientific work and to heap insults upon his Jesuit detractors. Lastly Newton, in 1687, dropped his work on the *Principia* to lead the fight at Cambridge against James II's attempt to appoint a Catholic Fellow. Iliffe shows that "hatred of Catholicism permeated every aspect of Newton's life."

In the sixth essay, Robert Markley places Newton's historical efforts into the context of the "universal histories" of the world that were being written from the late Renaissance onward. Such "universal histories" portrayed ancient history as a lost Golden Age and, reflecting the loss of social and economic stability in the early modern world, pessimistically portrayed the present age as decadent, corrupt, and incapable of sustaining itself physically, economically, or morally. In contrast to the world-view assumed within this tradition, Markley shows that Newton, in his own version of universal history, was committed to looking for a pure beginning which could be recaptured or reconstituted through religious reform and natural science.

In the seventh essay, Matt Goldish analyzes the most finished of Newton's manuscripts, *Of the Church.* This manuscript exists in two versions, one in the Bodmer Library in Geneva, the other in Jerusalem. Goldish shows that Newton's careful examination of ancient religion, Judaism, and early Church history stresses how original, pure religious doctrines became corrupted by additions of pagan ideas, gnosticism, kabbalism, and human errors. The early church was infected by metaphysical misinterpretations of doctrines and, finally, was taken over by Trinitarianism. Goldish also reveals the amazing scholarship which underlies Newton's analysis and underscores how Newton used his historical reconstruction of what Christianity should be to develop his basis for remaining within the Anglican Church while simultaneously opposing Latitudinarianism, deism, and the theosophy of Leibniz (which, for Newton, was nothing more than the emanationist metaphysics of the kabbalah all over again.)

In the eighth essay, Sarah Hutton examines what is usually brushed aside, Newton's interpretation of the Vials and the Trumpets in the Book of Revelation. Newton and Henry More had a falling out about how to interpret these texts and Newton felt it necessary to construct a diagram for Locke and Lady Masham so that they could understand his point. Hutton puts the discussion in terms of what the great exegete, Joseph Mede, had said about the relationship of Daniel, and about the symbolism therein, to the Book of Revelation. Both Newton and More broke with Mede on this issue. Hutton shows that Newton's interpretation of these texts was crucial for his Arianian theology and explains Newton's understanding of the crucially pivotal role of the non-divine Christ in Providential history. For Newton, the fights about these symbols was crucial in defining the essentials of the true faith.

In the ninth essay, Kenneth Knoespel shows us the fruits of his careful examination of one of the most puzzling of Newton's manuscripts, the *Theologiae gentilis origines philosophiae*, Yahuda MS 16 in Jerusalem. This unfinished writing, which Newton was working on for decades, has been seen lately as crucial for understanding Newton's thought. But the unfinished character, the mass of notes and corrections, has made it difficult to penetrate. Knoespel puts forth its structure and its relationship to the earlier work with a similar title by J. G. Vossius. Knoespel then traces Newton's picture of how ancient science and religion, so intimately connected, developed and sadly became corrupted. Knoespel suggests that "Viewed with the *Principia*, the *Origines* then provides historical justification for Newton's own momentous discoveries." The discoveries made by Newton in natural philosophy may be seen, from the perspective of this manuscript, as part of the rediscovery of ancient revelations.

In the tenth essay, Michael Murrin traces the history of the status of the Book of Revelation from ancient to early modern times. He points out that it was disowned by most of the early eastern Christian churches and only taken seriously in the West. During the Reformation it was at first disowned by Luther and only gradually became all important because it could be used to show that the Church of Rome was about to be overthrown. Some of the most important modern commentators on the Book of Revelation were Spanish Jesuits who connected its symbolism and content with current events and saw in it a map into the future. The moderate interpretations of these Spanish Jesuits were then rebutted by the 17th-century Millenarian Protestants, Joseph Mede, and his followers, Henry More, and Isaac Newton. In Murrin's analysis, Newton finally secularized its content and eliminated the possibility of genuinely prophesying when the events of the end times would occur.

In the eleventh essay, John Rogers treats the relationship between Newton's conception of empirical induction in science and its relation to his theology. Rogers shows that Newton did not get into the problem of trying to justify empirical induction against sceptical doubts on the subject because, for Newton, God was always active in the world and in making it possible for human beings to know enough about the world. Hume's type of puzzle about whether there could be any guarantee that the future will resemble the past could only be broached when and if the Newtonian God of total dominion vanished from the cosmos.

In the twelfth paper, my collaborator, James E. Force, seeks to locate Newton in the famous quarrel between the ancients and the moderns. In the 18th century, Newton was made commander-in-chief of the modernists. But Force shows that Newton himself strongly held that ancient wisdom was far superior to modern views except for those of the very few moderns, such as himself, who were busy recapturing the ancient wisdom. And Newton held this view about both science and religion. The true religion and the true science were known to the Patriarchs but was corrupted as pagans identified natural objects with historical figures and introduced human metaphysical views into theology. The corruptions of Christianity, especially from Athanasius onward, also corrupted the understanding of nature. Newton was concerned to find this ancient religious and scientific understanding in ancient monuments such as Solomon's Temple and in the circle of stones at Stonehenge. In understanding Newton's place in the controversy of the ancients and moderns, one can begin to see how the religious and the scientific Newton fitted together as the unified outlook of one man.

In the thirteenth, and last essay, Reiner Smolinski examines the actual theological physics of several of the leading Millenarians of the late-17th century and traces how they interpret what is supposed to happen at the end of the Book of Revelation in connection with their knowledge of modern physics. Newton's explanations are compared and contrasted with those of John Ray, Thomas Burnet, and the New England millenarian, Cotton Mather. Smolinski follows out, in fascinating detail, just how each of these thinkers seriously tried to figure out where the destructive fire would take place, how the elect would escape, where the New Jerusalem would come from, what its dimensions would be, etc. Newton is shown to be as concerned as these other millennial theorists about these problems. All of them were well aware of the latest developments in physics, geology, etc.

These thirteen essays show, we think, that the "other" Newton was part of a context, mostly lost by now, in which religious and scientific concerns went hand in hand. And for Newton, as well as many others of his time, the understanding of the Book of Nature and the Book of Scripture were complimentary exercises. We hope that the reader will see that one does not have to construct a Newton who had an internal switch by which he could turn off the scientific Newton and turn on the religious Newton, and vice versa. Instead some of us engaged in studying the vast corpus of Newton's theological and alchemical manuscript, can see the possibility that one man, a product of many intellectual and religious currents in the 17th century, could write great scientific works, great works in church history, in Biblical interpretation, etc., as part of one great enterprise, that of understanding man and his place in the grand scheme of God's creation.

Professor Force and I also hope that this collection of essays will lead to further conferences and collections of studies about Newton and his time and help us recapture the vision that animated his whole intellectual life.

ESSAY 1

Alchemy and Eschatology: Exploring the Connections between John Dee and Isaac Newton[1]

DEBORAH E. HARKNESS
University of California, Davis

In 1724/5, Isaac Newton explained to John Conduitt the key role he believed the comet of 1680 would play in the catastrophic end of the world.[2] Newton's understanding of the comet was informed not only by his natural philosophical views, but also his faith in a providential deity as described in holy scripture. While alchemical literature explained the vital role putrefaction played in generation, Newton believed that the ultimate restitution of nature and mankind foretold in the Bible could only come about after a cataclysmic event such as the one he expected when the comet of 1680 hit the sun. Until that time of spiritual and material regeneration, Newton aspired to increase human understanding of God's plan for the cosmos through what Betty Jo Dobbs described as "the knowledge of God's activity in the world."[3] Like many of his contemporaries, Newton believed that such knowledge could be increased by consulting ancient authorities who lived and worked when human understanding had been less corrupted by the consequences of Adam's Fall. Newton hoped that his search through ancient texts would yield a true religion and a true natural philosophy, both

[1] I am indebted to the participants in the Clark Library conference on "Newton and Religion" whose provocative and insightful comments greatly enriched this paper. This paper is dedicated to the memory of Betty Jo Teeter Dobbs, mentor and scholar, to whom I owe all my insights into the complicated figure of Newton and who helped me to see the complexity of John Dee.

[2] B. J. T. Dobbs, *The Janus Faces of Genius* (Cambridge: Cambridge University Press, 1991), p. 234; Simon Schaffer, 'Newton's Comets and the Transformation of Astrology', in *Astrology, Science and Society, Historical Essays*, ed. Patrick Curry, (Wolfeboro, NH: Boydell Press, 1987), pp. 219-43.

[3] Dobbs, *Janus Faces*, pp. 73-4 and 87.

J.E. Force and R.H. Popkin (eds.), Newton and Religion, 1–15.
© 1999 *Kluwer Academic Publishers.*

of which would help to increase human understanding of the natural world as well as the divine.[4]

Newton was not the first English natural philosopher to have such beliefs. At the end of the sixteenth century John Dee believed that the comet of 1577 was a sign of the imminent, catastrophic end of the world. Dee, working in his library, his study, and his alchemical laboratory, had become convinced that a restitution of nature was necessary to reverse the adverse spiritual and moral effects of the expulsion of Adam and Eve from Paradise. Dee's belief that he was living in a world crippled by decay and corruption encouraged him to look to the perfection of the heavens for guidance in his practice of natural philosophy. This guidance came not only in the form of comets and other celestial phenomena, but also in the form of angels, who visited Dee through the medium of an optical glass. Through his glass Dee received information about the role that natural philosophy would play in the ultimate restitution of all things. In addition, the angels conveyed God's language of creation, restored alchemy to its origins, and shared doctrines for a true religion which Adam had known, but which had been lost.[5]

Can Dee's interest in alchemy and eschatology help us to understand Newton's interest in the comet of 1680, in ancient texts, and in material and spiritual restitution? In this essay I will explore the intriguing connections between these two figures in the history of science, without systematically exploring their affinities nor suggesting that their distinctions are unworthy of consideration. Instead I will present John Dee's views on alchemy and eschatology as they are expressed in his minute records of conversations with angels to delineate a precursor to Newton's subsequent interest in alchemy, apocalypse, and restitution.[6] Too often dismissed by historians of science as a blend of religion and insanity, Dee's angel diaries contain a

[4] Dobbs, *Janus Faces*, p. 151.

[5] Deborah E. Harkness, "The Scientific Reformation: John Dee and the Restitution of Nature," Ph.D. diss., University of California, Davis, 1994; *John Dee's Conversations with Angels: Learning, Revelation and the End of Nature* (Cambridge: Cambridge University Press, forthcoming).

[6] The holograph transcripts of Dee's angel conversations are contained in a number of scattered and imperfect manuscripts at the British Library, London, and the Bodleian Library, Oxford: British Library, Sloane MS 3188-3189; British Library, Sloane MS 3191; British Library, Cotton Appendix MS XLVI, 2 vols.; British Library, Add. MS 36674; and Bodleian Library, Ashmole MS 1790. Printed editions of selections from the manuscripts are available and their reliability varies. Meric Casaubon was the first to print excerpts from the angel diaries under the title *A True and Faithful Relation* (London, 1659), but the work includes only conversations dated after 1583 and is not without textual inaccuracies. The seventeenth-century collector, Elias Ashmole, attempted to make corrections in the Casaubon edition. His corrections appear in his copy of the work, now Bodleian Library, Ashmole MS 580. The early spirit diaries, which date from 1581 to April 1583, have received much less attention. The earlier diaries were available only in manuscript until Christopher Whitby's careful transcript of *John Dee's Actions with Spirits* made them available to a wider audience. Whenever possible, I will refer to John Dee, *A True and Faithful Relation* ed. Meric Casaubon (London, 1659) and John Dee, *John Dee's Actions with Spirits*, 2 vols., ed. and trans. Christopher Whitby (New York: Garland, 1988).

wealth of information about his views on natural philosophy and religion: where they converge, overlap, support each other, and come into conflict. From the deciphering of eschatological signs embedded in the natural world to the sharing of revelations with the elect, Dee believed he played a critical, active role in the unfolding of God's plan for the cosmos.

John Dee began to ask God for heavenly assistance in natural philosophy around 1569 while deeply engaged in his alchemical and mathematical studies. He prayed for God to send him a "pious wise man and skilled philosopher" to help him with his studies.[7] Dee doubted whether any human could fulfill this request, so he suggested it might be more expedient for God to send "good spiritual ministers and angels ... who may fully and perfectly instruct me in the true and exact knowledge and understanding of your mysteries and great works (concerning all your creatures, their natural properties, and best employments)." Dee continued to pray for help until 1581 with little apparent success. In the final months of 1581, however, Dee's prayers were answered when the scryer Barnabas Saul began to see angels in a crystal stone. Dee, acutely aware of the importance of this development, began to keep detailed accounts of their activities. Today we have the records of over two hundred conversations with angels, most of which took place between 1581 and 1587 with the assistance of three different scryers: Barnabas Saul, Edward Kelly, and Bartholomew Hickman. Tantalizing gaps in the diaries suggest that these records, rich as they are, represent only a portion of a much larger body of texts now lost.[8]

Dee did not decide to contact angels suddenly—he came to the decision after years of studying ancient and medieval scholarship about the natural and supernatural worlds.[9] By the 1580s Dee had amassed the most extensive personal library in England. In his library Dee was able to come into contact with wisdom that had been handed down through the ages, sometimes in need of repair and restitution as every good humanist of the time knew, but in many cases still retaining traces of God's divine inspiration—the root of all earthly wisdom. Within Dee's library, treatises on military engineering and Greek literature jostled on the shelves with books on the Jewish cabala, alchemy, optics, and obscure ancient languages. As we know from the recovery efforts of Roberts and Watson, Dee did not simply own these books; like Isaac Newton, he *used* them.[10] In many cases evidence from

7 Dee, *Actions with Spirits* 2:5-6.

8 Harkness, "The Scientific Reformation," pp. 312-21.

9 Harkness, "The Scientific Reformation," pp. 62-209.

10 Newton's library and its important place in the development of his ideas has been studied by John Harrison. See John Harrison, *The Library of Isaac Newton* (Cambridge: Cambridge University Press, 1978). Newton did not own any of Dee's books, but Newton's active reading practices—most notably dog-earing—are similar. Harrison, especially pp. 1-27.

marginalia and annotations indicates Dee had read and carefully digested their contents, adding his own opinions as warranted.[11]

Dee's books provide valuable information about how he came to believe that contacting angels was not only possible, but desirable for a natural philosopher. Dee's interest in angels can be traced back as far as 1550 when he purchased a copy of Pompilius Azalus' *De omnibus rebus naturalibus*.[12] This work, though not primarily concerned with angels, does attempt to explicate all features of the natural world, including angels. Dee underlined and marginally annotated passages relating to the angels' functions as ministers of God in the natural world, including their role as guiding spirits. By 1560 Dee had acquired a comprehensive assortment of classical, Christian, and contemporary works germane to angelology such as Ficino's collection of neo-Platonic and Hermetic texts which contained treatises on spirits, Dionysius the Areopagite's "The Celestial Hierarchy" and "The Divine Names," Tommaso Rangoni's study of spiritual visitations, and complex works which combined Christian angelology with esoteric Jewish cabala such as Francesco Giorgi's *De harmonia mundi*, Agrippa's *De occulta philosophia*, Reuchlin's *De verbo mirifico*, and Trithemius' *De septem secundeis*.[13]

Not all of these works have been recovered, and not all those recovered are signed or annotated. Despite these limitations, however, it is clear that Dee read his books about angels very carefully, most notably the Dionysian corpus. Both Dionysius' "Celestial Hierarchy" and "Divine Names" reveal a number of connections between the information in Dee's library books and his angel conversations. Throughout the angel conversations, for example, Dee noted carefully the appearance of each angel as his scryers conveyed it to him.[14] In the "Celestial Hierarchy" Dee underlined Dionysius' descriptions of the angelic messengers, which emphasized the angels' many

[11] See Julian Roberts and Andrew G. Watson, *John Dee's Library Catalogue* (London: The Bibliographical Society, 1990), passim; Harkness, "The Scientific Reformation," passim; and William Sherman, *John Dee: the Politics of Reading and Writing in the English Renaissance* (Amherst: University of Massachusetts Press, 1995), pp. 29-100.

[12] Pompilius Azalus, *De omnibus rebus naturalibus* (Venice, 1544). Roberts and Watson, #134, now Emmanuel College, Cambridge 304.1.54.

[13] Dee's recovered copies are Dionysius, *De Mystica Theologica* (Rome, 1525), Roberts and Watson, #271, now Cambridge University Library H*8.22(c); Dionysius, *Opera*, comm. Jacques Léfèvre d'Étaples (Venice, 1556), Roberts and Watson, #975, now Magdalene College, Cambridge D.7.10; Marsilio Ficino, *Index eorum, quae hoc in libro habentur. Iamblichus, de mysterils Aegyptiorum . . .* (Venice, 1516), Roberts and Watson #256, now Folger Shakespeare Library BF 1501 J2 Copy 2 Cage; Johann Tritheim, *De Septem Secundeis* (Frankfurt, 1545), Roberts and Watson, #678, now Cambridge University Library Dd*4.5; Johann Trithemius, *Liber Octo Quaestionum, Quas Illi Dissolvendas Proposuit Maximilianus Ceasar* (Cologne, 1534), Roberts and Watson, #897, now Cambridge University Library H*15.9(f).

[14] See, for example, Dee, *Actions with Spirits* 2:71-8, and 94.

feet and their tendency to appear in the guise of animals such as a lion or a horse. Other annotations in Dionysius' works demonstrate Dee's familiarity with his organizational scheme for the celestial hierarchies and their relationship to the Deity, something that preoccupied Dee throughout his angel conversations.[15]

More complex annotations appear in the ample margins of Dee's books. In his copy of the "Divine Names," for example, Dee noted the name "Pele" in the margins against a passage that referred to the marvelous name of God which was above all other names.[16] The name "Pele" does not appear in the Dionysian text, but it does appear in an angel conversation which took place on 14 March 1582.[17] On that day the angel Michael revealed the design for a ring not seen on earth since the days of Solomon, which Solomon used for "divine works and wonders." The ring was inscribed with the name Pele, and Dee noted in the margins "See Reuchlin's book *De verbo mirifico* about, the name PELE."[18] 14 March 1582 proved to be a day full of bibliographic cross-references. Dee noted that he needed to look in Reuchlin not only for the name "PELE" but also for the name "NA" which the angel Michael told him was a divine name to use in times of trouble. The angel Uriel appeared briefly and told Dee that some called him Nariel, and Dee noted that this information could be found in a specific book and chapter of Agrippa's *magnum opus—De occulta philosophia.* Finally the angel Michael revealed a seal-referred to as the Sigillum Dei or "seal of God"—whose true name was "Emeth," Hebrew for "truth." Dee, his bibliographic abilities unwavering, gave two references for this information: one referring to Reuchlin's *De arte cabalistica*, and another to Agrippa's *De occulta philosophia.*

By 1579, however, Dee had entered into a period of intellectual crisis so profound that he began to doubt whether his library books could help him to achieve certain knowledge. In a prayer from that year Dee shared his despair with God and reminded Him of his long-held and fervent desire for "pure and sound wisdom and understanding" as well as his dedicated study of nature for "many years, in many places far and near, in many books and sundry languages." Finally, however, Dee admitted that he had reached a scholarly impasse and "could find no other way to true wisdom" without the

[15] Dionysius, *Opera*, p. 48v-54v.

[16] Dee's annotations in his copy of Dionysius, *Opera*, p. 146r: "Ac veluti ab omni eum divini nominis cognitione ab duce[n]tem dixisse cur de nomine me interrogas quod *est mirabile*. An vero istud non est *mirabile* nomen: quod est super omne nomen, quod est sine nomen, quod omne exsuperat nomen . . ." Against this passage, Dee noted "Pele."

[17] Dee, *Actions with Spirits* 2:31-3.

[18] Dee, *Actions with Spirits* 2:32. Unfortunately, Dee's powers of exact recall were faulty; the name does not appear in *De verbo mirifico*.

gift of divine revelation.[19] In this prayer Dee explains why he turned from his library to the heavens for answers to his most profound natural philosophical questions. Additional information from the collection helps further to clarify and contextualize this impulse. Specifically, the resurgence of apocalyptic sentiment in Europe in the 1570s and 1580s demonstrates the early modern belief in the end of days, and the presence of a sizable number of apocalyptic texts in Dee's library indicates his interest in the interpretation of eschatological signs.

Authors of early modern eschatological and apocalyptic treatises cited the decay of the earth, alterations in the heavens, the Reformation of the church, and strange natural occurrences to support their contention that the world was about to end.[20] In 1580, for instance, an Englishman named Francis Shakelton argued that the "constitution of the celestiall worlde, is not the same that it hath been in tymes paste, for ... the Sunne, is not so farre distant from us now, as it hath been heretofore."[21] Shakelton used the disturbing changes in the traditionally immutable heavens to suggest the existence of even greater instability in the imperfect sublunar world, where "every part... doeth feele some debilitie and weakenesse." Another author, Andrew Golding, confirmed that human decay and corruption went hand-in-hand with nature's alarming deterioration: "mannes nature [is] growing dayly more and more into decay with the perishing world nowe hasting [sic] to his ende, [it] is more subiecte too corruption, and less gyven to Godlynesse and vertue than ever it was."[22] Learning and knowledge were not exempt from decline, which Sheltco à Geveren pointed out was another "argument of the worldes consummation," since "al[l] good arts and learning...and Universities and schooles and scholasticall discipline... [have]

[19] Dee, *Actions with Spirits* 2:8-12.

[20] The literature on early modern apocalypticism and its antecedents is extensive. They include: Norman Cohn, *The Pursuit of the Millennium: Revolutionary Millenarians and Mystical Anarchists of the Middle Ages* (New York: Oxford University Press, 1970); Bryan W. Ball, *A Great Expectation: Eschatological Thought in English Protestantism to 1660* (Leiden: E. J. Brill, 1975); Richard Bauckham, ed., *Tudor Apocalypse: Sixteenth-Century Apocalypticism, Millenarianism, and the English Revolution* (Oxford: Oxford University Press, 1978); Ernest L. Tuveson, *Millenium and Utopia: A Study in the Background of the Idea of Progress* (New York: Harper and Row, 1964); Richard Foster Jones, *Ancients and Moderns: A Study of the Rise of the Scientific Movement in Seventeenth-Century England* (Berkeley and Los Angeles: University of California Press, 1965); Katharine R. Firth, *The Apocalyptic Tradition in Reformation Britain 1530-1645* (Oxford: Oxford University Press, 1979); C. A. Patrides and Joseph A. Wittreich, Jr., eds., *The Apocalypse in English Renaissance Thought and Literature* (Ithaca: Cornell University Press, 1984).

[21] Shakelton, *A blazyng Starre* (London, 1580), sig. Aiiiv-Avr, quoted in Jones, *Ancients and Moderns*, p. 24.

[22] Golding, quoted in Tuveson, *Millenium and Utopia*, p. 32.

almost in every place come to decay."[23] The attention devoted to the end of the world rose to a fevered pitch in the years leading up to 1583 when a grand conjunction took place in the constellation of Aries—the same configuration of stars and planets, it was thought, as had been present at the birth of Christ.[24]

Dee had been paying attention to other potentially eschatological signs since the 1570s. In 1572, he personally observed the strange new star which appeared in the constellation Cassiopeia, and then purchased nine European commentaries on its importance.[25] A few years later, in 1577, Dee spent three days at Windsor Castle advising Queen Elizabeth and her courtiers about the significance of a comet which had caused apocalyptic speculation and alarm.[26] In 1580 Dee experienced the strange earthquake which rocked London and noted his reactions to the event in his personal diary.[27] When a collection of letters discussing the earthquake was published by the poet Edmund Spenser and another English natural philosopher, Gabriel Harvey, Dee purchased the work for his library.[28] For Dee, however, the most significant eschatological sign came to him personally when his prayers were answered and angels finally appeared in his crystal stone to share their knowledge of the world and how it worked.

Though this broadly-based interest in eschatology and apocalypticism provides the context for Dee's belief in the decay of nature, it does not explain why he remained committed to the angel conversations for such a long time. The angels' revelations, when examined more closely, suggest

[23] Sheltco à Geveren, *Of the Ende of this Worlde, and the Second Commyng of Christ* (London, 1577), sig. Diiir; quoted in Tuveson, *Millenium and Utopia*, p. 45. Dee owned a copy of this work; for more information see Roberts and Watson, *John Dee's Library Catalogue*, #1727.

[24] For a detailed discussion of the grand conjunction, see Carroll Camden, "The Wonderful Yeere," in *Studies in Honor of De Witt T. Starnes*, ed. By Thomas P. Harrison (Austin: University of Texas Press, 1967), pp. 163-79.

[25] Dee's Library books concerning the new star were: Marcus Manilius, *Astronomica* (Basel, 1533), Roberts and Watson, #251, which contains his annotation on the new star; a work by Dee's landlord while he was in Prague conversing the angels and Rudolphine alchemists, Tadeas Hajek, *Dialexis de nova stella* (Frankfort, 1574) Roberts and Watson, #438; Theodorus Graminaei, *Erklerung oder Auszlegung eines Cometen* (Cologne, 1573) Roberts and Watson, #703; Jeronimo Muñoz, *Du nouveau comete de l'an 1572* (Paris, 1574), Roberts and Watson, #842; Georg Busch, *Die andere Beschreibung vondem Cometen welcher in 1572 Iar erschienen* (Willemberg?, 1573), Roberts and Watson, #1291; two copies of Cornelius Gemma and Guillaume Postel, *De peregrina stella 1572* (Venice, 1573) Roberts and Watson, #2137; Augustus, Elector of Saxony et al., *De publica poenitentia . . exorta peregrina stella 1572* (Wittemberg?, 1578) Roberts and Watson, #2217; Joannes Sommerus, *Refutatio scripti Petri Carolii editi Wittebergae* (Cracow, 1582) Roberts and Watson, #D20. More information on titles with Roberts and Watson numbers can be found in Roberts and Watson, *John Dee's Library Catalogue*.

[26] Dee, *Personal Diary*, p. 21. For an interpretation of this event, see Robert William Barone, "The Reputation of John Dee: A Critical Appraisal," Ph.D. diss., The Ohio State University, 1989, p. 51.

[27] Dee, *Personal Diary*, p. 22-3.

[28] Edmund Spenser and Gabriel Harvey, *Letters Touching the Earthquake* (London, 1580). For information on Dee's copy see Roberts and Watson, *John Dee's Library Catalogue*, #1720.

that they held Dee's attention because eschatology was combined with natural philosophy. Almost every conversation referred in some way to a complex combination of ideas that resonated with natural philosophical and theological implications: the redemption of language through a reintroduction of the divine language of creation; the restoration of alchemy to its original, pure state which would facilitate the material redemption of nature; and the revelation of a "true cabala of nature" which not only perfected the Jewish cabala, but also provided an exegetical tool specifically designed to meet the challenges of God's most complicated and corrupt text, the "book of nature."[29]

The cornerstone of the angels' redemptive plan was the revelation of God's language which had been used to create the world as described in the book of Genesis. The angels began to convey the divine language to Dee in March 1582. Their revelations included an alphabet (similar to the Hebrew alphabet) which consisted of twenty-one regular characters and one aspirated character, the proper names for each character, and the numerical values of each character. The angels also revealed five holy books in the divine language which the angels said would serve as scripture in the redeemed world.[30] The most important conversations concerned with the divine language, however, involved the true names of God, the angels, and parts of the created world. A complete knowledge of these true or essential names had been a goal of early modern occult philosophers and Christian cabalists such as Pico della Mirandola, Johannes Reuchlin, and Henry Cornelius Agrippa because the knowledge of a true name gave the *magus* or cabalist either direct dominion over that thing or, in the case of God and the angels, access to supernatural power.[31] Dee was understandably excited when he was given the true names of God and multitudes of angels along with the proper names of geographical regions and parts of the heavens, for this represented a lexicon of considerable power. The Adamic antecedents of the names only heightened Dee's excitement, for the angels told Dee that Adam, while in Paradise, used the divine language to communicate with God and the angels, and to bestow correct and essential names on all parts of the creation. This level of communication gave Adam his dominion over nature.[32] Adam lost his mastery of the divine language and the power that accompanied it, according to the angels, when he disobeyed God's

[29] Harkness, "The Scientific Reformation," pp. 394-488.

[30] Dee, *Actions with Spirits*, 2:237; Dee, *True and Faithful Relation*, p. 61.

[31] See Brian P. Copenhaver, "Léfèvre d'Étaples, Symphorien Champier, and the Secret Names of God," *Journal of the Warburg and Courtauld Institutes* 40 (1977), pp. 189-211; Charles Zika, "Reuchlin's *De verbo mirifico* and the Magic Debate of the Late Fifteenth Century," *Journal of the Warburg and Courtauld Institutes* 39 (1976), pp. 104-38.

[32] Dee, *True and Faithful Relation*, p. 92.

commandments and was expelled from Paradise. After the expulsion, Adam managed to retain only fragments of the divine language, which provided a basis for the development of Hebrew, but Adam's limited understanding of the language of creation had been lost in the Flood.[33]

The angels' delivery of the divine language represented an important, first step in the larger program of cosmic restitution. Because nature itself had decayed, Dee also needed to foster the material redemption of the cosmos. The outline for this transformation was provided in alchemical secrets he received from the angels. Alchemists had been asserting for centuries that their art attempted to perfect metals by hastening natural cycles of generation, degeneration, and regeneration in the mineral kingdom. In the alchemical process metals progressed from a chaotic mix of materials, through differentiation or separation, decay, and death, until the substance finally reached a state of redemption and perfection in the elusive Philosopher's Stone—a marvelous substance that could cure all sickness, render the alchemist immortal, and restore nature to its perfect state.[34] The alchemical redemption of matter was predicated on death and decay—just as Dee's angel conversations and the redemption they promised was predicated on the belief that the world had undergone its own decay and was fast approaching "death" in the long-awaited Apocalypse.

Some scholars, such as the Siennese metallurgist Vannoccio Biringuccio (1480-1539?), believed that the lofty goal of material perfection espoused by alchemists could not be achieved "unless someone should find some angelic patron."[35] Dee had such patrons, of course, and they conveyed to him new and uncorrupted information about alchemical processes and materials.[36] Through the conversations Dee discovered, for example, that "Dlafod" was the true name for sulfur, and that "Audcal" was mercury in the divine language. The angels gave allegorical explanations of the processes through which alchemical materials had to be subjected before they became "Dar" or the Philosopher's Stone. Allegorical parables involving traditional alchemical symbols brought more detailed and complex alchemical ideas to life. Pageants of angels carrying symbols of the metals and the elements promised Dee a greater mastery of the material world once he had decoded the angels' arcane messages.[37] Finally, the angels shared their alchemical "*medicina dei*" with Dee, a "medicine of God" which the angels promised

33 Dee, *Actions with Spirits*, 2:333.

34 Betty Jo Teeter Dobbs, *Alchemical Death and Resurrection: The Signficance of Alchemy in the Age of Newton* (Smithsonian Institution, 1990), p. 15.

35 Vannoccio Biringuccio, *The Pirotechnia of Vannoccio Biringuccio*, ed. and trans. Cyril Stanley Smith and Martha Teach Gnudi (New York: Dover, 1990), p. 35.

36 See, for example, Dee, *True and Faithful Relation*, pp. 387-8.

37 Dee, *Actions with Spirits*, 2:147-57.

would heal the natural world of its sicknesses and imperfections. As the angel Gabriel told Dee, the *medicina dei* was the end-result of restored alchemy and was a key ingredient in the restitution of nature.[38] Dee's apocalyptic interests were thus firmly tied to his alchemy. Once again, Dee was not alone; many alchemical theorists, including Paracelsus, believed that the redemption of metals which took place in alchemical transformation would occur with the redemption of humanity and the natural world, and usher in the age of Apocalyptic rebirth.[39] In Dee's conversations angelology, apocalypticism and alchemy collapse into a single narrative designed to present one message concealed among layers of imagery and symbolism: that Dee was instrumental in the world's progress towards material and spiritual restitution. This message was not delivered once, but repeatedly.[40]

Linked to the angelically-restored alchemy and the process of cosmic restitution was the "true cabala of nature." The cabala of nature was based upon the divine language and supported the restored alchemy because it furthered the interpretation of the sacred "book of nature." The emphasis on exegesis which the cabala of nature shared with traditional cabala was based on cabalistic techniques of *gematria, notarikon* and *temurah,* each of which appears in Dee's angel conversations. One angel, for instance, explained to Dee that each word the angel uttered actually contained sentences of untold wisdom—a reference to *notarikon*.[41] Dee also noted in the diaries his "diverse talks and dyscourses of letters"—*temurah*—a subject which he had already considered in his *Monas hieroglyphica* (1564).[42] Frequent references were made in the angel conversations to the numerical values of various words in the divine language, or *gematria*.[43] Dee needed these exegetical methods to help make the "book of nature"—a book "written" or created with the divine language—legible after centuries of decay and corruption. The angels told Dee that the cabala of nature would enable him to access a different, theurgic level of the cabalistic art that would facilitate the restitution of nature by drawing perfect, celestial influences down into the imperfect world. The theurgic or magical implications of the cabala of nature were most fully represented in the angel conversations when Dee was

[38] Dee, *True and Faithful Relation*, p. 251.

[39] Robin Bruce Barnes, *Prophecy and Gnosis: Apocalypticism in the Wake of the Lutheran Reformation* (Stanford: Stanford University Press, 1988), p. 176.

[40] For additional interpretation of these messages, see Deborah E. Harkness, "Shows in the Showstone: A Theater of Alchemy and Apocalypse in the Angel Conversations of John Dee," *Renaissance Quarterly* 49 (1996).

[41] Dee, *Actions with Spirits*, 2:224.

[42] Dee, "A Translation of John Dee's 'Monas hieroglyphica' (Antwerp, 1564), with an Introduction and Annotations," ed. and trans. C. H. Josten, *Ambix* 12 (1964), pp. 84-111, esp. p. 209; Dee, *Actions with Spirits*, 2:351.

[43] Dee, *Actions with Spriits* II: 351; Dee, *True and Faithful Relation*, pp. 92-3.

given the secret, true names of the forty-nine angels responsible for governing the earth who were to assist Dee in his alchemical work. When the names of the forty-nine governors of the earth were subjected to *notarikon, temurah* and *gematria* the angel Uriel promised that "the Fowntayne of wisdome ... [will be] opened. Nature shal be knowne: Earth with her secrets disclosed. The Elements with theyr powers judged.[44] With the help of the angelic governors Dee would "work in the quieting of the estates, in [the] lerning of wisdome, in [the] lerning of wisdome, ...as well as in the depths of waters, Secrets of the Ayre,...[and] in the bowells of the Earth."[45]

The angels stated that Dee, equipped with the divine language, restored alchemy, and the cabala of nature, would bring all levels of the cosmos back into perfect communication and would also purify the earth.[46] Dee took these responsibilities seriously, and he carefully noted the words of the angel Michael who promised that he would one day open the gates of the New Jerusalem so that Christ could enter and the age of peace, prosperity, and perfection could begin.[47] Theologically these were significant promises because they gave Dee an important role in the fulfillment of Biblical prophecies about the end of the known world and human history and the beginning of a new era and a new, perfect world. The promises also had a natural philosophical resonance because they depended on Dee's ability to learn from his angelic teachers and then observe, interpret, and redeem nature. At every level of Dee's angel conversations there is a layering of theological and natural philosophical ideas so dense that it becomes difficult to establish when religion ends and the natural philosophy begins.

There is no doubt that Dee's strategy for expanding human knowledge through conversations with angels was unique. The underlying beliefs which supported this strategy, however, were not. The information in Dee's angel diaries may increase our understanding of Dee and his contemporaries, especially Paracelsus and Postel. Furthermore, the diaries may also shed light on the religious and scientific climate of Newton's time because of Meric Casaubon's decision to publish selections of Dee's diaries in 1659. Casaubon's edition ensured that more of Newton's contemporaries knew of the conversations with angels than had Dee's contemporaries. Casaubon had elaborate goals for his volume; he hoped it would convert atheists who did not believe in the spiritual world, moderate advocates of prophecy and revelation such as the Anabaptists, and encourage readers to appreciate the power of prayer. In addition, Casaubon considered the diaries a persuasive

[44] Dee, *Actions with Spirits*, 2:124.

[45] Dee, *Actions With Spirits*, 2:124-5.

[46] Dee, *True and Faithful Relation*, p. 92.

[47] Dee, *True and Faithful Relation*, pp. 59-60.

weapon to use against people who consulted astrologers and conjurers—as well as those who read about such dangerous ideas in books.[48] Apparently, Casaubon's edition did not have the desired effect; late seventeenth-century responses to the work are numerous and suggest that the seventeenth century was no less interested in the occult features of the natural and supernatural worlds than Dee.

Manuscript diaries now in the British Library, for example, attest to the interest in Dee and his angels that can be found in the seventeenth century.[49] These manuscripts record the activities of a group who gathered between 24 July 1671 and 18 December 1688 to converse with angels through a crystal, just like Dee and his scryers. The similarities do not end there. The dates of the last conversations suggest that the group may have initially decided to contact the angels after they read Casaubon's edition and discovered there was some question in Dee's mind as to whether the year of cosmic restitution was actually 1588 or 1688. In addition, the angels contacted were identical to Dee's, and two of Dee's most idiosyncratic angels—Madimi and Galuah—make frequent appearances. There is little evidence to suggest that the practitioners were deeply engaged with the problems of natural philosophy, however. Instead, the participants were primarily concerned with their financial and personal futures rather than the future of the world.

The antiquarian Elias Ashmole (1617-1692) was intensely interested in Dee's angel conversations as well as his alchemical work. In 1672, after years of studying Dee and amassing manuscript remains belonging to the natural philosopher, Ashmole came into possession of a cache of Dee's earliest angel diaries not known to Casaubon.[50] Ashmole's interest in the diaries was alchemical as well as biographical. In Ashmole's alchemical compilation *Theatrum Chemicum Britannicum* (1652) he wrote that the Philosopher's Stone, when subjected to even higher and more mystical processes, yielded "vegetable," "magical," and "angelic" stones.[51] The most precious stone, the angelic stone, was a substance so subtle it could not be

48 Casaubon, "Preface," in Dee, *True and Faithful Relation*, sigs. [Hlr-v].

49 See British Library, Sloane MSS 3624-3628, passim. The group included a scryer "E. R[orbon]," and two other regular participants: "R. O." who appears to be the figure occupying Dee's role in the conversations, and another referred to "t times as "E. C." and occasionally as "Brother Collings."

50 Ashmole obtained the diaries from "Widow Jones," who purchased a wooden chest from a "parcell of the Good of Mr. John Woodall Chirurgeon." The chest contained Dee's earliest angel diaries in a secret compartment. The Jones' kitchen maid destroyed some of the manuscripts when she used the pages to line pie-plates. See Ashmole's introduction to the manuscript in Dee, *Actions with Spirits*, 2:2-4. Finally, the diaries passed from Ashmole's hands into the collections of the British Library through Sir Hans Sloane, although the exact delineation of this transfer is not clear. Christopher Whitby discusses the murky chain of events that might have led from Ashmole to Sloane. See Whitby, *Action with Spirits*, 2:3-7.

51 Elias Ashmole, "Prolegomena," *Theatrum Chemicum Britannicum* (London, 1652), sig. A4v.

seen, felt, or weighed, but only tasted. Ashmole described the stone in some detail:

> It hath a *Divine Power, Celestiall,* and *invisible,* above the rest; and endowes the possessor with *Divine Gifts. It* affords the *Apparition of Angells,* and gives a power of conversing with them by *Dreames* and *Revelations:* nor do any *Evil Spirits* approach the *Place* where it lodgeth. Because it is a *Quintessence wherein there is no corruptible Thing: and where the Elements* are not *corrupt, no Devil can stay or abide.*[52]

Ashmole also remarked that Hermes had knowledge of the angelic stone which he refused to share with anyone, and that the only other people to possess the angelic stone were Moses and Solomon.[53] Ashmole's "angelic stone" possessed many of the same properties of Dee's optical glass through which he saw angels, and no doubt contributed to the antiquarian's interest in Dee's manuscripts.

Natural philosophers and practitioners of the new science were not immune to the attractions of Dee's angel conversations. One of the first natural philosophers to express her views in print was Margaret Cavendish, Duchess of Newcastle (1623-73.) In her natural-philosophical fantasy *The Description of a New World, Called the Blazing World* (1666) Cavendish (thinly-veiled as the Empress), invited spirits to talk with her about the state of natural philosophy in her homeland. In particular the Empress asked whether anyone had discovered the secrets of the Jewish cabala. "Several have endeavored it," the spirits replied, "but those that came nearest ... were one Dr. Dee, and one Edward Kelly."[54] Despite the angels' subsequent condemnation of the two men and their conversations, the Empress proceeded to talk extensively with the spirits about the occult properties of the natural world and the intricacies of the cabala.[55]

Other prominent seventeenth-century natural philosophers also had views on the diaries and Dee's methods, namely Robert Boyle (1626-1691) and Robert Hooke (1635-1703.) Boyle, for example, was "tempted" to try his hand at scrying as had Dee.[56] Boyle confessed to several brushes with the art, including one involving a cleric who offered to show him "strange

[52] Ashmole, "Prolegomena," sig. Bv.

[53] Ashmole, "Prolegomena", sig. B2r. I was unable to find any mention of this stone in the texts of the Hermetic corpus; see the translation by Brian P. Copenhaver, *Hermetica: The Greek "Corpus Hermetica" and the Latin "Asclepius"* (Cambridge: Cambridge University Press, 1992), passim.

[54] Margaret Cavendish, *The Blazing World and Other Writings* ed, Kate Lilley (Harmondsworth: Penguin, 1992), p. 166.

[55] Cavendish, *Blazing World*, pp. 166-87.

[56] Michael Hunter, "Alchemy, Magic, and Moralism in the Thought of Robert Boyle," *British Journal for the History of Science* 23 (1990), pp. 387-410. I am indebted to Margaret G. Cook of the University of Calgary for bringing this incident to my attention.

representations" in a glass of water.[57] Though Boyle did not accept the cleric's services, he would later feel "the greatest Curiousity he ever felt in his life" when a gentleman brought him "an Ordinary double convexe glasse" capable of displaying spirits.[58] Boyle overcame his intense desire to look for spirits in the glass, and considered this refusal the greatest achievement of his life. Boyle's conviction that communication between a natural philosopher and the spiritual levels of the cosmos facilitated the genesis of the Philosopher's Stone—one of his ambitions—no doubt made the temptation (and his triumph over it) even more significant.[59] Robert Hooke's references in his Cutlerian Lecture "Of Dr. Dee's Book of Spirits" to the "many Discourses" overheard among the members of the Royal Society indicate that Boyle was not alone in his interest, for the angel conversations were a popular topic of discussion among the leading natural philosophers of the period.[60] Hooke referred to the angel conversations as "Dr. Dee's Delusion"' and rebuked Casaubon for publishing the diaries despite their value as weapons against atheists, "Enthusiasts, who altogether depend upon new Revelations," and those misguided enough to seek the counsel of witches, conjurers, and astrologers.[61] Even Hooke found something of interest in the diaries, however, arguing that the conversations were a form of cryptography—one of his passions—in which Dee had enciphered valuable political intelligence for Elizabeth in "feigned Stories."[62]

Though there is no evidence to suggest that Newton knew of Dee's diaries, the attention of Boyle, Hooke, and Cavendish does indicate that Dee's angel conversations had a certain resonance in the natural philosophical community of the late seventeenth century. Much had happened, however, since Dee engaged in the conversations: the mechanistic theories of Descartes, challenges to the status of astrology, additional Biblical scholarship. We might ask ourselves if Newton would have felt that angels were evidence of the "vegetable spirit"?[63] Would Dee's

57 Hunter, "Alchemy, Magic, and Moralism," p. 390.

58 Hunter, "Alchemy, Magic, and Moralism," p. 391.

59 Hunter, "Alchemy, Magic, and Moralism," pp. 391 and 396-7.

60 Robert Hooke, *The Posthumous Works* (London, 1705), p.205. For a discussion of Hooke's interest in and aversion to Dee see John Henry, "Robert Hooke, the Incongruous Mechanist," in *Robert Hooke: New Studies*, ed. Michael Hunter and Simon Schaffer, (Woodbridge, Suffolk: The Boydell Press, 1989), pp. 149-80.

61 Hooke, *Posthumous Works*, pp. 203-4.

62 Hooke, *Posthumous Works*, pp. 206-7. Though subsequent scholars have attempted to support Hooke's contention that Dee was a spy and cryptographer in Elizabeth's service, adequate proof has not been uncovered. A similar line of argumentation has been levelled against Trithemius's *Steganographia*, in which Dee was interested. See Nicholas Clulee, *John Dee's Natural Philosophy: Between Science and Religion* (New York: Routledge, 1988), pp. 136-7.

63 Dobbs, *Janus Faces*, p. 39.

enthusiastic and unqualified embrace of prophecy have disturbed Newton?[64] Did Newton's belief that "there is no way ^(wthout revelation)^ to come to ye knowledge of a Deity but by the frame of nature" leave room for serious consideration of Dee's angelic conversations?[65] These remain questions for further study. The publication of Dee's angel diaries in 1659 requires that we consider the information they may provide about the alchemical and eschatological ideas of the sixteenth century, and the impact these ideas may have had on the views of Isaac Newton.

[64] Dobbs, *Janus Faces*, p. 80 and Chap. 6.

[65] Newton, quoted in Dobbs, *Janus Faces*, p. 151.

ESSAY 2

Newton and the Rosicrucian Enlightenment

ALLISON P. COUDERT
Arizona State University

1. THE YATES THESIS

The aspect of the Yates thesis that has become the point of so much contention can be summed up as follows: modern science originated as the result of a new and optimistic view of human nature which fostered the belief that man had the ability to improve both himself and the world in which he lived. This optimistic assessment of man's nature first emerged in the writings of the late Florentine Neoplatonists and was carried into the seventeenth century through the subterranean channels of occult philosophy. The very idea that man could change his environment for the better and harness the powers of nature to his own advantage had its roots in the magical world of Renaissance Hermeticists, and the twin concepts of progress and reform, which are the hallmarks of modern science, emerged from the grandiose schemes of Renaissance magi. In Yates' view the *Rosicrucian Manifestos* of the early seventeenth century were perfect expressions of the new and exhilarating view of human potential and prowess that made the scientific revolution possible. With their call for the "Universal and General Reformation of the whole world" and their conviction that creation can be brought back to the state in which Adam found it, the *Rosicrucian Manifestos* provided a bridge between Renaissance Hermeticism and modern science. On the basis of this evaluation of their importance, Yates suggested the word "Rosicrucian" should enter the vocabulary of serious historians to describe the kind of activist, reforming mentality that paved the way for modern science.[1]

[1] Frances A. Yates, "The Hermetic Tradition in Renaissance Science," in Charles S. Singleton, ed., *Art, Science, and History in the Renaissance* (Baltimore: Johns Hopkins Press, 1968), p. 263.

J.E. Force and R.H. Popkin (eds.), Newton and Religion, 17–43.
© 1999 *Kluwer Academic Publishers.*

Yates' notion of a Rosicrucian Enlightenment has received serious criticism on the grounds that her evidence was scanty, hypothetical, and her generalizations unwarranted.[2] The idea that good science could come out of bad occultism clearly affronted some scholars.[3] The problem with this position, however, is that it fails to explain why so many thinkers in the early modern period clearly did belong in both the occult and the scientific camp, a point made repeatedly by many historians in recent years.[4] Mounting evidence has conclusively shown that the ideas of the major thinkers in the scientific pantheon were influenced by occult/ and especially alchemical theories. Consequently, it has become less and less tenable to separate good science from bad occultism. This was the basic conclusion reached by the panelists who assembled at the Folger library in 1982 to examine the Yates thesis. The majority of contributors concluded that while Yates' claim for Hermeticism as the decisive force was exaggerated, her basic insight into the manifold connections between occultism and science in the early modern period is beyond dispute. The most trenchant deconstruction of Hermeticism in this volume appears in Brian Copenhaver's essay, "Hermes Trismegistus, Proclus, and the Question of a Philosophy of Magic in the Renaissance." While not disputing the impact that magic had in fostering the notions of manipulation and control so central to the development of modern science, Copenhaver persuasively argues that the real source of magic theory was not Hermeticism but Neoplatonism. He therefore cautions those scholars who dismiss magic as

[2] Brian Vickers, "Frances Yates and the Writing of History," *Journal of Modern History* 51 (1979), pp. 287-316. For more measured criticism see Charles Webster, *The Great Instuaration: Science, Medicine and Reform, 1626-1660* (London: Duckworth, 1972), p. 576; J. E. McGuire, "Neoplatonism and Active Principles: Newton and the *Corpus Hermeticum*," in J.E. McGuire and Robert S. Westman, *Hermeticism and the Scientific Revolution* (Los Angeles and Berkeley: University of California Press, 1977); Paolo Rossi, "Hermeticism, Rationality and the Scientific Revolution," in M. L. Righini Bonelli and William R. Shea, eds., *Reason, Experiment, and Mysticism in the Scientific Revolution* (New York: Science History Publications, 1975), pp. 247-73. For a positive assessment of the Yates thesis, see D. P. Walker, *The Ancient Theology* (London: Duckworth, 1972), pp. 254-63; P. M. Rattansi, "Reason in Sixteenth and Seventeenth-Century Natural Philosophy," in Mikulaš Teich and Robert Young, eds., *Changing Perspectives in the History of Science: Essays in Honour of Joseph Needham* (London: Heinemann, 1973), pp. 160-5.

[3] Mary B. Hesse, "Hermeticism and Historiography: An Apology for the Internal History of Science," *Minnesota Studies in the Philosophy of Science* 5 (1971); Brian Vickers, "Analogy and Identity," in Brian Vickers, ed., *Occult & Scientific Mentalities in the Renaissance* (Cambridge: Cambridge University Press, 1984; *idem*, "On the Function of Analogy in the Occult," in Ingrid Merkel and Allen G. Debus, eds., *Hermeticism and the Renaissance: Intellectual History and the Occult in Early Modern Europe* (Washington: Folger Books, 1988), pp. 265-92. For a critique of Vickers' attempt to separate magic and science, see Patrick Curry, "Revision of Science and Magic," *History of Science* 23 (1985), pp. 299-325.

[4] For a good overview of this issue (with appropriate bibliography) see G. MacDonald Ross, "Occultism and Philosophy in the Seventeenth Century," in A. J. Holland, ed., *Philosophy, its History and Historiography* (Dordrecht: D. Reidel, 1985), pp. 95-115; Simon Schaffer, "Occultism and Reason," *Ibid.*, pp. 117-43; and Ross' "Reply to Simon Schaffer," *Ibid.*, pp. 145-7.

unimportant in the formation of modern science because they reject the importance of Hermeticism and suggest that they take into account all sources of Renaissance magic—Neoplatonism, Plato, Aristotle, Galen, Aquinas, and Albertus Magnus.[5]

While the participants in the Folger seminar accepted the Yates thesis along general lines, other scholars have argued that the optimistic view of human nature attributed by Yates to Hermeticism was derived from earlier and different sources. Charles Trinhaus, for example, thinks that patristic theology was a far more important and earlier source for this new optimism.[6] Following Garin, he believes the hermetic vision of deified man derived from the medieval Latin translation of the *Asclepius* was firmly in place in patristic and pre-scholastic medieval thought, and he seconds Garin's contention that the foundation for an optimistic assessment of human nature lay in two key Christian concepts: the Genesis image of man as created in the image of God and the doctrine of the Incarnation.[7] Trinkaus sees Augustine as a particularly important source in fostering the optimistic view of human potential characteristic of humanists and Renaissance Neoplatonists.[8]

To my mind there are several problems with Trinkhaus' analysis. In order for Augustine to have been an important source for the optimistic assessment of human nature characteristic of the early humanists, they would have had to have read his works very selectively, virtually ignoring the extreme pessimism of his later writings.[9] The early humanists could, of course, have been selective in their reading of Augustine. Nevertheless, enormous feats of mental acrobatics would be required to ignore the profound pessimism characteristic of Augustinian anthropology. William Bouswma brings out this pessimism in the contrast he draws between the Augustinian and Stoic view of man. While Stoics believed men contain a divine spark or seed, identified with reason, which allows them to comprehend and adapt themselves to the divine order, Augustine and later Augstinians stressed human depravity and insisted on the complete inability

[5] Copenhaver's paper confirmed the conclusions reached by J. E. McGuire in "Neoplatonism and Active Principles." A. J. Festugière also describes the eclectic nature of early Hermeticism in *La Rèvèlation d'Hermes Trismègiste*, 4 vols. (Paris: J. Gabalda, 1944-54). Charles Schmitt described Hermeticism as a not very distinctive subspecies of Neoplatonism in "Reappraisals in Renaissance Science," *History of Science* 16 (1978), pp. 200-14.

[6] Charles Trinkhaus, *In our Image and Likeness: Humanity and Divinity in Italian Humanist Thought*, 2 vols. (Chicago: University of Chicago Press, 1970), 2:476-508, 519-24.

[7] *Ibid.*, 1:187-8.

[8] *Ibid.*, 2:463-4, 501.

[9] Peter Brown, *Augustine of Hippo* (New York: Dorset Press, 1986; first published 1967), p. 395. On Augustine's pessimism, see also Aviad M. Kleinbrg, "De Agone christiano: The Preacher and his Audience," *Journal of Theological Studies*, NS 38 (April, 1987), pp. 16-33.

of man to rely on his own powers for salvation.[10] Bouwsma does, however, agree with Trinkhaus in finding a positive element in Augustine's rejection of determinism. Man may be depraved but he is not bound to cosmic laws as in Stoicism; a measure of freedom is therefore provided for human action and creativity.[11]

But even if one agrees that Augustine allowed a measure of freedom to man, this freedom seems to me utterly different from the kind of freedom envisioned by the Florentine Platonists. For in Augustine, as in all orthodox patristic and Christian writings, there is the inescapable fact that man must rely on grace for salvation. It is precisely this idea that is virtually eliminated in the Florentine Platonists and the numerous sixteenth and seventeenth-century proponents of occultism. The opening paragraph of Pico's *Oration on the Dignity of Man* describing man's ability to become whatever he wishes, even to the point of divinity, is too well known to need quoting. Pico places absolutely no limitations on man. He has only to "aspire" and to "will," both key words in the *Oration*, and he will find himself "inferior to... [the angels] in nothing. "[F]or if we will... ," says Pico, "we can."

Pico's emphasis on man's ability to control his own destiny without any external help appears in Ficino as well. Ficino describes man's striving to "become God" as entirely "natural," in the same way that flight is to birds; man requires no external, divine assistance to become perfect in this world.[12] Like Pico, Ficino quotes the *Asclepius* passage describing man's ability to become anything he wishes, and he quotes Hermes Trismegistus to the effect that man is "a great miracle." [13] One might argue that even this exalted vision of man's potential remains essentially Christian inasmuch as Ficino is careful to say that it is God who invests the power of perfectibility in human beings. But the very fact that this power is innate obviates the need for either Christ or the Catholic Church, and in taking such a radical position Ficino is anything but orthodox. The sense of sin that is so remarkable in Augustine seems to be entirely absent in Ficino, and in this he follows the Neoplatonists far more than any Christian. As R. T. Wallis has said of Plotinus, "[he] lacks any sense of sin or of the need for redemption. . . . Our true self is eternally saved and all that is required is to wake up to this fact, a process requiring self-discipline, but perfectly within the soul's own

[10] William Bouwsma, "The Two Faces of Humanism: Stoicism and Augustinianism in Renaissance Thought," in William Bouwsma, *A Usable Past: Essays in European Cultural History* (Berkeley: University of California Press, 1990), pp. 25-6.

[11] *Ibid.*, p. 27.

[12] *Theologia Platonica*, 2:247. Cited in Trinkhaus, *In Our Image and Likeness*, 2:487; see also 2: 491.

[13] *Theologia Platonica*, 2: 256-8. Cited in *Ibid.*, 2: 489-90.

power."[14] The idea that the individual has the ability to take complete charge of his own salvation as well as the world's, which appears to me to be a prominent feature of both Pico's and Ficino's philosophy, is, I would argue, problematic in terms of the Christian concepts of original sin and the atonement. As Hans Blumenberg has pointed out, the future envisioned by the modern idea of progress is a product of immanent development, not of some transcendental, external divine intervention (*deus ex machina*). Such an idea of progress could not, therefore, have developed from orthodox Christianity but only in opposition to it. [15]

What led Frances Yates to stress the importance of Hermeticism in the development of modern science was her work on Giordano Bruno. But while she stressed the mystical and magical side of Bruno's thought, scholars are now emphasizing the progressive scientific aspects. Hilary Gatti describes this new approach:

> Some of the most recent work on Bruno shows a marked reaction against such a primary emphasis on his mysticism and magic. His Copernicanism has been re-examined and revalued. His interest in, and knowledge of the scientific enquiries of his times, such as the study of comets, has been underlined. In general, it is precisely where Bruno breaks away from the Neoplatonic magi to establish a new cosmic vision that he is attracting the attention of many scholars today.[16]

Gatti does not deny the hermetic roots of Bruno's extraordinary vision of human potential, but she, along with other scholars, considers them (and Neoplatonic sources in general) less important in shaping Bruno's thought than pre-Socratic, Pythagorean, Stoic, and Epicurean sources.[17] In her opinion the following quotation, which on the face of it seems so clearly "Hermetic," more directly echoes Lucretian cosmology:

> Now behold the man who has fended the air, pierced the sky, journeyed amongst the stars, travelled beyond the margins of the world, dissolved the imaginary barriers of the first, eight, ninth, tenth, and other spheres, if it had been possible to add others through the reports of vain mathematicians, and the blind visions of false philosophers. Thus using every sense and reason, with the key of passionate enquiry, he has opened those cloisters of truth which it is in our power to open, laid bare veiled and hidden nature: he has

[14] R. T. Wallis, *Neoplatonism* (London: Duckworth, 1972), p. 90.

[15] Hans Blumenberg, *The Legitimacy of the Modern Age*, trans. Robert M. Wallace (Cambridge, Mass.: The MIT Press, 1983; first published in German in 1966).

[16] Hilary Gatti, *The Renaissance Drama of Knowledge: Giordano Bruno in England* (London & New York: Routledge, 1989), pp. 49-50.

[17] Gatti, private communication. See also, Michele Ciliberto, *Giordano Bruno* (Rome: Laterza, 1990); Giovanni Aquilecchia, *Giordano Bruno* (Rome: Instituto della enciclopedia italiana, 1971).

> given eyes to the moles, illuminated the blind who are unable to fix their eyes and focus her image in the many mirrors which on all sides reflect it to him.[18]

What Gatti and more recent scholars have shown is that this vision was thoroughly compatible with science, something that even Frances Yates was unwilling to claim. As Yates said at the end of her book on Bruno, "The procedures with which the Magus attempted to operate have nothing to do with genuine science. The question is, did they stimulate the will towards genuine science and its operation?"[19] Gatti concludes that it was not Bruno's Hermeticism so much as his indebtedness to rediscovered Greek and Roman scientific texts that led him to demand complete freedom of thought as a necessary condition for scientific investigation. He was executed for this demand, not for some hypothetical attempt to institute a magical-Hermetic philosophy. Bruno's unorthodox religious views—he had sympathies with Arian doctrine, denied the immaculate conception and Virgin birth, rejected the worship of saints, and advised his fellow monks to study the early Church Fathers—went hand and hand with his unorthodox cosmology.

Another aspect of the Yates thesis that has been questioned is her assertion that Hermeticism was responsible for orienting men towards an active rather than a contemplative life and for stressing the power of the will in shaping human affairs. According to Trinkhaus the idea that contemplation represented the highest achievement of human beings was only true of one medieval tradition, the Thomist, and not even for all of that. The early humanists had at their disposal plenty of patristic and medieval sources emphasizing the importance of the will and the value of human activity.[20] While there is some truth in this, the positive evaluation of the active life one finds in Ficino and Pico and their numerous followers seems to me to be qualitatively different from anything in patristic, medieval, or early humanist writings. Ficino offers a vision of man as a gnostic savior, something that is inconceivable in Augustinian theology:

> The human mind vindicates to itself a right to divinity not only in forming and shaping matter through the method of arts, as we have said, but also in transmuting the species of things by command, which work is indeed called a miracle, not because it is beyond the nature of our soul, when it is made an instrument of God, but because, since it is something great and rarely done, it generates admiration. Here we marvel that the souls of men dedicated to God rule the elements, call upon the winds, force the clouds to rain, chase away fogs, cure the diseases of human bodies and the rest. These plainly were done in certain ages among various peoples, as poets sing, historians narrate, and those who are the most excellent of philosophers, especially the Platonists, do

[18] Quoted in Gatti, *Ibid.*, p. 32.

[19] Frances Yates, *Giordano Bruno and the Hermetic Tradition* (London: Routledge, 1964), p. 449.

[20] Trinkaus, *In Our Image and Likeness*, 2:499-500.

> not deny, the ancient theologians testify, above all Hermes and Orpheus, and the later theologians also prove by word and deed.[21]

One of the reasons man is so potentially powerful is that he is a master of the mechanical arts. Ficino's high estimation of the mechanical arts comes out clearly in several passages. He praises man's accomplishments in the mechanical arts as one of his defining characteristics, and he mentions Daedalus and Icarus in a positive way, which as far as I know was not done in the Middle Ages.[22]

Although Lynn White, Jean Gimpel, Lewis Mumford, William Eamon, Elspeth Whitnery, and others argue that the Middle Ages provided the foundation for the modern work ethic and faith in technological progress,[23] the contemplative life was still valued more highly than the active life, and the mechanical arts were tainted by their association with manual labor and concern with worldly ends. The Aristotelian core of Scholasticism accentuated the division between the liberal and the mechanical arts by valuing mental work more than manual.[24] During the Middle Ages a distinction was drawn between wondrous nature and her pale imitator, art, a distinction which further contributed to a negative attitude towards manual labor. The pejorative view of the mechanical arts was enshrined in the etymology deriving the term from the Greek word for adultery (*moicheia*) on the grounds that the mechanical arts trick and deceive. This etymology was widespread in the Latin West.[25] Given such evidence, one is forced to conclude, I believe, that Ficino's exalted view of man's ability to use his inborn talents, his knowledge, and his skills in the mechanical arts to shape and change his environment marked a radical revision of medieval attitudes.

Such a conclusion is supported by the condemnation of worldly wisdom and especially of curiosity that runs like a leitmotif through the medieval

[21] *Theologia Platonica*, 2:229. Cited in Trinkaus, *Ibid.*, p. 486.

[22] *Theologia Platonica*, 2:223-5. Cited in Trinkaus, *Ibid.*, pp. 482, 483-4.

[23] Lynn White, "Cultural Climates and Technological Advance in the Middle Ages," in *Medieval Technology and Social Change* (Oxford: University of Oxford Press, 1962), pp. 217-53; Jean Gimpel, *The Medieval Machine: The Industrial Revolution of the Middle Ages* (New York: Holt, Rinehart, and Winston 1976); Lewis Mumford, *Technics and Civilization* (New York: Harcourt, Brace & World, Inc., 1963); William Eamon, "Technology as Magic in the late Middle Ages and the Renaissance," *Janus* 70 (1983), pp. 171-212; *idem*, *Science and the Secrets of Nature: Books of Secrets in Medieval and Early Modern Culture* (Princeton: Princeton University Press, 1994); Elspeth Whitnery, *Paradise Restored: The Mechanical Arts from Antiquity through the Thirteenth Century*, Transactions of the American Philosophical Society, 80 (Philadelphia: American Philosophical Society, 1990).

[24] George Ovitt, Jr., *The Restoration of Perfection: Labor and Technology in Medieval Culture* (New Brunswick, NJ: Rutgers University Press, 1987).

[25] William Newman, "Technology and Alchemical Debate in the Late Middle Ages," *Isis* 80 (1989), pp. 423-45.

period.[26] Ginzburg emphasizes the effect that Jerome's mistranslation of Romans 11.20 had in turning curiosity into a sin. Jerome translated Paul's injunction "be not high-minded" as "noli altum sapere," which was consistently interpreted to mean "do not seek to know high things."[27] No one was more critical of human curiosity than Augustine, who defined it as one of the three forms of the vice "concupiscence": lust of the flesh, lust of the eyes, and worldly ambition. Curiosity was "lust of the eyes,"[28] and Augustine links it directly with both the sin of pride and the Fall. Furthermore, curiosity was the source of heresy and the black arts. The association of curiosity with sin, pride, heresy, and magic became commonplace in medieval thought. The idea that curiosity could all too easily lead to magic was enshrined in the Faust Legend. Steiner denies that the Faust legend was a product of the Reformation and particularly of Lutheran thought. He provides ample evidence to show that it fits in splendidly with the medieval Church's condemnation of curiosity.

Faust was only one of many figures damned for their curiosity in the medieval period.[29] Icarus, Daedalus, Prometheus, and Proteus were others condemned either for aiming to high, or, as in the case of Proteus, for adapting too easily to changing circumstances. The Hermetic image of man as marvelously protean, the image that exhilarated Pico and Ficino,[30] had little appeal for those who feared and distrusted change, viewing the world ideally as a static hierarchy where individuals assumed, and were expected to keep, the occupations and roles into which they were born. The positive view of Icarus as the prototype of the scientist who dares to know comes out in Bruno, who, as Gatti has pointed out, shared the Neapolitan poet Tansillo's admiration for Icarus' daring, even though it led to death.[31]

[26] Eugene F. Rice, Jr., *The Renaissance Idea of Wisdom* (Cambridge: Harvard University Press, 1958), Chap. 1.

[27] Carlo Ginzburg, "High and Low: The Theme of Forbidden Knowledge in the Sixteenth and Seventeenth Centuries," *Past and Present* 73 (1976), pp. 28-41.

[28] *Confessions*, Bk. 5.3.3. For a discussion of patristic and medieval attitudes towards curiosity, see Arpad Steiner, "The Faust Legend and the Christian Tradition," *Proceedings of the Modern Language Association* 54 (1939), pp. 391-404; William Eamon, *Science and the Secrets of Nature: Books of Secrets in Medieval and Early Modern Culture* (Princeton: Princeton University Press, 1994), pp. 59ff; E. Peters, "'Libertas Inquirendi' and the Vitium Curiositatis in Medieval Thought," in George Makdishi, Dominique Sourdel, and Janine Sourdel-Thomine, eds., *La notion de liberté au Moyen Age: Islam, Bysance, Occident* (Paris, 1985), pp. 89-98.

[29] For a provocative reading of Marlowe's Dr. Faustus as illustrating the conflict between Renaissance optimism and Christian pessimism, see Hilary Gatti, *The Renaissance Drama of Knowledge*, Chap. 4.

[30] Pico compares man to a chameleon and to Proteus in his Oration. He describes himself as Protean. See A. Bartlett Giammatti, "Proteus Unbound: Some Versions of the Sea God in the Renaissance," in Peter Demetz, Thomas Greene, and Lowry Nelson, Jr., eds., *The Disciplines of Criticism: Essays in Literary Theory, Interpretation, and History* (New Haven: Yale University Press, 1968), pp. 437-75.

[31] Gatti points out that Bruno commented on Tansillo's poem (*The Renaissance Drama of Knowledge*, p. 87).

Francis Bacon also rejected the traditional interpretation of the Icarus myth in a passage that echoes Bruno:

> Icarus chose the better of the two; for all defects are justly esteemed more depraved than excesses. There is some magnanimity in excess, that, like a bird, claims kindred with heavens: but defect is a reptile, that basely crawls upon the earth.[32]

Bruno rejects two other *loci classici* for the dangers of forbidden knowledge, the Genesis story of the forbidden fruit and the myth of Prometheus. Those who willingly accept limits to intellectual inquiry are "asses" in his opinion because

> they remain unable to stretch out their hands ... like Adam to pluck the forbidden fruit from the tree of knowledge, and in consequence they remain without the fruits of the tree of life, or to hold out their hands like Prometheus (which is a metaphor of the same kind) to take the fire of Jove to light the lamp of rational ability.[33]

Given the accumulated evidence it would appear that the Middle Ages was not fertile ground for the optimistic evaluation of human potential and power so forcefully presented in the writings of the Renaissance Neoplatonists and their followers.

There was, however, one medieval source for such ideas, and that was alchemy. Nicholas Clulee has criticized Yates for underestimating the influence that medieval alchemy and particularly Roger Bacon had in fostering an optimistic view of human potential. Clulee rejects the very term "Hermetic magic," as essentially meaningless.[34] The idea that medieval alchemical writings are the place to look for a validation of technology and human potential has also been suggested by William Newman.[35]

Stephen Clucas also argues that a positive attitude towards technology and the mechanical arts—what he describes as the "will to operate"—can be found in the thirteenth century and earlier in the *ars notoria*. Clucas suggests that scholars chart the history of "the will to perform" and he looks to Augustine as an important proponent of the proper use of human agency: "This belief in the legitimacy of pious or inspired agency, which is close to the Augustine's doctrine of "legitimate use," confirms that the notion of operativity was not absent from medieval and Renaissance culture, prior to Yates' magico-scientific "operative" revolution—albeit it was heavily

[32] Quoted in *Ibid.*, p.88.

[33] *Ibid.*, pp. 87-8.

[34] Nicholas H.Clulee, *John Dee's Natural Philosophy: Between Science and Religion* (London: Routledge, 1988).

[35] William Newman, "Technology and the Alchemical Debate in the Late Middle Ages," *Isis* 80 (1989), pp. 424-5, 429.

circumscribed."[36] To my mind this is precisely the point. Not only was the medieval notion of "operativity" heavily circumscribed but it was also qualitatively different from the notion that developed from the time of the Renaissance Platonists onwards. And in fairness to Frances Yates, it should be remembered that both alchemy and the *ars notoria* were identified with Hermes Trismegistus. Alchemy was known as the "Hermetic Art," and Hermes Trismegistus was its reputed founder. Hermes is also mentioned with respect in the notorious *Picatrix*. Furthermore, although the hermetic view of man as a potentially divine gnostic savior was an aspect of alchemy from its inception, this idea was transformed and strengthened from the fifteenth to the seventeenth centuries.[37]

Yates' contention that Hermeticism provided the outlook and optimism conducive to the development of modern science is only one of several possible answers that have been given to the question of what caused the scientific revolution. The explanation most prevalent before hers was Protestantism. The thesis that Weber proposed linking Protestantism to the emergence of capitalism was adopted and adapted by many historians who concluded that science, like capitalism, was the product of a Protestant ethic and mentality. Merton's essay was the first in a long line of books and articles that argued for a Protestant and particularly a "Puritan" origin of modern science.[38] But while it is generally admitted that science did not develop to the same extent in Catholic as in Protestant countries after the middle of the seventeenth century, it seems less and less convincing to argue that the reasons for this lie in specific theological doctrines. After all, Copernicus, Kepler, Galileo, Mersenne, Pascal, and Descartes worked in a Catholic environment, and the discoveries made largely by Catholic and continental mathematicians paved the way for the extraordinarily accomplishments of Newton.

The difficulty of asserting a direct link between theology and science had the effect of turning historians a hundred and eighty degrees in the opposite direction, to the point that some claimed that science only emerged when religion issues were ignored. In the view of these historians, religious

[36] Stephen Clucas, "Wondrous Force and Operation: Renaissance Praxiology re-assessed," Paper delivered at the conference "Rethinking the Middle Ages and Renaissance," Center for Medieval and Renaissance Studies, Arizona State University, Tempe, Arizona, February 1995.

[37] Mebane makes this point in his *Renaissance Magic and the Return of the Golden Age: The Occult Tradition and Marlowe, Jonson, and Shakespeare* (Lincoln: University of Nebraska Press, 1989), pp. 210-11, note 27. Bernard Gorceix also argues that only during the sixteenth and seventeenth centuries, when alchemy moved in many cases out of the laboratory altogether, did it provide the foundation for the belief in progress and positive attitude towards change that lie at the heart of the modern scientific world view. See *La Bible des Rose-Croix. Traduction et Commentaire des trois premiers écrits rosicruciens...* (Paris: Presses Universitaires, 1970).

[38] I. Bernard Cohen, ed., *Puritanism and the Rise of Modern Science: The Merton Thesis* (New Brunswick: Rutgers University Press, 1990).

moderates and religious moderation laid the foundation for both the scientific revolution and the Enlightenment.[39] Trevor-Roper stated this position eloquently when he proposed that one look to Erasmus as the real progenitor of the enlightened scientific spirit with its combination of skepticism and a belief in the legitimacy of rational, critical inquiry.[40]

There is much to be said for this view. According to Thomas Sprat, the historian of the Royal Society, one of the stipulations of the Society was that religious issues should not be discussed by the members. But while moderation and an aversion for religious controversy may indeed have characterized the proponents of the new philosophy after it was in place, neither contributed to the new attitudes toward man, nature, and progress that provided the impetus for the scientific revolution itself. Erasmus was no Newton, nor even a Francis Bacon. And while he did advocate those aspects of mind essential to critical and methodical investigations of all kinds, there is no hint in his writing that man can and should transform his world as well as himself. It is virtually impossible to see Erasmus as the source for the kind of prophetic and millennial rhetoric one constantly meets in such thinkers as Bacon, Comenius, Hartlib, and the host of pansophic philosophers who did so much to popularize the view that science was the answer to human pain, discomfort, and distress. Even Boyle exhibits hints of enthusiasm when he argues that scientists are the true priests. In short, it is impossible to find in Erasmus' measured view of human potential the spirit of confidence that marks the work of seventeenth-century scientists. The belief that reason combined with a method provided a virtually infallible formula for scientific discovery and scientific progress became the accepted wisdom of the Enlightenment and subsequent centuries. And here I think we come back to Yates' thesis, for it is precisely this positive assessment of human potential that she stresses as so crucially important in providing the *sine qua non* for modern science. But perhaps instead of locating the source of this confidence in Hermeticism, or in the magical and occult traditions of the Renaissance, one might go a step further and find it in the revival of Pelagianism during the Renaissance. Friedrich Herr has suggested this in his book *Die Dritte Kraft*. Here he studies the attempts made by Humanists, frustrated by the latent Manichaeism of the Church and laity, to create a new Pelagian ethic.[41]

[39] Barbara Shapiro, "Latitudinarianism and Science in Seventeenth-century England," *Past and Present* 40 (1968). Lotte Mulligan, "Civil War Politics, Religion and the Royal Society," *Past and Present* 59 (1973).

[40] H. Trevor-Roper, *The Crisis of the Seventeenth Century: Religion, the Reformation and Social Change* (New York: Harper & Row, 1968), pp. 232-3.

[41] Friedrich Heer, *Die Dritte Kraft: Der europäische Humanismus zwischen den Fronten des konfessionellen Zeitalters* (Frankfurt am Main: S. Fischer Verlag, 1960). Trinkhaus refers to the

The revival of Pelagianism in the Renaissance is surely of great importance in the shaping of the later scientific mentality. But I would argue that the reason why Pelagianism came back into favor during the Renaissance was twofold: first, because of the rediscovery of classical text embodying Greek and Roman scientific traditions, and, second, because of the revival of various forms of Gnosticism.[42] In his now classic book *The Renaissance Idea of Wisdom*, Eugene Rice traces the way in which the ideal of wisdom became increasingly secular under the impact of the rediscovery and absorption of classical texts during the Renaissance and later centuries. These texts precipitated a confrontation between philosophy and theology, science and religion, and encouraged a reevaluation of the goal and potential of human reason. A more positive view of both nature and human nature emerged from this confrontation. Among these rediscovered texts were, of course, the *Corpus Hermeticum* translated by Ficino. Although Plotinus accused Gnostics of despising the sensible world (*Enneads*, II, 9), this was not true of all Gnostics. In recent years, especially as a result of the gnostic texts found at Nag Hammadi, certain Gnostics have been shown to have had a more optimistic world view, and this is the view that survived in the Hermetic texts translated by Ficino, in alchemy, in the Kabbalah, and, in Neoplatonism. In these sources one can find a "monadic gnosis,"[43] in which there was no chasm between man and God or need for a mediator between the two. Thus I would agree wholeheartedly with Trevor-Roper's suggestion that the origins of the Enlightenment are to be found in heresy.[44] The

---

"rhetorical Pelagianism" of the Humanists (*In Our Image and Likeness*, 2:633, 649) and suggests that hortatory writing of any sort presupposes some sort of Pelagianism.

42 In a recent article Joseph Dan argues against the use of the terms "Gnosticism" and "gnostic" on the grounds that they are too imprecisely used to be meaningful. See "Jewish Gnosticism?" *Jewish Studies Quarterly* 2 (1995), pp. 309-28. Michael Allen Williams makes the same point in *Rethinking "Gnosticism": An Argument for Dismantling a Dubious Category* (Princeton, NJ: Princeton University Press, 1996). It is therefore with some trepidation that I use the term in this article. I do so simply because I can think of no better term to describe the radical (and, from an orthodox Christian perspective, heretical) idea that individuals can gain the necessary knowledge for salvation through their own efforts, without the intervention of the Church or Jesus. I agree with Elaine Pagels that this idea, characteristic of gnostic Christians, survived as a suppressed current, re-emerging periodically, especially among radical sectarians in the seventeenth century. See Elaine Pagels, *The Gnostic Gospels* (New York: Random House, 1979), p. 150.

43 This term was used by Clement of Alexandria. See Pagels, *The Gnostic Gospels*, p. 31.

44 H. Trevor-Roper, *The Crisis of the Seventeenth Century*, pp. 232-3: "Thus, when we look into the religious origins of the Enlightenment we do not discover them in any one Church or Sect. They are to be found in both Churches and in several sects. What is common to the men who express such ideas is that all of them are, in some sense, heretical. That is, they either belong to dissident groups within their Churches or are themselves regarded as unorthodox. The orthodox Churches—Catholic, Lutheran, Anglican, Calvinist—look askance at them. Moreover, the heretical tradition which they share is not only independent of the Reformation from which it is so often supposed to have spring. It precedes the Reformation, and the Reformation, though it may at first have liberated it, has soon become a repressive movement, positively fragmenting and obstructing it. The intellectual tradition of scepticism, mysticism,

rediscovery of classical scientific texts and the revival of gnostic forms of thinking—whether they appear in the newly translated Hermetic texts, in alchemy, in Neoplatonism, or in the Kabbalah—led to the resurgence of the Pelagian heresy with its belief in the ability of man to determine his own destiny as well as that of the world's.

The gnostic part of my conclusion is not unique; it had been anticipated by Eric Voegelin over forty years ago. "[T]he essence of modernity, " he wrote in *The New Science of Politics*, "[is] the growth of Gnosticism." Voegelin's analysis of Gnosticism is, however, very different from mine. Writing from the perspective of an Austrian who had watched European democratic institutions collapse in the face of Fascism and Communism, Voegelin quite understandably saw only one side of Gnosticism, the intolerant side stemming from the notion of the "godded man," who, in Voegelin's view, was more often a bigot than a saint.[45] But in stressing its totalitarian potential, Voegelin completely misses the other side of Gnosticism, the side represented by Pico della Mirandola when he proclaimed the ability of every man to realize his divine potential and the side represented by Bacon, Comenius, Hartlib, and the members of the numerous scientific and philosophical societies committed to the ideal of men working together for the common good. The forms of Gnosticism revived during and after the Renaissance (Hermeticism, alchemy, and the Kabbalah, especially the Lurianic Kabblah) were characterized by the democratic and egalitarian belief that individual men are basically good and capable of determining their own destiny. All of these gnostic philosophies were predicated on the possibility of universal salvation and the restoration of the world to its prelapsarian state.[46] This is a very different philosophy from the totalitarian view which stresses the corporate identity of the individual and his subordination to society and the state.

---

critical scholarship, lay reason, free will, which was united in Erasmus was broken up and driven underground by the ideological struggles of the sixteenth and seventeenth centuries. What had once been a general movement within a united society, acceptable in the courts of princes and in the cathedrals of the established Church, became, under the impact of successive ideological struggles, a number of separate heresies, labeled with sectarian names and equally condemned by all right-minded members of the several religious establishments. In times of ideological peace, Olympian minds like those of Grotius or de Thou or Bacon would seek to reunite these ideas, to restore to them their original respectability, to develop them further. Once again princes and higher clergy would listen to them. But the return of religious war gave power to the radicals of orthodoxy; to the Calvinists who condemned Grotius in Holland, to the friars who condemned Galileo of Italy. The movement which might have been the orthodoxy of a united society became again heresies of divided Churches. The eighteenth-century Enlightenment, when it came, would be a reunion of all the heretics, the reintegration of a movement which religious revolution had arrested and transformed, but could not destroy."

[45] Eric Voegelin, *The New Science of Politics: An Introduction* (Chicago: The University of Chicago Press, 1952), pp. 147. Vogelin's assessment of Gnosticism is similar to that of Hans Jonas, "Gnosticism and Modern Nihilism," *Social Research* 19 (1952).

[46] Allison P. Coudert, *Leibniz and the Kabbalah* (Dordrecht: Kluwer, 1995).

2. ISAAC NEWTON AND THE YATES THESIS

Having arrived at the conclusion that the Yates thesis has validity in a modified form *if* careful attention is paid to the rediscovery and absorption of classical scientific texts and *if* the designation "Rosicrucian" is replaced by that of "gnostic," the question arises of how to square this modified Yates thesis with the theology and science of Isaac Newton. For Newton, the greatest scientist of his age, was decidedly hostile to Gnostics of any sort, and this included Rosicrucians. He condemned the *Fama* and the *Confession*, which he owned and read in Vaughan's translation, as "impostures," a strongly pejorative word in his vocabulary.[47] And as Frank Manuel and Matt Goldish have pointed out, Newton's criticism of Gnosticism lies at the heart of his conflict with Leibniz.[48]

While Newton was hostile to all varieties of gnostic thought, he was especially antagonized by the Jewish Kabbalah, which he considered a major source for the Gnosticism that had distorted the clear, straightforward teaching of early Christianity. As he says, "If the theology of the Cabbalists be compared to that of the Gnostics it will appear that the Cabbalists were Jewish Gnosticks and the Gnosticks were Christian Cabbalists."[49] Since Matt Goldish has done a thorough job of explaining and illustrating Newton's hostility to the Kabbalah, I need only review a few points here.

Newton believed that primitive Christianity had been contaminated by three sources: The Kabbalah, Platonism, and Gnosticism. He looked for the source of all three heresies in Egypt and variously suggested that Kabbalists learned their odious doctrines from Plato or vice versa.[50] He objected to Platonists, Kabbalists, and Gnostics on similar grounds. All three were responsible for introducing metaphysics into theology, thereby distorting the simple teaching of primitive Christianity. Newton insists that "the Scriptures were given to teach men not metaphysics but morals."[51] By postulating emanationist theories of creation and by denying creation *ex nihilo*, Platonists, Gnostics and Kabbalists had the further unfortunate effect of

[47] Frank Manuel, *The Religion of Isaac Newton* (Oxford: Clarendon Press, 1974), pp. 45-6.

[48] *Ibid.*, p. 75. Matt Goldish, "Newton and the Kabbalah," in James E. Force and Richard H. Popkin, eds., *The Books of Nature and Scripture: Recent Essays on Natural Philosophy, Theology, and Biblical Criticism in the Netherlands of Spinoza's Time and the British Isle of Newton's Time* (Dordrecht: Kluwer, 1994), pp. 98-9 and passim.

[49] Yahuda MS 15.7, fol.127r, at the Jewish National and University Library, Jerusalem. Quoted by permission. Cited in Goldish, "Newton and the Kabbalah," p. 95. Agrippa had also seen the affinity of the Kabbalah and Gnosticism. See Moshe Idel, *Kabbalah: New Perspectives* (New Haven: Yale University Press, 1988), pp. 5-6. Jacques Basnage had as well in his *History of the Jews* (English edition, 1708, pp. 76-7).

[50] Yahuda MS 15.7, fol. 138r; MS Bodmer, Chap. 4, p. 4. Cited in Goldish, "Newton and the Kabbalah," pp. 93-4.

[51] Yahuda MS 15.7, fol. 190r. Cited in Manuel, *Religion*, p. 72.

undermining the supreme power of the one omnipotent creator. They consequently turned what had originally been a monotheistic religion into pagan polytheism.[52] Newton considered gnostic myths of emanation degrading to God because they described God in human and, even worse, sexual terms.[53] Furthermore, by hypostasizing divine mental processes, they created a wholly unscriptural family of consubstantial gods.[54] The infiltration of false gnostic doctrines into primitive Christianity had dire consequences not only for religion but for science as well. Newton believed that polytheism went had in hand with bad science, while monotheism was the source of good science dedicated to finding the simple, unifying cause of all natural phenomena. Just as there was only one all-powerful divine Creator and Father, there was only one cause for all the disparate phenomena in the created world. The law of simplicity was valid for both the interpretation of nature and scripture.[55]

Although Newton never mentions Spinoza (as far as I know), it seems probable that his hostility to the Kabbalah was colored by the charge made by several of his contemporaries that Spinoza was a Kabbalist. Newton owned Jacques Basnage de Beauval's *Antiquitez judäiques, or Remarques critiques sur la République des Hébreux* (in both the French edition of 1713 and the abridged English edition of 1708.) This work contained a substantial critique of Spinoza, associating his ideas with the Kabbalah. Basnage accuses Spinoza of accepting the kabbalistic belief that there was only one unique substance, God, and that all other created beings were merely modifications of this substance.[56] Leibniz was also aware of the charge that Spinoza was a Kabbalist. He mentions this in his *Theodicy*, which Newton also owned.[57] Henry More also drew a parallel between Spinoza's philosophy and the Kabbalah, which I will come back to because I think More's ideas influenced Newton. José Faur thinks that the General Scholium of the *Principia* was aimed squarely at Spinoza.[58] This may have been the case, but the scholium could just as well have been directed against Gnostics and Kabbalists in general on the grounds of their unwarranted anthropomorphizing of God. As Newton says:

[52] Yahuda MS 15.7, fol. 137r; Goldish, "Newton and the Kabbalah," p. 93.

[53] Yahuda, MS 15.7, fol. 109v; Manuel, *Religion*, p. 73.

[54] Yahuda MS 15.7, fol. 108v. Cited in Manuel, *Religion*, p. 96.

[55] Yahuda MS 1. 1, fol. 14r. Cited in Manuel, *Religion*, pp. 48-9.

[56] Richard H. Popkin, "Spinoza, "Neoplatonic Kabbalist?" in Lenn E. Goodman, ed., *Neoplatonism and Jewish Thought* (Albany: State University of New York Press, 1992), pp. 387-409. For the connection drawn by Basnage between Spinoza and the Kabbalah, see the French edition, pp. 149-50 and p. 297 in the English edition.

[57] *Theodicy*, ed. E. M. Huggard (New Haven: Yale University Press, 1952), p. 79.

[58] José Faur, "Newton, Maimonides, and Esoteric Knowledge," *Cross Currents* (Winter 1990), p. 531.

> he is all similar, all eye, all ear, all brain, all arm, all power to perceive, to understand, and to act; but in a manner not at all human, in a manner not at all corporeal, in a manner utterly unknown to us. As a blind man has no idea of colors, so have we no idea of the manner by which the all-wise God perceives and understands all things. He is utterly void of all body and bodily figure, and can therefore neither be seen nor heard nor touched; nor ought he to be worshipped under the representation of any corporeal thing. We have ideas of his attributes, but what the real substance of anything is we know not ... much less, then have we any idea of the substance of God.[59]

All this evidence illustrating Newton's hostility to the Kabbalah confirms Goldish's conclusion that both Hutin's depiction of Newton as a "Christian Kabbalist" and the idea that he drew on the Kabbalah for his idea of space and time are unfounded.[60]

As far as I can ascertain, Newton derived his knowledge of Gnostic heresies primarily from the Church Fathers. The catalogue of what is left of his library shows that some 28% of the books dealt with theological subjects, and a large part of these with Judaism and early Christianity.[61] He owned Irenaeus' *Adversus hereses* and uses many of the same arguments against the Gnostics presented in this work. For example, Irenaeus accused Gnostics of talking nonsense and proposing "fictitious doctrines." They are "sick" and "foolish" (I. 16. 3). Their exegesis of scripture is arbitrary and absurd, and their teachings are borrowed from philosophers, not from Scripture. They hypostasized mental processes and used human analogies excessively. Irenaeus believed that gnostic heresies were introduced into Christianity by Simon Magus (II. 23.2), a point made by Newton as well. Irenaeus, like Newton, stresses the uniqueness of God as the creator of the world and father of Jesus Christ (I. 10.1). And he rejects the distinction Gnostics make between "learned" theologians" and simple believers. Newton also owned the *Opera omnia* of Epiphanius, who stresses the connection between heresy and libertinism. Epiphanius was particularly critical of gnostic speculation about divine and cosmic genealogies. There is a measure of prurience in his description of these that may have helped to color Newton's own highly charged description of Simon Magus as a libertine. [62] Newton also possessed the works of other Church Fathers

[59] *Ibid.*, p. 531.

[60] Serge Hutin, "Note sur la création chez trois kabbalistes chrétiens anglais: Robert Fludd, Henry More et Isaac Newton," *Kabbalistes Chrétiens* (Paris, 1979), pp. 149-56. Brian Copenhaver, "Jewish Theologies of Space in the Scientific Revolution: Henry More, Joseph Raphson, Isaac Newton and their Predecessors," *Annals of Science* 37 (1980).

[61] John Harrison, *The Library of Isaac Newton* (Cambridge: Cambridge University Press, 1978), p. 59.

[62] Yahuda MS 15.3, fol. 53r. Cited in Manuel, *Religion*, p. 73. It would be interesting to know exactly what Newton's reaction was to Epiphanius. For while Newton would have agreed with his hostile

critical of the Gnostics, for example, Justin, Eusebius, Tertullian, and Theodoretus.

While Newton's knowledge of Gnostics and Gnosticism came primarily from the writings of the Church Fathers, he owed his knowledge of the Kabbalah to Knorr von Rosenroth's *Kabbala denudata* (1677, 1684), a multi-volume work of kabbalistic texts translated into Latin with commentaries. Richard Popkin says that Newton had been given a copy of this work by Knorr's collaborator Francis Mercury van Helmont.[63] According to Harrison some fifteen pages of this work were turned down by Newton, and there are "several other signs of dog-earing," which suggests that Newton read the work fairly carefully.[64] Knorr's object was to present the Latin-reading public with portions of the *Zohar*, the most important source of kabbalistic thought. But in order to make the *Zohar* more accessible and comprehensible, Knorr included other treatises deriving from the Lurianic Kabbalah of the sixteenth century and commentaries written by Jewish Kabbalists with strong Neoplatonic sympathies. He accompanied these with introductions, explanations, and treatises written by himself, his collaborator Francis Mercury van Helmont, and the Cambridge Platonist Henry More. More's contribution to the *Kabbala denudata* is extremely important, I would argue, because of the influence I believe it had on Newton's view of the Kabbalah.

Newton's relationship with the Cambridge Platonists in general and Henry More in particular has been a subject of debate. Burtt, Rattansi, McGuire, Dobbs, and Copenhaver have stressed their affinity, while Manuel remains quizzical.[65] While it is true that Newton rejected More and Cudworth's *Principium Hylarchicum*, or Spirit of Nature, on the ground that there could be no intermediary between God and matter because the two were utterly distinct, there are many places in Newton's published and unpublished writing where this distinction is blurred and where something

---

assessment of gnostic heretics, Epiphanius was an ardent supporter of Nicea, which, of course, Newton attacked as the source of the Trinitarianism he unequivocally, though clandestinely, rejected.

63 Popkin, "Spinoza," p. 404, n. 20.

64 Van Helmont knew both Henry More and Ezekiel Foxcroft, who was at Cambridge during the 1660s and part of the 1670s. Foxcroft was also friendly with More, and he shared Newton and van Helmont's alchemical interests, translating Andreae's *Chemical Wedding*. Popkin says that the pages of the *Kabbala denudata* remained uncut, which seems to contradict Harrison's description of the condition of the book.

65 E. A Burtt, *The Metaphysical Foundations of Modern Physical Science*; J. E. McQuire and P. M. Rattansi, "Newton and the Pipes of Pan"; J. E. McQuire, "Neoplatonism and Active Principles: Newton and the Corpus Hermeticum," in Robert S. Westman and J. E. McGuire, *Hermeticism and the Scientific Revolution*; B. J. T. Dobbs, *The Foundations of Newton's Alchemy: or, The Hunting of the Greene Lyone* (Cambridge: Cambridge University Press,1975), pp. 102ff.; Brian Copenhaver, "Jewish Theologies of Space in the Scientific Revolution"; Manuel, *Religion*, p. 72

akin to More's "Spirit of Nature" appears.[66] Newton also rejected innate ideas, a cornerstone of Cambridge Platonism. But once again, the situation is not completely clear-cut in as far as Newton seems to have believed that the mind possessed certain innate faculties and was therefore not a *tabula rasa*.[67] Given these differences, there were, however, fundamental areas of agreement between Newton and More, and, as I suggested earlier, there appear to me to be echoes of More's criticism of the Kabbalah in Newton's unpublished writings.

More's criticisms of the Kabbalah appeared in six treatises published in the *Kabbalah denudata*. They were therefore accessible to Newton. By the time More finished his last kabbalistic critique in 1675, he had read Spinoza and was profoundly disturbed by the pantheistic implications he discovered in both philosophies. In fact, More criticizes the Kabbalah and Spinoza in almost identical terms, on the grounds that neither made a clear distinction between matter and spirit. Consequently, both fell into the heresy of pantheism, anthropomorphism, and polytheism. If God had not created the world *ex nihilo,* out of a substance utterly separate and different from himself, then the world would somehow have emanated from God. Created entities would be nothing less than divine fragments, an appalling notion for More because it suggested than "God can become stones, dung, a little louse, a toad, a devil, etc."[68] The idea that God could somehow be divided was incomprehensible to More since in his view divisibility was *the* property that distinguished matter from spirit. In the preface to his *Immortality of the Soul* (1659), he coined the term "indiscerpibility" to characterize this essential property of matter. Koyré and Cohen have pointed out that Newton uses the word "indiscerpible" in his college notebook of 1661, where he specifically refers to More.[69]

There is a further way that More's criticism of the Kabbalah anticipated Newton's. Like Newton, More believed that the essence of Christian belief consisted of simple and straightforward moral precepts, not abstruse philosophical opinions. While it was perfectly legitimate for Christians to engage in philosophical speculation, such speculation was not to be

66 Although Newton disclaimed any connection between his concept of force and More's "Spirit of Nature, the two were similar in that both penetrated matter, were indivisible, and admitted degrees of intensity. See J. E. McGuire, "Force, Active Principles, and Newton's Invisible Realm," Ambix 15 (1968): 154-208.

67 G. A. J. Rogers, "Locke, Newton and the Cambridge Platonists on Innate Ideas," *Journal of the History of Ideas* 40 (1979), pp. 191-205.

68 More, "Fundamenta Philosophiae, sive Cabbala Aeto-Paedo-Melissaea," *Kabbala denudata,* 1, 2, p. 298.

69 Alexander Koyré & I. B. Cohen, "Newton and the Leibnitz-Clarke Correspondence," *Archives Internationales d'histoire des Sciences* 15 (1962), pp. 63-126. The authors also point out that the term "indiscerpible" was "silently corrected" to "indiscernible" in the French and English editions of the Leibniz-Clarke correspondence edited by Robinet and Alexander, but that this correction makes no sense.

considered an essential part of the faith. Those people who turned such speculation into various doctrines and formed sects around them elicited More's anger and contempt.[70] As we have seen, Newton insisted that "The Scriptures were given to teach men not metaphysics but morals." In his approach to scripture, he relied on literal interpretation first. As James Force, Richard Popkin, and José Faur have pointed out, Newton distinguished between doctrines suitable for the majority of Christians, which he called "Milk for Babes," and more abstruse doctrines, which he describes a "Strong Meats for Elders," phrases that come from Maimonides, whose work had a great influence on Newton.[71] These latter doctrines cannot be considered binding; they are speculative and conjectural and should never be matters for quarrels or excommunication.[72] More made a similar distinction between a literal and philosophic interpretation. Like Newton he believed that scripture had been written to accommodate the understanding of ordinary men, but this did not mean that there were not more noble truths, which those with the gift of interpretation might discover. This was the rationale behind More's own essay into the field of Kabbalah. He wrote his *Conjectura Cabbalistica* (1653) with the express purpose of exposing this double interpretation of scripture. More's goal was to show how the first three chapters of Genesis were compatible with "modern" science, particularly that of Descartes.[73] More wrote the *Conjectura Cabbalistica* before he had any real knowledge of authentic kabbalistic texts. The two Jewish authorities he cites most frequently, and presumably the only ones he actually read, are Philo and Maimonides, neither of whom were Kabbalists. Thus both More and Newton were indebted to Maimonides, and More's interest may have encouraged Newton's.

More and Newton's acceptance of an exoteric and esoteric interpretation of scripture was linked to another shared assumption, namely, the existence of a *Prisca Theologia* or *Prisca Sapientia*, which had been corrupted over the succeeding centuries. They both believed that through their respective studies of the biblical text and especially of the biblical prophecies in the book of *Daniel* and *Revelations*, they had rediscovered this *Prisca Theologia*. As we have seen, More equates this with Cartesianism in his

[70] More, *Annotations upon the two foregoing Treatises, Lux Orientalis. . . and the Discourse of Truth* (London, 1682), pp. 147-8.

[71] These phrases are also in Hebrews 5:12-14. Faur, "Newton, Maimonides, and Esoteric Knowledge." Popkin, "Newton's Biblical Theology and his Theological Physics," in P. B. Scheurer and G. Debrock, eds., *Newton's Scientific and Philosophical Legacy* (Dordrecht: Kluwer, 1988), pp. 81-97. Force, "Newton, The Lord God of Israel and Knowledge of Nature," *Jewish Christians and Christian Jews*, ed. Richard H. Popkin and G. M. Weiner (Dordrecht: Kluwer, 1994), pp. 131-58.

[72] Force, "Newton, the Lord God of Israel and Knowledge of Nature," in Weiner, *Jewish Christians and Christian Jews*, pp. 142-4 and passim.

[73] More, *Conjectura Cabbalistica* (London, 1653).

Conjectura Cabbalistica. Newton was convinced his scientific theories had been anticipated by the ancients. [74] What is especially interesting is that both More and Newton thought they had rediscovered Pythagorean philosophy, which they believed represented one of the earliest and most authentic traditions of the original *Prisca Theologia*. Newton found both the notion of atoms and gravity in Pythagoras. Dobbs suggests that Newton may also have found in Cicero's and Justin Martyr's description of Pythagorean doctrine confirmation of his conviction that God is an all-pervasive spirit infusing the universe.[75] David Gregory described the changes Newton intended to make in the *Principia* to make the affinity of his ideas with the ancients even clearer.[76]

I have taken time to illustrate the similarities between Newton and More's thought because I believe these can prove useful in bringing the discussion back to the Yates thesis and my contention that the thesis becomes more applicable if one substitutes Gnosticism for Hermeticism. As we have seen, both Newton and More take a very dim view of the Kabbalah and gnostic theories of emanation. On the face of it, this fact should preclude both men from taking any part in a so-called "Rosicrucian Enlightenment." But I am not so sure that this is the case.

Richard Popkin and Sarah Hutton have discussed how important prophecy was for Newton.[77] Popkin has gone as far as to describe prophecy and millenarianism as a "Third Force" in the scientific revolution. He has convincingly argued that one way of dealing with the skeptical crisis affecting so many intellectuals was to turn to the certainty generated by the correct interpretation of biblical prophecies, particularly that of the book of Daniel, according to which knowledge would increase as the millennium approached. [78]

Newton was not modest about his own endeavors at prophetic interpretation. Manuel, Popkin, Force, and Hutton have illustrated Newton's meticulous scrutiny of the book of *Daniel* and the *Apocalypse*. They provide a description of his working method, with his sustained effort to produce a

[74] J. E. McGuire and P. M. Rattansi, "Newton and the Pipes of Pan," *Notes and Records of the Royal Society* 21 (1966), p. 112.

[75] B. J. T. Dobbs, "Alchemy and Newton's Principle of Gravitation," in P. B. Scheurer and G. Debrock, eds., *Newton's Scientific and Philosophical Legacy* (Dordrecht: Kluwer, 1988), pp. 64ff.

[76] McQuire and Rattansi, "Newton and the Pipes of Pan," p. 110.

[77] Sarah Hutton, "More, Newton, and the Language of Biblical Prophecy," in James E. Force and Richard H. Popkin, eds., *The Books of Nature and Scripture: Recent Essays on Natural Philosophy, Theology, and Biblical Criticism in the Netherlands of Spinoza's Time and the British Isles of Newton's Time* (Dordrecht: Kluwer, 1994), pp. 39-53; Popkin, *The Third Force in Seventeenth-Century Thought* (Leiden: E. J. Brill, 1992).

[78] Popkin, "The Third Force in 17th Century Philosophy: Skepticism, Science and Biblical Prophecy," *Nouvelles de la Republique des Lettres* 1 (1983), p. 54.

dictionary of equivalents for the images, symbols, and allegories in both texts, together with his attempt to place these symbols and allegories in the framework of ancient history. Newton was convinced of the soundness of his method. He refused to accept other interpretations, as More found out to his amusement, or perhaps chagrin.[79]

Henry More was equally convinced of the correctness of his prophetic interpretations, which was why he and Newton reached a stand-off in their discussion of the seven trumpets and vials. The prophetic character of More's thought and the certainty that this engendered is revealed by the quotations he places on the title pages of various of his works. For example, on the title page of his collected theological works in Latin, he quotes *Daniel* 12:10: "many shall be purified, and made white, and tried; but the wicked shall do wickedly and none of the wicked shall understand; but the wise shall understand." He quotes another passage along the same lines, *Ecclesiasticus*, 1: 39: "It is not so with the man who applies himself, / And studies the Law of the Most High ./ He searches out the wisdom of all the ancients, / And busies himself with prophecies." This passage is only comprehensible when one realizes that it comes directly after a passage describing the impediments an active life place in the way of acquiring knowledge.[80] However skeptical he appears at times,[81] More's choice of quotations reveals the moral and prophetic dimension of his epistemology. The beginning of *Ecclesiasticus* supplied the certainty that lies at its heart: "All wisdom comes from the Lord, / and remains with him forever/ . . . But he supplied her liberally to those who loved him" (1:1,10). This certainty is not grounded on philosophical argument but on divine illumination. It seems clear to me that More did believe himself to be so illuminated. How can we otherwise understand those ecstatic passages in his writings, of which the following is a sample?

> For God doth not ride me as a Horse, and guide me I know not whither myself; but converseth with me as a *Friend*: and speaks to me in such a *Dialect* as I understand fully, and can make others understand, that have not make Shipwrack of the Faculties that God hath given them, by *Superstition* or *Sensuality* ... for God hath permitted to me all these things; and I have it

[79] Marjorie Hope Nicolson, ed., *The Conway Letters* (Oxford: Clarendon Press, 1992; rev. edition, ed. Sarah Hutton), p. 478. Manuel provides a suggestive psychological explanation for Newton's deep certainty that he was one of the select few able to comprehend fully the mystery of both the book of nature and scripture (*Religion*, pp. 19-20).

[80] "A scribe attains wisdom through the opportunities of leisure, / And the man who has little business to do can become wise. / How can the man who holds the plow become wise, / Who grives oxen, and guides them at their work, / And whose discourse is with the sons of bulls?/ He sets his mind on turning his furrows, / And his anxiety is about fodder for heifers. /It is so with every creaftsman and builder... ."

[81] R. H. Popkin, "The 'Incurable Scepticism' of Henry More, Blaise Pascal and Søren Kierkergaard," *The Third Force in Seventeenth-Century Thought*, pp. 203-21.

under the Broad Seal of Heaven. Who dare charge me? God doth acquit me. For He hath made me full lord of the Four elements; and hath constituted me Emperour of the World. I am in the *Fire of Choler*, and am not burn'd; in the *Water of Phlegm*, and am not drown'd; in the *Airy Sanguine*, and yet am not blown away with every blast of transient Pleasure, or van doctrines of Men; I descend also into the sad *Earthly Melancholy*, and yet am not buried from the sight of my God. I am, Philalethes, an *Inhabitant of Paradise* and *Heaven* upon Earth—I sport with the Beast of the Earth; the Lion licks my Hand like a spaniel; and the Serpent sleeps upon my Lap and stings me not. I play with the fowls of Heaven; and the Birds of the air sit Singing on my fist. All these things are true in a sober Sense. And the *Dispensation* I live in, is more *Happiness* above all Measure, than if thou couldst call down the moon so near thee by thy *Magick* Charms that thou mayst kiss her, as she is said to have kiss'd *Endymion*; or couldst stop the course of the Sun; or which is all one, with one Stamp of thy foot stay the Motion of the Earth.[82]

This rhapsodic passage bring to mind the Hermetic soliloquy delivered by *Nous* to Hermes Trismegistus and the Hermetic vision of man as the microcosm with the ability to understand the macrocosm because it lies within himself.[83]

The kind of prophetic certainty that More and Newton possessed also appears in Francis Bacon. Charles Whitney has discussed at length the "prophetic character" of Bacon's thought. He convincingly claims that by its very nature prophecy is revolutionary, that it tends to subvert existing institutions and ideologies by proposing a vision of a new beginning inexorably tied to moral and material rejuvenation. The "Great Instauration" that Bacon envisioned would eradicate the effects of the Fall and bring man back to an almost prelapsarian perfection. Such is Bacon's self-described hope: "Francis of Verulam. . . thought all trial should be made, whether that commerce between the mind of man and the nature of things. . . might by any means be restored to its perfect and original condition."[84] He believed that the loss of man's "state of innocency" and "his dominion over creation"..."can in this life be in some part repaired; the former by religion

[82] Richard Ward, *Life of Henry More* (London, 1711), pp. 89-90.

[83] On More's enthusiasm, see Craig A. Staudenbauer, "Galileo, Ficino and Henry More's *Psychathanasia*," *Journal of the History of Ideas* 29 (1968), pp. 565-78; Arlene Guinsberg, "Henry More, Thomas Vaughan and the late renaissance magical tradition," *Ambix* 27 (1980), pp. 36-57; Robert Crocker, "Mysticism and enthusiasm in Henry More," in Sarah Hutton, ed., *Henry More (1614-1687): Tercentenary Studies* (Dordrecht: Kluwer, 1990), pp. 1-17; Allison P. Coudert, "Henry More, the Kabbalah, and the Quakers," in Richard Kroll, Richard Ashcraft, and Perez Zagorin, eds., *Philosophy, Science, and Religion in England, 1640-1700* (Cambridge: Cambridge University Press, 1992), pp. 31-67.

[84] *Works*, 4: 8. Cited in Charles Whitney, *Francis Bacon and Modernity* (New Haven: Yale University Press, 1986), p.24.

and faith, the latter by arts and sciences."[85] Bacon stressed the religious and moral roots of knowledge and clearly felt that he was in some very profound sense favored by God. As he says, ". . . if any man think that he knoweth any thing, he knoweth nothing yet as he ought to know. . . . But if any man love God, the same is known of him."[86] Paolo Rossi stresses the humility of Bacon's approach to learning. He does this in a effort to draw a distinction between Bacon's valid scientific work and the grandiose claims and schemes of Renaissance magicians.[87] But, as Whitney points out, there is something suspect about the "titanic modesty" of these and other similar statements, especially when one remembers that Bacon's saw himself as *sui generis* and his work as unprecedented. In the following passage, for example, Bacon sets himself apart and suggests, just as More did, that his insights are the product of a kind of divine intoxication:

> I may say then of myself that which one said in jest ... "it cannot be that we should think alike, when one drinks water and the other drinks wine." ... I pledge mankind in a liquor strained from countless grapes, from grapes ripe and fully seasoned, collected in clusters, and gathered, and then squeezed in the press, and finally purified and clarified in the vat. And therefore it is no wonder if they and I do not think alike.[88]

Like More and Newton, Bacon is aware of the novelty of his ideas, yet he claims that they perfectly reflect ancient wisdom. They are, he says, "quite new, totally new in their kind: and yet they are copied from a very ancient model, even the world itself and the nature of things and of the mind."[89] According to Whitney's insightful analysis of statements such as these, Bacon's humility is like that of the prophets, whose overweening certainty arises from their conviction that they are the special recipients of divine knowledge. [90]

Bacon's goal, "To write an apocalypse or true vision of the footsteps of the Creator imprinted on his creatures,"[91] appears to me to be exactly what Newton attempted and accomplished in the *Principia* with his mathematical demonstration of gravitational forces. Newton's *Principia,* like Bacon's "Great Instauration," was as much a religious as a scientific work: the goal of both was to provide empirical evidence of divine Providence. By uncovering secondary causes, Bacon believed that he could offer infallible

[85] *Works*, 4:247-8. Cited in *Ibid.*, p. 25.

[86] *Works,* 3:264, 266. Cited in Whitney, *Francis Bacon*, p. 39.

[87] Paolo Rossi, "Hermeticism, Rationality, and the Scientific Revolution."

[88] *Works*, 4:109. Cited in Whitney, *Francis Bacon*, p. 83.

[89] *Works*, 4:11. Cited in *Ibid.*, pp. 82.

[90] *Ibid.*, p. 105.

[91] *Works*, 4:33. Cited in Whitney, *Francis Bacon*, p. 27.

proof for the existence of God.[92] Both David Kubrin and John Henry have suggested that Newton's goal was also to establish the existence of secondary causes on experimental and mathematical grounds, for secondary causes must themselves have a cause, and this could only be God. Gravity was just such a secondary cause. Henry's contention that Newton conceived of gravity as an agent or active principle that God had added either to matter or to the aether (depending on the different phases of Newton's scientific thinking) is convincing. According to Henry, Newton did not object to the idea that gravity might be a property of matter, only to the idea that it was an *essential* property of matter. Were it an essential property, then God would be superfluous; matter would be entirely self-activating, and the floodgates would be flung open to atheism.[93] Newton, of course, leveled precisely this same charge of atheism against both against Cartesianism and Gnosticism.

It appears to me that Newton was entirely aware of his accomplishments—how could he not have been given the adulation of his contemporaries? And it also appears to me that in this awareness, he exhibited the kind of optimism characteristic of the millenarians and pansophists of the Rosicrucian Enlightenment. Manuel denies this. One does not find in Newton's writing anything resembling John Wilkins's or Joseph Glanvill's enthusiasm for the proliferation of utilitarian inventions. Newton's scrutiny of nature was directed almost exclusively to the knowledge of God and not to the increase of sensate pleasure or comfort. Science was pursued for what it could teach men about God, not for easement or commondiousness.[94]

It is certainly true that Newton did not engage in the kind of optimistic reveries characteristic of many of his contemporaries, including Leibniz.[95] And there is also the added factor that Newton accepted the idea of a cyclical cosmos, anticipating that this world would be destroyed and other worlds created, hardly an optimistic notion on the face of it. But the idea that scientific discovery could contribute to the restoration of prelapsarian perfection, which was the hallmark of the Rosicrucian thinking described by Frances Yates, appears in Newton. Newton's vision is more moderate, or perhaps one should say, more moderately put, but the grand claim that a restoration is possible is clearly there, as is the implicit claim that Newtonian science will be the direct cause of this restoration:

[92] *Works*, 3:267-8. Cited in *Ibid.*, p. 29.

[93] John Henry, "'Pray Do Not Ascribe that Notion to Me': God and Newton's Gravity," in James E. Force and Richard H. Popkin, eds., *The Books of Nature and Scripture: Recent Essays on Natural Philosophy, Theology, and Biblical Criticism in the Netherlands of Spinoza's Time and the British Isles of Newton's Time* (Dordrecht: Kluwer, 1994), pp. 123-47.

[94] Manuel, *Religion*, p. 48.

[95] Coudert, *Leibniz and the Kabbalah*, pp. 153ff.

> If Natural Philosophy in all its parts, by pursuing this method, shall at length be perfected, the bounds of Moral Philosophy will be also enlarged. For so far as we can know by Natural Philosophy what is the first cause, what power He has over us, and what benefits we receive from Him, so far our duty towards Him, as well as that towards one another, will appear to us by the Light of Nature. And no doubt, if the worship of false gods had not blinded the heathen, their moral philosophy would have gone farther than to the four cardinal virtues; and instead of teaching the transmigration of souls, to worship the sun and moon and dead heroes, they would have taught us to worship our true Author and Benefactor, as their ancestors did under the government of Noah and his sons before they corrupted themselves.[96]

Newton phrases this in hypothetical terms. But we know that he believed he had rediscovered the *Priscia Theologia* or *Priscia Sapientia* by discovering the laws of gravity. Natural philosophy therefore had been perfected and time would do the same for moral philosophy. Once men began to worship God correctly, which they could now do because they could finally philosophize correctly, the days before Noah and his sons "corrupted themselves" would once again return.

3. Conclusion

As we have seen, Newton rejected Gnosticism and the Kabbalah—which he considered a form of Gnosticism—because both taught "metaphysical" doctrines at odds with the morality of primitive Christianity and true science. Yet, as we have also seen, Newton shared the gnostic vision of a future time of scientific and religious renovation expected by so many of his contemporaries and set forth in Francis Bacon's clarion call for the renovation of knowledge. There are clear indications that Newton felt he had a special gift for interpreting God's word in both the book of scripture and of nature, that, in fact, he considered himself one of God's anointed. It also seems clear that Newton was not immune to the millennial hopes of so many of his contemporaries, and that whatever ideas he had about a cyclical cosmos in the long term, he envisioned some kind of return to a prelapsarian perfection in the short term. Thus he subscribed to many of the progressive ideas of his contemporaries, who like him conceived of progress in terms of the return to, or a renewal of, an ideal past. Charles Webster has described the way our modern idea of progress developed from the belief in the return of an Edenic past.[97] McGuire and Rattansi ask if Newton was behind the

[96] Isaac Newton, *Opticks* (New York: Dover Publications, 1952), pp. 405-6. Cited in McGuire and Rattansi, "Newton and the Pipes of Pan," pp. 122-3.

[97] Charles Webster, *The Great Instauration* (London: Duckworth, 1972).

times because he looked backwards rather than forwards.[98] But I would argue that Newton, like his contemporaries, was Janus-faced; they could look in both directions, but still anticipate improvements and advances never described in an ancient context. Leibniz believed in the *Prisca Theologia* just as much as Newton did, but he could envision submarines, airplanes, and all kinds of things absent from ancient texts.[99] For that matter, so could Roger Bacon and Leonard da Vinci. Past perfect and future perfect only began to be separated during Newton's lifetime. It would take the eighteenth century to accomplish this separation, at a cost well-documented by Romantics and conservatives of every subsequent generation.

What I propose in conclusion is that the Yates thesis is valid up to the point I suggested earlier. The remarkable upsurge in confidence and optimism Yates documents among the Renaissance Neoplatonists and which she attributed to Hermeticism, did exist; but could better be attributed to the rediscovery of classical scientific texts and to the revival of various forms of Gnosticism. If this is the case, then Newton should be beyond the pale, so to speak, because of his outright rejection of Gnosticism. But taking a lead from Moshe Idel, who has convincingly argued that Gnosticism is an integral part of Judaism, I suggest that Gnosticism is an integral part of Christianity as well. Idel argues that juxtaposing rationalist Maimonides with mystical Kabbalists, or to go back far earlier, and juxtapose Rabbinical Judaism to Kabbalah, is a mistake.[100] The two were in constant tension, which perhaps is only understandable. A similar tension appears in Christianity. For every theologian stressing the utter transcendence of God, there is a mystic who claims to have achieved union with the divine and even to have become divine. The Judeo-Christian doctrine that man was made in God's image spawned all kinds of heretical, gnostic, and Pelagian doctrines which, like Hydra heads, emerged and reemerged through all the centuries, down to Newton's own and beyond. In this sense Trinkhaus is entirely right to emphasize the biblical and patristic sources for the idea of man's greatness and ability. But these ideas were marginalized by the orthodox Church.[101] Thus, there was a religious antecedent that paved the way for the concentration on man and his potential in Eighteenth century thought, but it sprang from unorthodox roots. In one way or another, and from a wide variety of classical and gnostic sources, ideas arose minimizing or denying the orthodox Augustinian view that man could not be redeemed by his own efforts but only through Christ. Although Newton did not receive

[98] McGuire and Rattansi, "Newton and the Pipes of Pan," p. 127.

[99] P. Wiener, *Leibniz: Selections* (New York: Charles Scribner's Sons, 1951), p. xxi.

[100] Idel, Kabbalah; *idem*, "Rabbinism versus Kabbalism: On G. Scholem's Phenomenology of Judaism," *Modern Judaism* 11 (1991), pp. 281-96.

[101] Pagels, *The Gnostic Gospels* (New York: Random House, 1979).

these ideas of man's power and potential from self-declared Gnostics and Kabbalists, he found similar ideas in the scriptures themselves, in the Greek Church Fathers, and particularly in the prophetic books he spent so much time decoding. And he absorbed these ideas while living in an age when gnostic heresies advocating man's potential flourished.

ESSAY 3

From Paracelsus to Newton: The Word of God, the Book of Nature, and the Eclipse of the "Emblematic World View"

JAMES J. BONO
State University of New York at Buffalo

> I do not assume the irrelevancy of Newton's pursuit of an ancient, occult wisdom to those great syntheses of his that mark the foundation of modern science. The Janus-like faces of Isaac Newton were after all the production of a single mind, and their very bifurcation may be more of a modern optical illusion than an actuality. Newton's mind was equipped with a certain fundamental assumption, common to his age, from which his various lines of investigation flowed naturally: the assumption of the unity of Truth. True knowledge was all in some sense a knowledge of God; Truth was one, its unity guaranteed by the unity of God. Reason and revelation were not in conflict but were supplementary. God's attributes were recorded in the written Word but were also directly reflected in the nature of nature. Natural philosophy thus had immediate theological meaning for Newton and he deemed it capable of revealing to him those aspects of the divine never recorded in the Bible or the record of which had been corrupted by time and human error.[1]

The unity of Sir Isaac Newton's various esoteric and exoteric studies, the fundamental intersection—even convergence—of what we would regard now as discordant discourses, and the grounding in religion of Newton's singular vision of Word and nature are foundational assumptions of much recent and revisionist Newtonian scholarship. This essay embraces the general spirit of the late Betty Jo Teeter Dobbs's remarks in the above epigraph, while offering them for two strategic reasons. First, as an assurance to the reader that the subject of this paper does have some pertinence to Newton, even if its focus (and the limits of my scholarly

[1] Betty Jo Teeter Dobbs, *The Janus Face of Genius: The Role of Alchemy in Newton's Thought* (Cambridge University Press, 1991), pp. 5-6.

J.E. Force and R.H. Popkin (eds.), Newton and Religion, 45–76.
© 1999 *Kluwer Academic Publishers.*

horizons and competence) situates him on the periphery of the essay's field of vision: for the Word of God and the Book of Nature do define metaphoric contexts for the narrative expression of Newton's "vision," or cultural project. Second, to suggest that certain specific features contributing to the general spirit of Dobbs's portrait of Newton need to stand out in sharper relief: what meanings might "the assumption of the unity of Truth" have for the sixteenth and seventeenth centuries? How, and to what extent, was such "Truth" accessible to humans? What were the relationships between God, His Word, and nature? In what senses were God and His Word seen to be "reflected" in nature? And finally, how was theological meaning encoded in nature and "retrieved" by human interpreters?

I propose to approach Newton indirectly by surveying some of the terrain that surrounds, and perhaps leads up to, his own redescriptions of the boundaries of science and religion. Specifically, I want to see if we can bring some of the above features that Dobbs finds so prominent a part of Newton's double-faced genius into sharper relief by casting our gaze upon the role played by refashioned understandings of the Word of God and the Book of Nature in the eclipse of the "emblematic world view" now associated with the Scientific Revolution.

1. The Emblematic World View

The emblematic world view can now be regarded as an important component of the historiography of the Scientific Revolution. This accomplishment is largely the work of William Ashworth, Jr., whose essay on this subject characterizes a way of thinking about the natural world that acquired a good deal of legitimacy in the sixteenth century.[2] While Ashworth is particularly telling in capturing the core meaning and importance of the emblematic world view, his attempts to explain its collapse are less satsisfying. Indeed, as I shall suggest, part of the problem is that this emblematic world view needs to be situated in a broader intellectual and cultural context before we can begin to account for its historical transformation. Before I begin by summarizing Ashworth's notion of the emblematic world view, let me first indicate the direction and larger purpose of the argument I shall be making.

First, I shall argue that we must consider what lies "beyond" the emblematic world view. I mean this in two senses: (1) I want to show that Ashworth's "emblematic world view" is part of a *larger* set of *interpretive*

[2] William B. Ashworth, Jr., "Natural History and the Emblematic World View," in David C. Lindberg and Robert S. Westman, eds., *Reappraisals of the Scientific Revolution* (Cambridge: Cambridge University Press, 1990), pp. 303-32.

practices for reading the "Book of Nature" in this period. "Practices" and "reading" are central here: how did natural historians, physicians, and natural philosophers *engage* animate nature? What tools did they have at their disposal and how did they define (or redefine) them? (2) I also want to examine Ashworth's claims regarding the *end* of the emblematic world view: what lies "beyond" it? This is especially important in the present context, since Newton operates, in Ashworth's scheme, well outside the chronological limits of the emblematic world view. Yet, Ashworth's account of the latter's demise leaves us in a quandry when considering where to situate and how to account for Newton's complex, "Janus-faced," view of nature.

I shall also invoke the example of Paracelsianism in order to illustrate how we need to look "beyond the emblematic world view" in order to situate and understand the set of interpretive practices Ashworth discusses *and* relate them to complementary (and conflicting) practices among early modern students of nature, medicine, and the life sciences.

I shall, moreover, explore the metaphor of the "Book of Nature" as deployed in the sixteenth century as key to the question of the timing and causes of the end of the emblematic world view. Specifically, I propose the central importance of shifting cultural—especially Biblical—narratives to shifting understandings of the nature and meaning of this divine "book of nature." Here we shall see that the trope of nature as God's "text" forges a strong link between natural things and language: (1) because figuring Nature as a text raises the question of the *proper language* for reading and hence deciphering nature and natural things; and this, in turn, raises the question of the relationship of this "language of Nature" (the divine creative language?) to human languages; and (2) because the very legacy of the creation of Adamic man—conjoining perfect knowledge of language [the *names* of the Edenic creatures] with mastery of natural things—and his subsequent Edenic, postlapsarian, and post-Babylonic history underscores the common search for knowledge, wisdom, and dominion through recapturing the lost unity of the Adamic language *and* of Edenic natural forms.

Thus, the significance in postlapsarian history of diversity [variety and difference] and unity to both nature and language link the latter together culturally and historically. Their linking provides a key to dating, and understanding causes for, the decline of the emblematic world view. That is, by attending to *the shifting valuations of originative unity versus difference and variety* in the study of languages in the early-modern period we can better grasp the movements that led to the eclipse of that search for a seamless text of interconnected *meanings* (unity in diversity) in the "Book of Nature" that forms the context for the decline of Ashworth's "emblematic world view."

The Emblematic World View: Ashworth's Thesis

> I hope to show, even in this brief reappraisal, that when we look at natural history through contemporary eyes, we see an entirely different world from ours, a world where animals are just one aspect of an intricate language of metaphor, symbols, and emblems. This "emblematic world view," as I choose to call it, was the single most important factor in determining the content and scope of Renaissance natural history. (p. 305)

According to William Ashworth, Jr., historians of Renaissance natural history have failed to ask the right questions and, as a result, have far too frequently drawn a skewed picture of their subject. In place of an account that sought to uncover the full range of practices employed by sixteenth-century naturalists and to understand the meanings that they discovered in the natural forms that constituted the objects of their gaze, modern historians have instead looked "for the roots of modern zoology and botany" (p. 304). As a result, they have championed examples of natural history, such as Belon and Rondelet, that appear closest to the descriptive, anatomical, and classificatory preoccupations of a much later age. A towering, and unavoidable, figure such as Conrad Gesner, on the other hand, "is lauded for his attempt to gather firsthand information and for his illustrations; he is chided for his humanist fondness for philology and for his lack of any critical sense" (p. 304).

Ashworth counters this dominant historiographical trend by first noting the exceptional contents of natural histories such as Gesner's, where an essay on the peacock, for example, would record not simply its description, but also its habits, characteristics (observed and reputed, from myriad sources), the etymology of its name, its association with colors, stones, rivers, etc. named after it, its mythological associations, and much, much more (p. 306). As Ashworth suggests, "Gesner believed that to know the peacock, you must know its associations—its affinities, similitudes, and sympathies with the rest of the created order" (p. 306). Second, Ashworth charts the various kinds of resources that the Renaissance naturalist would draw upon to flesh out his account of such associations. These included, according to Ashworth, "hieroglyphics," "antique coins and Renaissance medals," "Aesopic fables," "classical mythology," "adages and epigrams," and "emblems and devices" (pp. 307-11). It is from the growing use of emblems that Ashworth coins his name for this style of natural history. A result of this trend was to expand the "web of associations" that the naturalist would provide for each animal or plant, so that Gesner's article, at eight pages, was soon outdone by Ulisse Aldrovandi's, which devoted thirty-one pages to the peacock (p. 313).

Here, then, is Ashworth's own characterization of the perspective practiced by Renaissance natural historians:

> The emblematic world view is, in my opinion, the single most important factor in determining late Renaissance attitudes toward the natural world, and the contents of their treatises about it. The essence of this view is the belief that every kind of thing in the cosmos has myriad hidden meanings and that knowledge consists of an attempt to comprehend as many of these as possible. To know the peacock, as Gesner wanted to know it, one must know not only what the peacock looks like but what its name means, in every language; what kind of proverbial associations it has; what it symbolizes to both pagans and Christians; what other animals it has sympathies and affinities with; and any other posssible connection it might have with stars, plants, minerals, numbers, coins, or whatever.... The notion that a peacock should be studied in isolation from the rest of the universe, and that inquiry should be limited to anatomy, physiology, and physical description, was a notion completely foreign to Renaissance thought (p. 312).

Ashworth and the "End" of the Emblematic World View

According to Ashworth, "the demise of emblematic natural history was a crucial part of the development that we call the Scientific Revolution. It was not simply an aftermath of Descartes and the mechanical philosophy but an independent, and perhaps even broader, cultural shift that had profound consequences for the evolution of seventeenth-century science."[3] I am inclined to agree with Ashworth's assesment. I would, however, stress that the peculiar practices involved in his emblematic natural history represent but a portion of more extensive "technologies" for reading the Book of Nature as symbolic and as unified by an interconnected network of analogy and resemblance. This qualification becomes significant once we ask what changes produce the "demise of emblematic natural history."

Ashworth himself sees Joannes Jonston's *Historia naturalis* (1650-1653) as a "watershed publication" marking, if not the end of monumental publishing projects, the end at least of emblematic natural history. Jonston's discussion of the peacock, Ashworth notes, constitutes by Renaissance standards an insubstantial two pages, with little more than description.[4] What is Ashworth's explanation for this dramatic change?

Ashworth notes that attempts to write natural histories of animals found in the new world had to confront the singular circumstance that "the animals

[3] Ashworth, "Emblematic World View," p. 305.

[4] Ashworth, "Emblematic World View," pp. 317-8. Joannes Jonston, *Historia Naturalis*, 6 vols. (Frankfurt, 1650-1653).

of the new world had no known similitudes."[5] Nature, or at least a part of it, seemed to have escaped the Renaissance network of symbols and meaning. Here, Ashworth appears to be onto a significant development, but he fails to press this line of inquiry far enough. Indeed, his essay leaves the impression that the simple profusion of new natural forms and the absence of ready-made emblematic associations for them was too wide a gap for the older symbolic mode of reading nature's book to bridge. I would argue, by contrast, that the assimilation of novelty, of difference, and even of vast and at first sight bewildering variety was a commonplace and strength of the emblematic, mythographic, and narrative traditions that Ashworth invokes as constituents of his emblematic world view. Thus, the tradition of examining and recuperating myths of the pagan gods extended well beyond Boccaccio and could attempt to embrace the mythographic structures and beliefs of native Americans, just as the culture, habits, political organization, and social structure of the new world natives presented challenges to theologians, lawyers, and political theorists who were nonetheless fully capable of responding without immediately jettisoning belief in the unity of human nature and humankind.[6] The problem then is not the absence of preexistent classical or postclassical emblematic associations, as Ashworth supposes, but rather when and why many Europeans abandoned habitual efforts to assimilate variety and difference to a preexistent divine architecture of resemblance and sympathetic/symbolic unity.

Another blow to the emblematic world view, for Ashworth, was the emergence of critical assessment of claims contained in the various fables, proverbs, myths, and emblems that surrounded particular animals in works by Gesner, Aldrovandi, and others. This critical perspective he traces to the new "antiquarianism and the quest for historical truth" of the seventeenth century.[7] Unfortunately, Ashworth's claims regarding the role played by northern European antiquarianism, with its attempts to use artifacts as "vital historical clues," though highly suggestive, remain insufficiently developed as an argument. In particular, *if* there is a new and critical attitude toward historical evidence evinced by the use of artifacts,[8] Ashworth does not give a convincing argument for its emergence. Of course, he is quite correct to point to the impressive development of "great museum collections" in the early seventeenth century as a significant event in the evolution of natural

[5] Ashworth, "Emblematic World View," p. 318.

[6] See Anthony Pagden, *The Fall of Natural Man: The American Indian and the Origins of Comparative Ethnology* (Cambridge: Cambridge University Press, 1982); and Anthony Pagden, *European Encounters with the New World* (New Haven, CT: Yale University Press, 1993).

[7] Ashworth, "Emblematic World View," pp. 319-22.

[8] Ashworth, "Emblematic World View," p. 321.

history, and undoubtedly this had an effect on fostering a comparative and critical spirit.[9] But one suspects that this critical spirit was not as fully developed, nor as internally consistent and uncontested, as Ashworth seems to imply. Certainly, in his list of antiquarian scholars one finds names, such as that of Jan Goropius Becanus, who remain wedded to the central importance of etymology for uncovering divine wisdom through attempts to rediscover and study the originary, Adamic language.[10]

Finally, Ashworth, in my opinion, comes closest to articulating a line of inquiry that can begin to make sense of the "demise" of the emblematic world view in a brief section on Sir Francis Bacon. There, he notes that "Bacon's rejection of the notion that the natural world is a divine language, encoded by God, is almost certainly related to his views on human language." Furthermore, he draws from Bacon's views implications that run directly counter to the assumptions undergirding the emblematic world view: "If words have no hidden meanings, why should nature? If the language of man is arbitrary, can there be a language of nature at all? How can the Book of Nature shed light on God's plan, if the language of that book is devoid of meaning?"[11] These are important questions, not least for any attempt to situate Sir Isaac Newton's own reading of the Book of Nature. Unfortunately, Ashworth fails to follow his own lead. He fails, that is, to explore the relationship of his emblematic world view to the variety of ways—changing ways, I would add—that the Book of Nature was understood as a code or language in the early modern period. For Ashworth, Bacon simply comes too late, and his influence on the naturalists he highlights in his essay is nonexistent. As we shall see, however, there is more to the story of language, the Book of Nature, and the eclipse of the emblematic world view than Ashworth's narrow and negative focus on Bacon imagines.

[9] *Ibid.* On early-modern museums, collecting, natural history, and scientific culture, see Bruce T. Moran, ed., *Patronage and Institutions: Science, Technology, and Medicine at the European Court, 1500-1750* (Rochester: The Boydell Press, 1991), especially Paula Findlen, "The Economy of Scientific Exchange in Early Modern Italy," pp. 5-24; Paula Findlen, "The Museum: Its Classical Etymology and Renaissance Genealogy," *Journal of the History of Collections* 1 (1989), pp. 59-78; Paula Findlen, "Museums, Collecting and Scientific Culture in Early Modern Italy," Ph.D. diss., University of California, Berkeley, 1989; Oliver Impey and Arthur MacGregor, eds., *The Origins of Museums: The Cabinet of Curiosities in Sixteenth- and Seventeenth-Century Europe* (Oxford: Clarendon Press, 1985); and Adalgisa Lugli, *Naturalia et mirablia. Il collezionsimo enciclopedico nelle Wunderkammern d'Europa* (Milan, 1983).

[10] See John Francis Eros, "Diachronic Linguistics in Seventeenth-Century England, with Special Attention to the Theories of Meric Casaubon," Ph.D. diss., University of Wisconsin, 1972.

[11] Ashworth, "Emblematic World View," p. 323.

2. Beyond the Emblematic World View: Other Hermeneutical Practices and Symbolic Exegesis of the Book of Nature

How can we understand the eclipse of the emblematic world view? We need, first, to see emblematic natural history as part of a larger network of cultural beliefs, practices, and narratives. We must look beyond the emblematic world view to a set of related beliefs and practices in order to grasp their common underpinning and, hence, related fortunes.

Here, I would like to stress that the very resources that Ashworth identifies with emblematic natural history—hieroglyphics, mythography, emblems, and devices—are but contingent and highly specialized technologies for reading the Book of Nature and producing (its practitioners would say discovering) a kind of divinely-originating meaning. The assumptions that such meaning exists in the Book of Nature, that it is accessible to human beings, and that technologies for reading that Book and decoding such meanings are available for humans to employ are fundamental not only to Ashworth's emblematic natural history, but to a number of rather different, and at times ideologically opposed, hermeneutic strategies in Renaissance natural philosophy and medicine.

For example, the well known sixteenth-century academic physician, Jean Fernel, who authored comprehensive "textbooks" of virtually the whole of Galenic medicine, sought in his natural philosophical and medical works to uncover the divine, and divinely-originating meaning, in the Book of Nature. Fernel, trained at Paris, juxtaposed scholastic elements of analysis with an exegetical approach to texts and nature in his medical and philosophical works written while a professor in the faculty of medicine at the University of Paris. Such textually-based exegesis, I have argued elsewhere,[12] prescribed a hermeneutics of scientific practice that focused upon the interpretation of language and texts as bearers of a lost, but recoverable, Adamic and divine understanding of nature and things. Exegesis as a technology for reading nature assumes that language itself—deriving ultimately from the most antique and "pure" of human languages—contains a kernel of pristine wisdom mirroring the divine knowledge that found concrete expression in the things contained in God's Book of Nature. Thus God's Word, unfolding as his myriad Works in that Book, ensures human access to the truths of nature through language and texts themselves.

Alternatively, the close connection between God's Word, the trope of language, and the Book of Nature found in the very notion of exegesis also led—by transference to "things" as the "pages" upon which God stamped

[12] On Fernel, see James J. Bono, "Reform and the Languages of Renaissance Theoretical Medicine: Harvey versus Fernel," *Journal of the History of Biology* 23 (1990), pp. 341-87; and Bono, *The Word of God and the Languages of Man: Interpreting Nature in Early Modern Science and Medicine*, vol. 1, *Ficino to Descartes* (Madison: University of Wisconsin Press, 1995), Chap. 4.

his mysterious, veiled symbols (or "signatures") in creating His Book of Nature—to a related exegetical hermeneutics applied directly to that Book. Within this important Renaissance hermeneutics of nature, the analogical, symbolic universe of the middle ages was concretized. That is to say, the "abstract symbolism" of the middle ages was transformed into a concrete "symbolic literalism" in which material things themselves came to be seen as symbolic and as constituting an elaborate tropological network linked together by intersecting, metaphorically conceived, relationships in which the distinction between the "metaphorical" and the "literal" was continually effaced.[13] Such a hermeneutics of nature bears strong resemblances to what Foucault characterized as "similitude" and "resemblance."[14]

In Renaissance efforts to uncover the Truth contained in the Book of Nature, the relative emphasis placed upon language and texts as repositories for a "hidden" Adamic, divine "meaning" on the one hand, and, alternatively, upon the symbolic traces of the divine Word stamped upon things themselves, depended upon how individual natural philosophers, physicians, and their heterodox compatriots negotiated complex cultural narratives. Specifically, the actual deployment of linguistic (textual) exegetical strategies and symbolic exegetical strategies (and the relationship between them) depended upon how individual practitioners situated their attempts to read God's Book of Nature within larger, often recast, narratives of the Fall, the confusion of tongues, and the Pentecost.

Paracelsus may serve as an example.[15] Man, says Paracelsus, "received from God in Paradise the privilege of ruling over and dominating all other creatures, and not of obeying them."[16] Although relinquishing this "right" as

[13] I have discussed 'abstract symbolism' and 'symbolic literalism' in James J. Bono, "Medical Spirits and the Medieval Language of Life," *Traditio* 40 (1984), pp. 91-130, esp. pp. 100-01. This distinction is applied to the metaphorical discourse of Renaissance medicine and its tendency toward slippage from the metaphorical to the literal in Bono, "Harvey versus Fernel," and *The Word of God*, Chap. 4.

[14] Michel Foucault, *The Order of Things: An Archaeology of the Human Sciences* (New York: Vintage, 1973). For criticism of Foucault's view of the Renaissance, see George Huppert, "'Divinatio et Eruditio': Thoughts on Foucault," *History and Theory* 13 (1974), pp. 191-207. See also Wolfgang Harms, "On Natural History and Emblematics in the Sixteenth Century," in Allan Ellenius, ed. *The Natural Sciences and the Arts*, Acta Universitatis Upsaliensis, Figura Nova, vol. 22 (Uppsala: Almqvist & Wiksell, 1985), pp. 67-83; and Harms, "Bedeutung als Teil der Sache in zoologischen Standardwerken der frühen Neuzeit (Konrad Gesner, Ulisse Aldrovandi)," in *Lebenslehren und Weltenwürfe in Übergang von Mittelalter zur Neuzeit* (Gottengen: Vandenhoeck & Ruprecht, 1989), pp. 352-69.

[15] The following discussion of Paracelsus is from Bono *The Word of God*, pp. 130-7, where original texts and references are cited in full.

[16] Paracelsus, *Die 9 Bucher De Natura Rerum [1537]*, in *Samtliche Werke*, vol. 11, ed. Karl Sudhoff (München; Berlin: R. Oldenbourg, 1928), p. 378. The translation is from Paracelsus, *The Hermetic and Alchemical Writings of Aureolus Philippus Theophrastus Bombast of Hohenheim, Called Paracelsus the Great*, ed. Arthur Edward Waite, 2 vols. (Boulder: Shambhala, 1976), 1:174. Where necessary, I have modified this English version with my own translation from the original German found in Sudhoff's text.

a consequence of the Fall,[17] humans still retain freedom even in the face of an unruly nature. Paracelsus wishes to make something of this freedom, and to do so he must unveil both humankind's strengths and weaknesses. Postlapsarian, Christian "man's" assets and deficits may be reckoned by attending to Adam's prelapsarian powers and the legacy of the fall. Here Paracelsus invokes the *ars signata* as link between the paradisical and fallen Adam. This *ars signata* "teaches the true names to give to all things."[18] Underscoring its significance, Paracelsus tells us that:

> Adam our first father had complete knowledge and perfect understanding of these names. For directly after the Creation he gave to all things their own proper and specific names. He gave to each of the animals,—and also to the trees, roots, stones, ores, metals, waters and to the fruits of the earth, water, air and fire—its own special name. And as Adam then christened them with their names, so was God pleased to ordain them. For their names were based upon real foundations, not upon their pleasant appearances, but rather upon a predestined art, namely the signatory art [*kunst signata*]. And for that reason, Adam was the first practitioner of this art of signs [*signator*].[19]

Although Paracelsus attributes to Hebrew (obviously in light of the belief that Adam spoke that language in the Garden of Paradise) the ability to capture the nature of things through their Hebrew names,[20] that capacity of language does not concern him. Rather, what remains significant for Paracelsus is the conviction that Adam's ability to give all creatures their proper names was dependent upon our first parent's capacity to penetrate the mere appearances presented by natural things to uncover their real natures. Indeed, Paracelsian natural philosophy turns upon a vision of man, not as dwelling in the house of language, but as restless sojourner in the world of things. For postlapsarian man must become, like Adam, a *signator*: a practitioner of the *ars signata*, one who operates within and on Nature, fathoming signs and manipulating the hidden virtues of things.

This, Paracelsus hints, is what the wise man (*der weis man*) does.[21] His wisdom comes, not from his intellect, not from his ability to read texts like a slavish scholar, but from his ability to transform himself into a new Adam through his direct engagement with nature. The nature of that engagement

[17] Paracelsus, *De natura rerum*, p. 378.

[18] Paracelsus, *De natura rerum*, p. 397: "die kunst signata leret die rechten namen geben allen dingen." *Hermetic and Alchemical Writings*, p. 188.

[19] *Ibid.* I have substantially modified the English version provided in Waite's edition.

[20] Paracelsus, *De natura rerum*, pp. 397-8.

[21] Paracelsus, *De natura rerum*, p. 378. Paracelsus contrasts the wise man to the bestial man (*einen viehischen menschen*). One rules the stars, the other is ruled by them. Implicit in this passage is the notion that postlapsarian man, even though he has lost Adam's free, easy, and immediate dominion over nature, nonetheless may win back some measure of his lost control. To do so, he must revive and master the Adamic *kunst signata*.

ultimately depends upon Paracelsus's understanding of the relationship of the *verbum Dei*, the divine and creative Word, to the world and to man. That relationship is at the very heart of what Paracelsus calls the *ars signata* which turns upon the Paracelsian notion of *signatura*—the signatures of natural things.

Thus the Paracelsian account prompts us to ask how it is that Adam, as the first practitioner of this "art," can give things their real names. Let us examine the passage concerning Adam's naming of all creatures more closely. Our Paracelsian author states that Adam's naming reflects his "complete knowledge and perfect understanding." What is the source of this understanding? Here we encounter a knot in the Paracelsian narrative, for there seem to coexist two contesting themes. On the one hand, Adam appears to choose names freely, if aptly; on the other hand, the names themselves are grounded in what the story tells us is a "predestined art." This "knot" can, I believe, only be loosened, not untied. Indeed, the tension it embodies is, in my view, essential to the Paracelsians' vision of man in the created universe. While emphasizing the creatureliness of humans—their dependence upon God and, indeed, the fallen nature that clouds the unaided human intellect—Paracelsus and his followers for the most part draw back from reducing humans to a creature utterly devoid of the freedom to act, to control nature, and to influence its own destiny.

Adam—prelapsarian man—freely chooses the names that he gives to all creatures in the Garden of Eden. Yet, this freedom does not mean that the names they are given are in any sense arbitrary. Language, or at least the Adamic language, is not a product of use, a human social construct, in Paracelsus's view. Free from all sin, Adam's language bears a close relationship to the *verbum Dei*. God has not directly imposed this language upon Adam. Rather, God provides Adam with a tool that, given his pure and unfallen nature, he is then able to use correctly. This tool—the signatory art [*kunst signata*]—is not itself knowledge of the true names of things. Rather, its proper use requires that Adam turn directly to nature—to God's creatures—where he may then read those "signs" that mark each natural thing as a natural kind. That is to say, Adam's perfect understanding of the names of things depends upon his ability to grasp, through direct experience, signs that reveal to the able practitioner of the *ars signata* the innermost secrets, hidden properties, occult virtues, and hence true natures of things. By freely and accurately reading such signs, Adam exercises his privilege as unfallen *imago dei* and microcosm to rule over all other creatures. He also comes to enjoy knowledge of those very elements of things that are based, or patterned, upon the Divine plan or ideas that are the blueprint for all creatures. As a result, the names spoken by Adam—the Adamic language—is one that "pleases" God as the true, correct names that God then, in His power and freedom, ordains as such. The Adamic names thus bear a direct and univocal relationship to the Word of God.

Postlapsarian man has lost full knowledge of the signatory art that brought Adam perfect understanding of things. But, while his intelligence is clouded, man's freedom enables him to choose the path of wisdom or of bestiality. Choice of the former path leads humans to the *ars signata* and from thence to nature's secrets. How is this path opened to humans in Paracelsus's view?

Not through language. Not through mere textual exegesis and etymologies. Paracelsus will not abide humans playing cat and mouse with language and texts. The way to the Word, to the effects of the Divine *Logos* upon nature, is through nature itself. But humans—postlapsarian man—must approach nature properly, in the proper spirit. Man's freedom—limited though real—must seek the proper path and encounter nature piously and, hence, with the aid of God. For "Adam" represents not only humankind in paradise; rather "Adam," as Kurt Goldammer insists, "c'est l'homme au sens théologique, c'est-à-dire tout d'abord un etre *perdu* et *reprouvé*, un *prisonnier du péché*."[22] This prisoner of sin nonetheless retains a certain liberty and can consequently avail himself of whatever help God provides him in his fallen state. And it is precisely the divinely-implanted light of nature (*lumen naturae*) that enables humans, Goldammer asserts, "à surmonter le *status corruptionis*."[23]

But it is not such an easy task, this overcoming of man's fallen nature.[24] While the key may be the light of nature, access to that light requires preparation. Given that preparation, however, nature and its secrets lay open before all of mankind. Man must prepare himself for God's illumination in order that the light of nature may awaken his intellect to the correct interpretation of the signs God's created Nature has stamped upon things. Humans must learn, through the *ars signata*, those practices that allow them to be bathed in the radiance of the *lumen naturale* in his direct engagements with Nature. In short, humans must turn, with piety, toward nature to encounter in the concreteness of lived experience both the grace of divine insight and the light with which nature reveals its own innermost secrets. In his conceptions of man, nature, experience, and God's "signatures," Paracelsus points the way to a hermeneutics of nature that promises to

[22] Kurt Goldammer, "La conception paracelsienne de l'homme entre la tradition théologique, la mythologie et la science de la nature," in *Sciences de la renaissance, viiie Congrès International de Tours* (Paris: Vrin, 1973), p. 248.

[23] *Ibid.*

[24] Goldammer, "La conception paracelsienne," provides a complex understanding of Paracelsus's divided and tension-filled view of man, at once limited and filled with possibilities. See especially, pp. 248-51 and 258.

complete the narrative of Adam: of the "old," once privileged and now fallen, Adam and the "new" regenerate Adam to come.[25]

Why, for Paracelsus, is it essential that man, this prisoner of sin, must turn to direct and concrete experience of nature? How can such experience open the human mind to the "light of nature?" The answer to these questions perhaps lies in what would have seemed a paradox to Paracelsus. For it was precisely what is most accessible, ready-at-hand, and visible that provides the key to unlocking the most intimate secrets of nature hidden to human reason. Aristotelians, Galenists, and other postlapsarian students of nature ignored what was under their very noses to pursue fictitious entities raised up by their clouded, but prideful, reason. While it is true for Paracelsus that the properties, powers or natures that give rise to the activity of things in nature are occult—hidden from our direct apprehension—, it is equally the case that Nature has left evident "traces" of such animating properties in the visible texture of things themselves. Thus, as Paracelsus was so fond of saying, "It is the exterior thing alone that gives knowledge of the interior; otherwise no inner thing could come to be known."[26] Postlapsarian man has ignored the external signs of the internal nature of things; by his failure to engage things directly, he has failed to penetrate the very secrets of nature. His failure, then, is at least in part a failure to learn how humans must read the visible book of nature placed before them by our Creator.

Paracelsus's "doctrine of signatures" is then both an account of the relations between the inner and outer aspects of concrete things in nature, and a hermeneutics that teaches fallen man how he is to read things, to read Nature itself. The themes of nature's signatures and of our dependence for knowledge of the hidden, inner nature of things upon external "signs" that signatures (re)present to man, pervade virtually all of Paracelsus's work. In different ways, he tells us that the understanding of signs is indispensible to the knowledge and manipulation of nature. "Nothing," he asserts, "exists without its signature." Indeed, there is nothing produced by nature that it has not marked by a sign of what is inside it. But such signatures have completely fallen out of use, forgotten entirely by fallen, error-ridden man.

[25] On the new Adam, Christ, God's Word, and the light of nature, see Paracelsus's *Astronomia Magna, oder die ganze Philosophia Sagax der grossen und kleinen Welt [1537-38]*, in *Samtliche Werke*, vol. 12, ed. Karl Sudhoff (Munich; Berlin: R. Oldenbourg, 1929), p. 398. On Paracelsus and theology also see Hartmut Rudolph, "Kosmosspekulation und Trinitätslehre, Weltbild und Theologie bei Paracelsus," in *Paracelsus in der Tradition*, in *Salzb. Beitr. Paracelsus- Forschung*, ed. S. Domandl, 21 (1980), pp. 32-47; and Hartmut Rudolph, "Einige Gesichtspunkte zum Thema 'Paracelsus und Luther'," *Archiv für Reformationsgeschichte* 72 (1981), pp. 34-54.

[26] Paracelsus, *Paragranum [1529-1530]*, in *Samtliche Werke*, vol. 8, ed. Karl Sudhoff (Munchen: Otto Wilhelm Barth, 1924), p. 97: "Allein die eussern ding geben die erkantnus des inneren, sonst mag kein inner ding erkant werden."

Hence it is imperative that "those who wish to depict natural things must grasp their signs and understand the same through their signatures."[27]

Paracelsus urges that humans become adept at reading signs. Nature imprints all things that it creates with an outward mark, a sign or signature. Each sign enables one who knows how to read them to know what it is that a given thing—a plant, an animal, a mineral, even the stars—contains within it. This "inner" thing controls, shapes, and activates that which we recognize by its external form and properties in nature. But it is the internal, occult, and secret properties of a thing that constitutes its very nature, that is the key to our understanding its "thingliness" as a creature of God, and that contains those very "virtues" that empower it to act according to its nature. Understanding of the latter—an understanding that may only be grasped experientially—enables man to tap into or extract the powers that lay hidden in nature and thence to control, transform, channel, and transmute them. Thus, Paracelsus envisions a society in which those who are privileged by their experience—by the expertise that grounds itself in intimate experience of nature—can use their expert knowledge to fathom the secrets of nature and use such hidden virtues to improve the lot of humans on earth. Thus, for example,

> The expert practitioner of the art of signs [*signator*] may recognize by means of the signature the virtue inhabiting each material being—that which is in herbs and in trees, in sensible and in insensible things. For consequently such expert signators discover a great many medicaments, remedies and other powers in natural things. And whosoever does not note the power and efficacy of a plant from its signature, that person does not comprehend what he writes. Indeed, he writes like a blind person who does not understand what he writes.[28]

Postlapsarian man, blinded by his fall and turned away from things to the contentious world of mere human words, does not comprehend what he writes and does not gain access to the hidden nature of things through his vain "sciences." The order prescribed by God in His creation does not permit such access except to those who practice a different "science"—one that relentlessly follows the tracks ordained by God in Nature that links the external signs of things back to their inner virtues and actions.[29] This new Paracelsian "science"—the true knowledge of natural things—reads in the

[27] Paracelsus, *Von Den Naturlichen Dingen [1525?]*, in *Samtliche Werke*, vol. 2, ed. Karl Sudhoff (Munich; Berlin: R. Oldenbourg, 1930), p. 86.

[28] Paracelsus, *Astronomia Magna*, p. 173.

[29] See Massimo Luigi Bianchi, *Signatura rerum. Segni, magia e conoscenza da Paracelso a Leibniz* (Rome: Edizioni dell'Ateneo, 1987), p. 62.

visible signatures of things their origins and meaning in the invisible and spiritual nature of their inner constitution.[30]

Visible and invisible, material and spiritual, concrete signature and symbolic order of beings: all find themselves inextricably woven together in the texture of God's creation. God's authorship of the Book of Nature, like Christ's voice speaking of divine mysteries in the languages of man, animates the mere material traces of the letters He has stamped upon this world. In fact, the very light of nature that Paracelsus extolls throughout his works is but nothing, were it not for the fact that it came from God. God's Word is the source of our illumination through the light of nature, just as our very nourishment and life itself comes "not from the earth, but from God through His Word."[31] Where the blind Galenist or Aristotelian sees but mere accidents—a mere cloak of passive matter to be cast aside in the profound search for *entia rationis*—the divinely illumined Paracelsian sees living symbols of Nature's hidden truths. Thus Paracelsus rails against the mere "bookish" medical learning of a Galen, Avicenna, Mesue, or Rasis. The door they hold open to their so-called art of medicine leads to but a blind alley. Only nature itself offers humans true access to the secret healing powers in things. The true door opening upon the art of medicine is that produced by the "light of nature" alone. Against the false, prideful, and pagan, books of man, one must oppose the only true book: "that which God Himself has given, written, devised, and established."[32] It is to this book alone that man must turn if he is to uncover, through the light of nature, the occult properties and secret virtues deep within natural things. Those who ignore God's book, who foresake the *lumen naturae*, are condemned to wander in the world as if in a labyrinth without hope of escape.[33]

The example of Paracelsus attests to the powerful role played by the metaphor of the Book of Nature and its deployment through highly localized rereadings of traditional Judeo-Christian biblical narratives in emerging strategies for interpreting nature in the late Renaissance. Such rereadings authorized particular modes of human access to the "Word of God" contained in nature's text.[34] The priority given to texts in exegetical technologies championed by those like Fernel, though based upon common

[30] Paracelsus, *Astronomia magna*, p. 177.

[31] Paracelsus, *Astronomia magna*, p. 396: "Also ist die speis und das leben nicht von der erden, sondern von got durch sein wort;" and, p. 397, "So das ist, so ist das natürliche liecht nichts, sondern es muss aus got gehen, dan so ist es genug."

[32] Paracelsus, *Labyrinthus medicorum errantium [1537-38]*, in *Samtliche Werke*, vol. 11, ed. Karl Sudhoff (Munich; Berlin: R. Oldenbourg, 1928), p. 169.

[33] Paracelsus, *Labyrinthus medicorum*, p. 170: "Was concordirt in das liecht der natur, das bestet und hat kraft. was aber in das nit concordirt, das ist ein labyrinthus der kein gewissen eingang noch ausgang hat."

[34] For details see Bono *The Word of God*, especially Chap. 3.

assumptions about the "Word of God" and the "Book of Nature," contrasted sharply with the exclusive priority given to experience in Paracelsus's symbolic technology for the exegesis of "things."

Emblematic natural history combines both technologies in so far as it weaves together the meanings encountered in texts (myths; emblematic sayings; etc.) with the experiential decoding of symbols embedded in natural things. Belief that "divine" meaning—a meaning that transcends phenomenal differences and exhibits their roots in the "complicated" unity of God's being and plan for the world—can be uncovered through language alone generally rests upon the special assumption that there is an intimate, if corrupted and occult, relationship between the languages of human textual traditions and an originary divine, "Adamic," language, the Word of God. Alternatively, belief in a divine meaning embedded in things themselves rests on the assumption that God's creative Word directly formed things and that, in so doing, the Word left traces, or signatures, on things that humans, properly aided and motivated, can learn to read as a symbolic language of nature.

Both these assumptions lead to strategies for reading the Book of Nature—at one extreme relying chiefly on texts; at the other, on "things" as bearers of symbolic meaning—, and both find their legitimation in the reworking of narratives regarding Adam, the Fall, the Tower of Babel and confusion of tongues, and, as we shall see later, the Pentecost, that authorized various relationships between the Word of God, the "languages of man," and the Book of Nature. Radically altered narratives and shifting interpretations of key biblical episodes—together with resulting shifts in fundamental assumptions about the relationships between the "Word of God," the "languages of man," and the "Book of Nature"—are then, fundamental to charting important changes in the technologies and strategies for reading nature in the Scientific Revolution of the sixteenth and seventeenth centuries.

3. The Eclipse of the Emblematic World View

Returning to Ashworth's emblematic world view, I would like to stress how the link between language theory and cultural narratives in the Renaissance provides a way of understanding its eclipse.[35] While it is quite likely that changed attitudes toward natural history and both botanical and zoological specimens were intimately related to transformations in the evaluation and interpretation of human artifacts, the grounds for such changes are not explored by Ashworth. Such radical revisioning of nature and history

[35] The following discussion is from Bono, *The Word of God*, pp. 179-92.

implied, and I would argue necessitated, radical recasting of fundamental narratives about Man, Nature, and language. Only within such a radically altered narrative frame could a discourse emerge that placed stress upon the critical *differences* distinguishing animals and plants, rather than on the correspondences and similitudes that bind them into an analogical and symbolic network. The latter network supposed a creation narrative in which God was the Judeo-Christian *Logos* whose Word gave visible expression to the archetypes or ideas of the divine mind. Nature was then the image of God and even a divine *language* that man must decipher. The question raised by the rise and fall of the emblematic world view recounted by Ashworth is, then, how did this story lose so much of its cultural force as authorizing narrative legitimating scientific practices?

Two developments contributed to this loss of narrative authority. First, the intimate and necessary connection between the Word of God and Nature came to be questioned and redefined. Second, language began to lose its status as originary source of knowledge through a gradual delegitimization of narratives that figured humans as privileged readers of the text of nature. As a result, the project of seeking knowledge of nature directly through deciphering the symbols imprinted upon things, or indirectly, through seeking the truth of things in language—that is, in etymology[36] or exegetical hermeneutics—lost its firm foundations.

In many ways the second development prefigured and prepared the way for the first, although it is not possible to insist upon any strict chronology. Not only did these developments overlap with one another, the very discourses and strategies authorized by conflicting cultural narratives survived and coexisted for much of the late sixteenth through at least the early eighteenth centuries. The point, however, is that Ashworth's account stops short of connecting the impact of the new "critical" antiquarianism with the groundwork prepared by both of these developments. Significantly, he does note that Francis Bacon's rejection of the emblematic world view and its approach to natural history was rooted in his firm denial that nature is in any way a mirror, image, or reflection of God, thus suggesting an important role for the first development mentioned above.[37] But Bacon's significance in the demise of the emblematic world view is not solidly connected by Ashworth to the role he claims for the new antiquarian spirit of criticism. And, more fundamently, the importance of Renaissance cultural narratives—and their gradual delegitimization—figuring man as enjoying privileged access to the "Word of God" through his Adamic inheritance (which allowed him to read the text of Nature) nowhere appears in

[36] As Marian Rothstein reminds us, "the very word 'etymology' [is] ... from *etumos*; true, real...," "Etymology, Genealogy, and the Immutability of Origins," *Renaissance Quarterly* 43 (1990), p. 332.

[37] Ashworth, "Emblematic World View," pp. 322-3.

Ashworth's account as a context integral to the emblematic world view and its decline.

In order to underscore the significance of this narrative context for an understanding of the hermeneutic strategies employed by natural history in this period, let me briefly turn to a text cited by Ashworth, Edward Topsell's *The Historie of Fovre-Footed Beastes* (1607).[38] For Ashworth, Topsell epitomizes the emblematic world view: "his 'Epistle Dedicatory' is a hymn to animals as symbolic images."[39] We should therefore find telling Topsell's preoccupation with the study of nature as integral to humanity's quest for salvation; a quest that, without the aid provided by the text of nature, may be doomed to failure by the curse of Babel.

> As an Interpretor in a strange country is necessary for a traveller that is ignorant of Languages (or else he should perish), so is knowledge and learning to us poore Pilgrims in this our Perigrination, out of Paradice, unto Paradice; whereby confused BABELS tongues are againe reduced to their significant Dialects, not in the builders of BABELL to further and finish an earthly Tower, but in the builders of IERVSALEM, to bring them all to their owne Countrey Which they seeke, and to the desired rest of soules.[40]

What is this knowledge and learning that serves as an interpreter, allowing us access to the language man needs in his journey away from his unfortunate fall out of a state of grace in the Garden of Paradise toward the Paradise of the soul's repose with God in heaven? Topsell would have us answer that it is knowledge and learning concerning the beasts that will reveal the one language that can guide us on our journey toward salvation. Indeed, Topsell's text insistently implicates and emplots human pursuit of natural knowledge within the grand narrative of possession of the Adamic language and its loss through Adam's tragic fall and the subsequent confusion at Babel.

Through his Fall and through Babel "man" has lost his way on his journey toward God and eternal salvation in Paradise. But, although he has lost the original Adamic language, postlapsarian and post-babylonic man is still heir to the Adamic legacy. That legacy—knowledge of the true names of all creatures and, hence, access to the creative "Word of God"—is his, if man will simply turn to the text of nature and read its divine message. Knowledge of that text—knowledge deepened by the etymological, mythological, and emblematic learning of ancient and modern authors—

[38] Edward Topsell, *A Historie of Four-footed Beastes* (London, 1607).

[39] Ashworth, "Emblematic World View," p. 316.

[40] Topsell, "Beastes," sig. A3.

serves as our interpreter on the journey to Paradise.[41] The knowledge once lost, is yet again available to man:

> Their [i.e., the beasts] life and creation is Devine in respect of their maker, their naming divine, in respect that Adam out of the plenty of his own devine wisdome, gave them their several appellations, as it were out of a Fountaine of prophesie, foreshewing the nature of every kind in one elegant & significant denomination, which to the great losse of all his children was taken away, lost, & confounded at Babel. When I affirm that the knowledg of Beasts is Devine, I do meane no other thing then the right and perfect description of their names, figures, and natures, and this is in the Creator himself most Devine, & therefore such as is the fountain, such are the streams yssuing from the same into the minds of men.[42]

The "names, figures, and natures" of animals are in God Himself; from God as source, they may flow "into the minds of men." Thus, the possibility of recapturing Adam's lost knowledge of the creatures, of repairing the effects of Babel and reconstituting a language that mirrors nature perfectly—that, in effect, closes the gap between the languages of man and the Word of God—is alive and very much present for Topsell. That possibility can be enacted through study of the very beasts themselves, for Topsell is convinced that God saved all animals from the flood by placing them on Noah's ark precisely to enable man access to Divine knowledge and the saving wisdom that may come with it. "Surely," Topsell, speaking of the animals saved by Noah, proclaims, "it was for that a man might gaine out of them much devine knowledge, such as is imprinted in them by nature, as a tipe or spark of that great wisdome whereby they were created."[43]

God's very Word, his Scriptures, "compare the Divell to a Lyon;...false prophets to Wolves;...Heretickes and false Preachers to Scorpions..." Man needs the knowledge God has "imprinted in" animals if he is not only to uncover, but also comprehend, His saving Word. For it is "cleare that every beast is a natural vision, which we ought to see and understand, for the more cleare apprehension of the invisible Maiesty of God." Thus it is that Topsell counsels direct investigation of all animals found in the book of nature, and that he brings to this task "what the writers of all ages have herein observed" and recorded in stories, images and emblems.[44] God is author of the

[41] Thus, note Topsell's insistent tone in setting forth, near the end of his "Epistle Dedicatory," the claims of the Book of Nature for bringing the counsel of God to his people: "For how shall we be able to speake the whole Counsell of God unto his people, if we read unto them but one of his bookes, when he hath another in the worlde, which wee never study past the title or outside; although the great God have made them an Epistle Dedicatory to the whole race of mankind," no pagination.

[42] Topsell, *Beastes*, in "The Epistle Dedicatory," no pagination.

[43] Topsell, *Beastes*, in "The Epistle Dedicatory," no pagination.

[44] Topsell, *Beastes*, in "The Epistle Dedicatory," no pagination.

universe, and humanity's task is to learn to read the divine text of nature: "but this sheweth that Chronicle which was made by God himselfe, every living beast being a word, every kind being a sentence, and al of them together a large history, containing admirable knowledge & learning, which was, which is, which shall continue, (if not for ever) yet to the world's end."[45] Topsell returns, then, to man and language. For him, natural history is the divine text written in the language of creatures that contains that "knowledge & learning" that Adam and his offspring lost through the fall and curse of Babel. In the very making of Ashworth's "emblematic world view" we therefore find a *re-making* of the grand Renaissance cultural narrative of man's fall and redemption. Natural history, as both the tool and product of reading God's text, enables humans access to their lost Adamic heritage, and with it, access to the saving *verbum Dei*. Renaissance natural history, the emblematic world view, and much of what I have elsewhere called the hermeneutics of symbolic exegesis have been emplotted in this story of man's fall and salvation.[46] The natural history of Gesner, Aldrovandi, Topsell and others is therefore part of that Renaissance quest for the wisdom of the Adamic language lost in the Babylonic confusion.

Let us now turn for examples of natural history to Paula Findlen's recent work, where we shall also find clues for understanding the eventual delegitimation of emblematic natural history. Ultimately, this concern will bring us back to consideration of changes in Renaissance language theory—particularly changes in those very cultural narratives that shaped attitudes toward language—and in the scholarly study of languages that were a crucial development in the decline of the emblematic world view and, more broadly, of symbolic exegesis.

Findlen's concern has been with natural curiosities that Renaissance naturalists and collectors increasingly viewed as "jokes of nature": *lusus naturae*. Her achievement has been to bring what appears to the modern as patently marginal to the forefront of the cultural and social system of early-modern science, particularly natural history. Jokes of nature—for example, nature's replication of natural forms, such as clouds, mountains, or faces, upon stones; shells in the shape of ears; the transformation of coral from a plant-like to a stone-like substance; fossils; or, flowers that assume the shapes of other natural creatures—are no mere oddities, nor simply the abberant imaginings of otherwise rational, even systematic, students of nature. Rather, they display fundamental assumptions about both nature and the practice of late Renaissance natural history.

[45] Topsell, *Beastes*, in "The Epistle Dedicatory," no pagination. Ashworth, "Emblematic World View," quotes this passage on p. 316.

[46] See Bono, *The Word of God*, for discussion of symbolic exegesis.

Nature is fundamentally playful and creative; it is, in short, poetic. As poetic, nature displays all the imaginative and creative powers associated with fiction. Nature can mimic; it can take natural forms and make them metamorphose into other shapes; it is inherently active, transformative, plastic. Because, at times, of "her" willful disposition, nature's playfulness may seem to subvert the very categories and kinds into which creatures are ordered. This aspect of the *lusus naturae* marks something of the independence of nature from the usually stable order of God's creation. Nature, through creative powers instilled in her, spills over with variety and invention and threatens to disrupt the very order of things. Yet, her playfulness also exhibits the workings of patterns, of archetypes in nature. Stones and plants mimic the forms of animals and humans; menstrual blood undergoes metamorphosis into toads. Such transformations and mirroring, then, display the analogical, metaphorical, and hierarchical structure of the divine system of nature. Its very playfulness was authorized by the way in which nature, as a divine language of things, was constituted by the replication of divine archetypes, vertically within different strata, and horizontally among its different "kingdoms and worlds."[47] Hence, the decline of nature's playfulness, as with the decline of the emblematic world view, stemmed from the delegitimization of this archetypal, symbolic vision of nature as divine language.

Within such a view, variety may be rampant. Yet, the overarching system of correspondences and analogies provide a way of recuperating variety and difference, of transforming them into so many instances of a unified language of things. The very image of God as creator unfolding and "explicating" the unity of His Being into the multiplicity of material things carries with it the complementary image of movement back toward unity. Unity becomes then the very ground for the possibility of diversity; and diversity—the multiplication, even the playfulness, of natural forms—becomes the occasion for the quest for an originative unity.

It is precisely the coincidence of this double movement, from unity to diversity and back again to unity, that the cultural narratives authorizing Renaissance language theory and legitimating the hermeneutics of symbolic exegesis in the study of nature enforce in their own domains and reinforce among each other. Consequently, we should view with great interest those movements across the grain of unity-in-diversity that erupt at the margins of Renaissance linguistics and language theory and that begin to complicate the story of language, both manifesting and instigating revision of the narrative of man's Adamic legacy.

[47] Paula Findlen, "Jokes of Nature and Jokes of Knowledge: The Playfulness of Scientific Discourse in Early Modern Europe," *Renaissance Quarterly* 43 (1990), p. 325.

Jean Céard enables us to see just such roots of change in sixteenth-century transformations of the story of Babel and the Pentecost. Céard, well known as a student of sixteenth-century natural sciences,[48] expresses fascination with contesting tendencies in Renaissance engagements with language and its status. Transfixed by the epic quest for the original Adamic language, the Renaissance nonetheless finds itself drawn to the sheer diversity of languages.[49] The story he relates illustrates how a guiding assumption such as that of an originary language can foster inquiry that, in turn, transforms the very terms in which language, and the ideal of an original and universal language, are understood.[50]

Céard recounts the Renaissance confrontation with the legacy of Babel and consequent search for a perfect, universal, Adamic tongue. He discloses the grounds of sixteenth-century obsessions with Hebrew, with etymologies, and with the lure that cabalistic attention to the form itself of letters, words, language exercised over the imaginations of contemporary scholars.[51] Yet, even this obsession constantly stumbled upon the effects of diversity within language, the inevitable presence of a multiplicity that contrasted with desired unity.

The question was, what to do about this variety within, and of, language? The recuperative power of Renaissance discourses are much in evidence in the challenge posed by the obvious variety of the languages of man. Every tendency exerted its force to find order, and therefore traces of a deeper unity, in the systematic study of language. Céard testifies to the imposition of the mental grid of Renaissance naturalists—also students of diversity—upon linguistics in the sixteenth century.[52] In particular Céard cites, as an almost archetypal example of this recuperation of variety to a hermeneutics that enforced unity as the ground of all interpretation, the great sixteenth-century naturalist and embodiment of the "emblematic world view," Conrad Gesner. Both a natural historian of the first rank and early student of comparative linguistics, Gesner was author of *Mithridates*, significantly subtitled, *De differentiis linguarum*, and the monumental *Historica animalium* also of 1555.[53] According to Céard, "dans les deux livres, Gesner met en oeuvre la meme conception de l'ordre et de la varieté."[54]

[48] See Jean Céard, *La Nature et les Prodiges: l'insolite au seizième siècle* (Geneva: Droz, 1977).

[49] Céard, "De Babel à la Pentecôte: la transformation du mythe de la confusion des langues au XVIe siècle," *Bibliothèque d'Humanisme et Renaissance* 42 (1980), pp. 577-94.

[50] *Ibid.*

[51] Céard, "De Babel à la Pentecôte," p. 578 ff. See also Bono, *The Word of God.*

[52] Céard, "De Babel à la Pentecôte," p. 581.

[53] Konrad Gesner, *Mithridates. De Differentiis Linguarum*, ed. Manfred Peters (Aalen: Scientia Verlag, 1974), reprint of 1555 Zurich edition.

[54] Céard, "De Babel à la Pentecôte," p. 581.

Linguistic diversity was, of course, a legacy of Babel. As such, one might dismiss the sheer variety of vernacular tongues as merely exhibiting the decadence and degeneration from an ideal language implicit in the notion of an historic confusion at Babel. Hence, one response to this linguistic variety might well be to deem its study ignoble. For Gesner, Céard notes, this conclusion fit the case of "barbaric" tongues, particularly Hungarian, which had nothing in common with languages that enjoyed a more noble pedigree. Just as Gesner the naturalist lavished scrupulous attention upon the viviparous quadrupedes whose affinities to man lent them a certain dignity, Gesner the linguist evinced preference for those vernaculars, and ancient tongues, dignified by relationship to the sacred languages of Hebrew, Greek, and Latin, "celles dans lesquelles s'exprime le verbe de Dieu."[55] For Gesner, barbaric languages like Hungarian, occupy the same ignoble place among languages in general as the lowly insects do among living things.[56]

Thus, while Gesner encouraged some attention to the variety of linguistic phenomena, his strongly hierarchical understanding of languages—flowing from his implicit adoption of a reading of the Babel narrative as suggesting decadence and degeneration from an ideal—placed limits upon the significance of such diversity. Other readings that looked to Babel as the origin of "linguarum varietas" were perhaps more receptive to details of such variation and their intrinsic significance.[57] Thus, for some the very diversity of languages might mask a deeper, hidden harmony.[58] As Claude Duret, author of massive tomes on language and on natural history,[59] revealed according to Cèard:

> Il y a "cinq differentes sortes d'escrire": les Asiatiques et les Africains écrivent de droite à gauche; les Européens, de gauche à droite; les Indiens, Chinois et Japonais, de haut en bas; les Mexicains, de bas en haut et, dans certains cas, en spirale. Il en conclut qu'ainsi sont exprimés "les secretz et mysteres de la croisee du Monde, et de la forme de la Croix, ensemble de la rotondité du Ciel et de la terre."[60]

Here we see the continuation of that impulse Céard noted earlier: the tendency promoted by belief in an original, Adamic language to embrace linguistic variety. In Duret's case, such variety leads back to a more fundamental, if sometimes occult, unity. But, as Céard strikingly contends,

[55] Céard, "De Babel à la Pentecôte," p. 583, and pp. 581-84 for this discussion in general.

[56] Céard, "De Babel à la Pentecôte," p. 583.

[57] Céard, "De Babel à la Pentecôte," p. 581.

[58] Céard, "De Babel à la Pentecôte," p. 585.

[59] Claude Duret, *Thresor de l'histoire des langues de cest univers (1613)* (Geneva: Slatkine Reprints, 1972), original publication: Cologny, 1613; Claude Duret, *Histoire admirable des plantes et herbes* (Paris, 1605).

[60] Céard, "De Babel à la Pentecôte," p. 585.

that willingness to embrace linguistic diversity authorized by the myth of Babel's originary view of language also led to radical revaluation of the relationship between linguistic variety and universality in the late sixteenth and early seventeenth centuries. If the encounter with variety prompted this revaluation, what extended and authorized it was another myth, another re-reading of Renaissance cultural narratives that gave prominence to the Pentecost as privileged, Christian context for repairing the effects of Babel.

First, linguistic variety was no longer simply a divine punishment and hindrance to man's reformation. Linguistic variety was, instead, a fact of human existence, and one within which humans learned to operate and to communicate. Céard presents a catalogue of opinions, each illustrating ways in which Renaissance thinkers came to domesticate linguistic diversity and to see in such variety opportunities for enobling "man's" status. Thus, for example, the Renaissance came to admire the polyglot, the person who forged bonds of universality, if not unity, with others through an appropriation of diverse tongues. Languages became a vehicle not only for communication, but also for acquiring the totality of human heritages—learning and wisdom.[61]

The surgeon Paré testifies to certain changes regarding language as well. While animals enjoy the ability to understand every individual member of their own species, man, according to Paré, displays his superiority to animals in his "dexterité d'apprendre toutes langues."[62] Céard sees in such views the beginnings of a new attitude toward language and the curse of Babel inflicted upon man: "Par cette 'dexterité' universelle, il échappe à la punition de Babel: l'unité perdue, il la reconstitue dans l'universalité."[63] Clearly, this step in itself does not constitute the end of the emblematic world view. Paré did not abandon a vision of nature as expression of the divine Word accessible to man. His approach to nature remains wedded to analogy and the semiotics of resemblance and a symbolic order.

But such revaluations of linguistic diversity did open a discursive space for the articulation and transformation of powerful Renaissance cultural narratives. And such transformations could have the effect of delegitimizing the very narrative and discursive foundations underwriting the hermeneutics of symbolic exegesis, including the emblematic world view. It is this aspect of Céard's examination of language and the myths of Babel and the Pentecost that are of such importance to our own story. His point is decisive for the historical changes we have been considering:

[61] Céard, "De Babel à la Pentecôte," see pp. 585 ff.

[62] Ambroise Paré, *Le livre des animavx et de l'excellence de l'homme*, ed. J. F. Malgaigne, in vol. 3 of *Oeuvres complètes d'Ambroise Paré*, 735-69 (Paris: J. B. Baillière, 1841), as quoted by Céard, "De Babel à la Pentecôte," p. 589.

[63] Céard, "De Babel à la Pentecôte," p. 589.

> Le mythe d'une langue unique et universelle cède la place à un autre mythe: celui d'une aptitude à parler toutes les langues du monde, à disposer de toutes les richesses du verbe. La langue universelle, c'est le 'thresor' de toutes les langues de cet univers. Dans cette abondance se refait la langue parfaite des origines.[64]

Even here, however, we must resist the temptation to proclaim a rupture in history. What we have is no more than an alternative articulation, a recast narrative whose terms are so familiar that they are always in danger of being reappropriated to a familiar discourse, reinscribed within the very narrative it attempts to discard. The alternative myth of speaking all tongues, does lead to an embrace of diversity and variety unparalleled in the more traditional quest for the traces of the lost, Adamic language. And yet that "new" myth remains haunted by the shadowy presence of a perfect originative language.

What Céard has exposed in sixteenth-century thought is not so much a rupture within discourse as its tensions and opposing valences. That does not mean that nothing has changed. Quite to the contrary, the exposure and articulation of these tensions only serve to heighten new possibilities for language theory and for the hermeneutics of nature. My point is simply that these new possibilities did not vanquish the old; rather they existed side by side, even indeed within the same text and discourse. This state of affairs was in fact productive of a veritable explosion of interest in the diversity and variety of nature. On the one hand, a Gesner could appreciate the diversity of languages and the variety of natural forms as worthy of scholarly and scientific investigation. Nature could be laid bare as a text upon which God had imprinted a rich array of forms, a veritable profusion of natural dialects of the unifying and creative Word of God. For Gesner, such diversity was contained within the larger frame of belief in a single, originary divine language of man and creation. Gesner's language theory and his practice as natural historian fit within the mainstream of the emblematic world view. His world, and that of his standard-bearers like Aldrovandi, was one that awaited symbolic exegesis.

On the other hand, one could also find in a Guillaume Rondelet a fascination with the variety of natural forms that resisted reinscription in the cultural text of a quest for an originative unity—a unity, not just of the specific forms themselves and the orders to which they belonged, but of the similitude of all forms that constitute the originative divine language of creation. It is precisely this symbolic, emblematic quality to the natural order, and to natural history, that Rondelet resists, and that has set him apart—along with the likes of Belon—among Renaissance naturalists in standard histories of the subject. Indeed, Ashworth's historiographic plea for

[64] Céard, "De Babel à la Pentecôte," pp. 589-90.

the centrality of his emblematic world view constitutes a reaction to what he sees as the anachronism of elevating the spare, descriptive natural history of a Rondelet to the standard against which Renaissance texts must be judged.[65] Rondelet should not, in fact, become such a standard. Nonetheless, the very century that produced a Gesner and an Aldrovandi also produced a Rondelet. They must, I think, be seen as emerging from the very possibilities for alternative discursive solutions to the problem of variety and unity that Céard's treatment of the myths of Babel and the Pentecost allow us to see in Renaissance language theory and natural history. Finally, we may regard a Paré as exhibiting a tolerance for linguistic diversity that finds itself easily recuperated by the belief in an originary *verbum Dei* authorizing a basically analogical, metaphorical, and hierarchical understanding of nature.

What Céard has uncovered is that opening up of a discursive space in the Renaissance that would, in time, widen into something like a chasm. That space allows for the articulation of a new linguistic ideal authorized by a subtlely recast narrative of cultural origins that never quite breaks with past, and one might even say dominating, narratives. The ideal is that of a universal language. The ideal itself is but a trope: a projective desire for a whole that will bind together the many fragmented parts constituting the languages of man. In this sense then, this new ideal generates itself out of a trope that it shares with its rival. But this new ideal threatens to destablize discourse by reemploting that trope in a narrative that alters its fundamental meaning. For the trope of a universal language now looks, not to the past, to the unity of Adamic innocence and perfection for its authorization and its model, but instead to the future. The metaphoricity of the Adamic language now becomes transformed into the synecdoche of an ideal universal language. The universal language is not something that has been, but rather an ideal to be created by humans out of the multiplicity of linguistic phenomena constituting the very languages of man.[66]

By the late sixteenth century, then, one can find the cultural narrative within which theories of language are emplotted broadened and diversified to allow for new possibilities in the understanding of the relationship between the Word of God and the languages of man. The story of the confusion of tongues at Babel may now be read through "le miracle de la Pentecôte."[67] The languages of man thereby gain the possibility of a new relationship to the *verbum Dei*: the very diversity that was once a sign of their corrupt, fallen status—a status to be overcome through supression of their individuality and alterity—now becomes the very material out of which a rich and proper universal language of things may be forged anew.

[65] Ashworth, "Emblematic World View," pp. 304-5.

[66] Céard, "De Babel à la Pentecôte," p. 592.

[67] Céard, "De Babel à la Pentecôte," p. 593.

But the prospect of such a universal language, as we have noted above, was not easily disengaged from the old ideal of an originative unity. Only a distant hope, this new ideal did help to open up the very scope of natural history and the investigation of natural phenomena. By allowing that the perfection and universality of truth—of true knowledge of nature—was not to be *rediscovered* through exegesis of texts and the book of nature, but rather was something to be *made* by mankind through our efforts and work, this new ideal—however tentatively and partially articulated—looked toward a new hermeneutics of nature. Just as the miracle of the Pentecost embraced the diversity of tongues and of mankind in an image of God's Apostles engaged in the work of creating a new Christian unity, those pious disciples of the Lord who choose to work upon God's Book of Nature can create a new order by de-in-scribing God's plan within the variety of his earthly creatures. This new Pentecostal narrative of man's fall and redemption hence pointed toward emphases upon human industry and the observation of nature in all its diversity—and away from a symbolic ordering of nature—that was to become a hallmark of the new science of Bacon, Galileo, Mersenne, Descartes, and Boyle.

Returning for a moment to Ashworth, the parallelism and, indeed, the intimate relationship between language theory and natural history, together with their mutual legitimation through the work done in recasting cultural narratives, must not be forgotten in attempts to explain shifting models for the practice of natural history. For, as in the case of the encyclopedic and comparative study of language in the sixteenth century, the sheer diversity of natural things that the discourse of encyclopedic, emblematic natural history uncovered (including the flora and fauna of the new world) tended to overwhelm and undermine the discourse of resemblance, correspondence, and symbolism and its underlying ideal of an originative unity. And yet, as was also the case in the study of language, this sheer diversity—the sheer weight of things and facts—was not of itself enough to transform the discourse of natural history. Instead, diversity in language as well as in nature had first to be reinscribed within transformed narratives. Only such new, legitimating narratives could work to delegitimize the relationship between the *verbum Dei* and nature that authorized the ideal and discourse of originative unity. By recasting stories of Babel and the Pentecost, "man" was now authorised to study diversity—both in language and in nature—within the context of a new relationship between the languages of man and the Word of God.

4. From Paracelsus to Newton: The Word of God and the Book of Nature "Beyond" the Emblematic World View

How do we get from Paracelsus to Newton? If we read Ashworth's account of the "demise" of the emblematic world view with this question in mind, we may be left with a number of impressions. We may surmise that the sheer weight of "new" and previously unknown things, and the lack of ready-made symbolic associations, led, by the mid-seventeenth century, to the utter devaluation of symbols generally and, more specifically, of the bookish search for knowledge pertinent to nature in the textual traditions of ancient biblical and pagan cultures. I have myself called attention to the fact that attacks by proponents of the "new science" such as Galileo, Bacon, and Sprat on "the uses of metaphor and fables in the study of nature implied that, unlike their opponents, they did not surround their descriptions of things with ever-expanding webs of meaning fashioned from words, myth, poetry, and symbols" and that "this distinction is in one sense true."[68] In another sense, however, this "fact" is too simple and contributes to the marginalization of Newton's "esoteric" interests (and their divorce from his "real science") in the heroic narrative of modern science.

This heroic narrative can be reinforced, unwittingly, by Ashworth's account of the demise of the emblematic world view, since it implies a nearly frictionless—virtually "natural"—transition to things that stand by, and for, themselves and that demand only to be described. Coming after the great divide—"beyond" the emblematic world view—, Newton's return to symbols (in some sense) and to ancient biblical and pagan texts fails to do what should now come "naturally." Is Newton simply unique, or even anachronistic?

The sense of anachronism diminishes, I think, once we view the "demise" of the emblematic world view as, instead, an "eclipse," and locate the transition not *simply* as one from "text to things" or from "nature emblematized to nature laid bare" (as its polemicists would have us think), but as the product of an essentially narrative reconfiguration of early modern European culture itself. Here, the work of narrative reconfiguration is fundamentally religious in content and purpose. Reconfiguring narrative relationships among the "Word of God," the "languages of Man," and the "Book of Nature" through reinterpreting key biblical stories allowed European culture to reposition and rebuild itself in the wake of confessional and sectarian divisiveness, reassertions of human frailty and Divine omnipotence, and, yes, the growing sense of the diversity of things and peoples.

[68] Bono, *The Word of God*, p. 274.

Viewed in this way, the eclipse of the emblematic world view need not be seen as an intrinsic part of a larger—and later—Enlightenment project of "reason" and "emancipation." Rather, we can regard it as yet another transformation of Western "bookish culture"; a transformation, that is, of the culture of the "Word." Narratively reconfigured, the "World" no longer "reflects" the divine Word as a mimetic image. It is now a contingent expression of an all-powerful Word, reflecting the Word's power, majesty, and providence, but not constrained to reflect its (incomprehensible) essence.

This account allows for the complexly negotiated relationships between "science" and religion that we find emerging in the latter half of the seventeenth century among figures like Boyle and Newton. Boyle's intense interest both in a descriptive, experimental "new" philosophy stripped bare of emblematic associations and in biblical language and the excellency of theology, manifest continued preoccupation with narratives of the "Word of God," the "languages of Man," and the "Book of Nature" that I have argued are essential to the eclipse of the emblematic world view.[69]

Newton, too, exhibits unending concern with the "Word" and the "Book of Nature," and, indeed, preoccupied himself with the work of reexamining the chronology and meaning of biblical and grand cultural narratives throughout his lifetime. As James Force, Kenneth Knoespel, Richard Popkin, and many of the scholars contributing to this volume have stressed in their own investigations,[70] Newton's voluminous unpublished works on the scriptures are not isolated preoccupations, but are, rather, deeply connected to his reading of the Book of Nature "as scientist." I would add, neither are they anachronistic, for they represent fresh attempts to situate and legitimate human inquiry into the Book of Nature: in a sense, they follow in the tradition of rereadings of narratives that provoked the eclipse of the emblematic world view.

While Newton's world is a world "beyond" the emblematic world view, it is nonetheless a world of symbols, biblical narratives, and prophecy. Just as Newton sought "Rules for interpreting ye words and language of Scripture"[71] and, as Sarah Hutton says, always leaned "towards the

[69] On Boyle, see two recent works: Rose-Mary Sargent, *The Diffident Naturalist: Robert Boyle and the Philosophy of Experiment* (Chicago: University of Chicago Press, 1995), especially Chap. 5, "Biblical Hermeneutics"; and Robert Markley, *Fallen Languages: Crises of Represenation in Newtonian England, 1660-1740* (Ithaca, N.Y.: Cornell University Press, 1993).

[70] For example, see Kenneth J. Knoespel, "Newton in the School of Time: The *Chronology of Ancient Kingdoms Amended* and the Crisis of Seventeenth-Century Historiography," *The Eighteenth Century: Theory and Interpretation* 30 (1989), pp. 19-41; and, James E. Force and Richard H. Popkin, *Essays on the Context, Nature, and Influence of Isaac Newton's Theology* (Dordrecht: Kluwer, 1990), and Force and Popkin, eds., *The Books of Nature and Scripture* (Dordrecht: Kluwer, 1994).

[71] This quotation from Yahuda Newton MS 1 is taken from Sarah Hutton, "More, Newton, and the Language of Biblical Prophecy," in Force and Popkin, *Books of Nature and Scripture*, p. 46.

literal,"[72] so, too, Newton departs from the emblematic world view in so far as he both sought symbols in nature as keys to the divine order and unity of nature and yet neither conceived such symbols as identifying God with his creation, nor as constraining God's freedom and power. For Newton symbols and the emblematic could themselves become occasions for mistakenly introducing idolatrous ideas that corrupted the meaning of the scriptures and undermined the truth of God's omnipotence and the utter contingency of his creation. Thus, he was committed both to reading scriptures and nature in a way that revealed their divine truth and meaning, while avoiding and denying these theological errors.

James Force has noted how Newton links the origins of "idolatrous gentile theology" with misappropriation of biblical figures like Noah and his sons (linking them with the planets) and with "encoding these esoteric doctrines in hieroglyphic symbols." Furthermore, "with the Egyptian conquest of her neighbors, including the Jews, these new idolatrous hieroglyphic symbols were taken as the cause of the polytheistic worship of the gods." Quoting a long passage from Newton's *Chronology of Ancient Kingdoms Amended*, Force shows Newton connecting this ancient polytheism with (in Newton's words) "this emblematical way of writing."[73]

I do not presume that Newton uses the term "emblematical" in the same sense we attach to Ashworth's "emblematic world view." Yet it may be useful to note that Newton, on the other side of the great divide separating emblematic natural history from modern science, assumes a certain vigilance with respect to the presence and operation of symbols in cultural narratives and in nature. In pointing to the close association between "this emblematical way of writing" and the error of polytheism, Newton wants to guard against any diminution of God's unique status. Similarly, the presence of symbols in nature can be misconstrued as necessitating *essential* links between natural things and God, thus limiting both God's freedom and power. As I have discussed elsewhere, under the influence of a radical Calvinist reading of signs in nature, what was once proclaimed by Paracelsus as a necessary congruence between the outer signature of a natural thing and its innermost essence came, in the seventeenth century, to be read by some of his protestant followers as a merely contingent and arbitrary link utterly dependent upon the will of God. Knowledge of such links, in turn, were themselves contingent upon God's grace.[74] Taken a step further, not only does the merely carnal and external sign stamped by God

[72] Hutton, "More, Newton, and the Language of Biblical Prophecy," p. 46.

[73] I am indebted to James Force for allowing me to see page proofs of his now published essay, "Samuel Clarke's Four categories of Deism, Isaac Newton, and the Bible," in Richard H. Popkin, ed., *Scepticism in the History of Philosophy* (Dordrecht: Kluwer, 1996), p. 58.

[74] See Bono, *The Word of God*, pp. 140-66, for discussion of Croll's Calvinist Paracelsianism.

on things become arbitrary, the very order and nature of physical things and systems themselves become merely contingent expressions of God's will rather than mirrors of His eternal and unchanging nature. "Newton," as James Force says, "...urges instead a cautious empiricism through which he recognizes the power of God to effect changes even in created natural law" (p. 63).

Although Newton has crossed the great divide separating emblematic natural history from modern science, his is still a culture of the Book. For, while Newton rejects the particular forms of "bookish culture" associated with earlier interpretive strategies for reading the text of Nature as symbolic and emblematic, his own work and practices remain embedded in the metaphoric contexts of the "Book of Nature" and the "Word of God." Unlike historiographical positions that assume a radical *cultural* disjunction between the new philosophy and premodern natural philosophy, it may be useful to think of Newton as *recreating* inherited metaphoric contexts, rather than simply rejecting them. Newton is neither Renaissance symbolic hermeneut, nor secular Enlightenment philosophe, although he is undoubtedly closer to the former than the latter.

Thus, Newton's fundamental assumption—the assumption of "true knowledge" as a "knowledge of God" and therefore of the "unity of truth"—must be understood in the context of his own reworking of the traditional tropes of the "Book of Nature" and the "Word of God." Within this Newtonian context, truth itself is reconfigured not as an occultly interconnected symbolic order of nature necessarily reflecting the essence and order of the divine *Mens*, but rather as the divinely established, if contingent, order of things and of history. These orders surely bespeak for Newton God's majesty, power, and other attributes, although they remain products of His will and therefore manifestations of His activity in the world. The truth that humans must work to uncover, then, is precisely the evidence of God's activity. Such truth was, for Newton, "one," as Dobbs suggested, but the assumption of the "unity of truth" differers markedly from that found in the emblematic world view and related technologies for reading the Book of Nature. The unity that Newton seeks is not that of an architectonic symbolic order based upon essentialized analogical, or metaphorical, links among things and their deeper meanings. Newton, as we've seen, seeks to go beyond the corruptions of mere man-made symbols and idolatrous, "emblematic" ways of writing. Rather, by enjoining humans to uncover evidence of God's activity in the world, Newton refigures "unity" as the unity of various parts making up the larger whole of God's two Books, of the universe as a whole. Yet, as Robert Markley has argued, Newton remains profoundly skeptical of traditional readings of these two books, precisely because of the corruptions and distortions introduced by human interpreters and transmitters of biblical and natural knowledge.[75] This narrative refiguring of the "Word" and the "Book of Nature" leads Newton to envision a difficult, incomplete, and perhaps unending task of interpretation that may best be considered an inevitable consequence of "man's" Fall. While enjoined to work—through the laborious processes of induction and of textual-historical criticism—, humans necessarily encounter the opacity of God's Word and World since they are mediated through the incomplete and, at times, distorting lenses of human languages and traditions.[76]

Betty Jo Dobbs understood this well. Let me then end, in Newtonian fashion, not with a definitive conclusion, but by a return to our beginning point, by implication an invitation to a new beginning. Dobbs has written of Newton's view that

> Human beings had once known how to worship God properly, and had understood the true religion. But humanity had slipped into idolatry...and the knowledge of true religion had been corrupted or lost. However, true religion might be restored. One comes finally to that powerful fountain of motivation that kept Newton at his furnace year after year...—the knowledge of God's activity in the world.
>
> ...His actions are His glory.... His actions, the "issues" of His will, are indeed
>
> > the only glory by which God manifests Himself to His creatures, and which His creatures are able to behold in Him, the reason why His creatures worship Him, and the life and soul of all worship we can give Him....
>
> The worship of God for His *activity* in the world—in creating it, preserving it, and governing it according to His will—that is the true religion. That is what Isaac Newton intended to restore through his study of God's activity in cosmology, alchemy, and history.[77]

Understanding Newton's use of symbols, beyond the emblematic world view, must begin with thorough appreciation of this narratively refigured Newtonian context.

[75] Markley, *Fallen Languages*, esp. Chap. 4, pp. 131-77; p. 144.

[76] See works by Knoespel and Markley cited earlier and their essays infra.

[77] Dobbs, *Janus Faces of Genius*, pp. 87-8, quoting Yahuda MS. Var. I, Newton MS 21, ff. 2-3.

ESSAY 4

"Acceptable to inquisitive men": Some Simonian Contexts for Newton's Biblical Criticism, 1680-1692[1]

JUSTIN A. I. CHAMPION
Royal Holloway College, University of London

1

In early April 1691, the Genevan born reformed theologian, Jean Leclerc (1657-1736), wrote to John Locke commenting upon a manuscript the latter had forwarded to him for consideration: "Des que j'aurai loisir, je traduirai ou en Latin, ou en Francois le petit *Historical Account* etc qui mérite de voir le jour." While Leclerc was happy to oblige Locke's request that the piece might be translated for a wider European audience, he commented further: "je croi poutant qu'il pourait être meilleur, si l'auteur avoit lu avec soin ce que M. Simon a dit du sujet, dont il parle dans la Critique de NT." On the 10/20 January 1692, Leclerc wrote again to Locke on the same subject: both dissertations had been translated into Latin carefully including the additions that Locke had sent "sur le passage de S.J." On the 1/10 of April, 1692, Leclerc wrote again to Locke expressing his surprise (and pique: he had, after all, translated the manuscripts into Latin) that the edition was to be withdrawn. In July, the correspondence between Leclerc and Locke over this particular matter ended with the former commenting that he would faithfully guard the manuscript until the author decided what to do with it.[2] Although Leclerc probably did not immediately know the author of the manuscript in

[1] I would like to thank Jim Force and Dick Popkin for organising the conference that provoked this paper. For their suggestions and insights I would also like to thank John Marshall, Robert Iliffe, Antony McKenna, Michael Hunter, and Scot Mandlebrote.

[2] See E. S. de Beer, ed., *The Correspondence of John Locke,* 8 vols., (Oxford: Clarendon Press), 4:Letters 1381, 1446, 1486, and 1511.

J.E. Force and R.H. Popkin (eds.), Newton and Religion, 77–96.
© 1999 *Kluwer Academic Publishers.*

question, Locke (and subsequent historians) knew that the pieces were composed by Isaac Newton.

Newton had a continuing and complex correspondence with Locke particularily concerned with matters of Biblical Criticism. They exchanged opinions and interpretation of difficult and controversial passages, news of scholarly works, and details of recently discovered manuscript versions of Scripture. It would not be too ambitious to suggest that they convened a continuing hermeneutical workshop: there is evidence that they collaborated in attempting to understand and examine the message and purpose of Scripture.[3] Perhaps the most ample piece of evidence for this continuing, intimate, and private critical relationship is the subject of Leclerc's correspondence with Locke—the letter sent to Locke on 14 November 1690, subtitled "An Historical Account of two notable corruptions of Scripture, in a letter to a Friend."[4] Newton wrote apologetically to Locke for burdening him with the manuscripts. It is clear that the composition had its genesis in an enduring discussion with Locke about the authenticity of I John 5:7 and I Tim. 3:16—Newton feared that "I have not only made you stay too long for them, but also made them too long by an addition." Still he hoped that, although the work was "swelled to ye bignesse you see," that they would prove "acceptable to inquisitive men." Unusually for Newton, he also instructed Locke to set in process plans for the translation and publication of the work: "I fear the length of what I say on both texts may occasion you too much trouble, & therefore if at present you get only what concerns ye first done in to French, that of ye other may stay till we see what success the first will have."[5] Indeed, as Newton continued to suggest, although he had sent Locke the only copy of his work, he wanted it preserved carefully "because I may perhaps after it has gone abroad long enough in French put it forth in English." By February of 1692, as his letter implies, Newton's confidence in the sensitivity of the translation and publication meant that he revised his intentions: "I am sorry," he lamented, "to heare there is news about them [the dissertations]." A note of desperation crept in to Newton's tone as he continued: "let me entreat you to stop their translation and impression as soon as you can for I designe to suppress them. If your friend hath been at

[3] See J. Marshall, *John Locke. Resistance, Religion and Responsibility* (Cambridge: Cambridge University Press, 1994).

[4] The dissertations are reprinted in *The Correspondence of Isaac Newton*, ed. H.W. Turnbull et al., 7 vols. (Cambridge: Published for the Royal Society at the University Press, 1957-77), 3: 83-144. The original of the first two letters are in the Bodleian Library Oxford MS New College 361/4 ff. 2-41: it includes the additions that Newton made referred to in the Leclerc letter above. See Locke, *Correspondence*, 4: Letter 1446. There are two copies of the third letter in the same archive MS New College 361/4 ff. 49v-68 and ff. 70-83.

[5] Newton, *Correspondence*, Letter 357, 3:82.

pains and charge I will repay it and gratify him."[6] What changed Newton's mind is difficult to be sure about: self-censorship so late in the process of publication suggests some incipient incident or context that made the publication of the dissertations particularly dangerous, compromising, or threatening. With good cause, both Locke and Newton were meticulous and obsessive about retaining the privacy of their theological and hermeneutical opinion, after all antitrinitarian belief was illegal and subject to statutory proscription.[7] It may be that Newton, especially given the ever increasing furore surrounding impious attacks upon the authenticity of Scripture in the 1690s, simply felt that his text would have dangerous resonances with other published accounts and that his anonymity could not be preserved successfully in face of undoubted hostility that would have received the book. Jean Leclerc was no stranger himself to controversy, suffering hostile vilification for his own Biblical criticism: indeed the publication and reception of the English abridgement of his polemic with Richard Simon in 1690, *The Five Letters Concerning Inspiration*, was the subject of intense and aggressive scrutiny.[8] Locke himself had corresponded with Limborch about the original French edition of Leclerc's *Sentimens* (1685) suggesting that, although it contained much merit, it was open to hostile and dangerous interpretation. Locke broadly approved of the discussion but was unhappy with the "harsh expressions" used by Leclerc, which were unsuitable for a public audience. Locke was aware that Leclerc's volume was provocative: "the moment I came on that passage in the eleventh letter I seemed to myself to hear cries of protest as if it meant the end of all religion. You know the ways of such people: the less they are able to refute some heterodox opinion, the more they are aroused, in their anxiety not to appear idle in God's cause, to outcries, accusations and calumnies."[9] Limborch, in reply, also noted that he had passed on Locke's criticisms to Leclerc: "He ought, I said, to have set forth distinctly what he attributed to the Apostles and what he denied to them, that so the point at issue in the controversy might be seen

[6] Locke, *Correspondence*, Letter 1465, 4:387.

[7] Marshall, *John Locke*, passim; See, for some insight into the dangers of publicity, James E. Force, *William Whiston: Honest Newtonian* (Cambridge: Cambridge University Press, 1985).

[8] Leclerc's abridgement which included the controversial eleventh and twelve letters of the *Sentimens* was attacked by Claude Lamothe, *The Inspiration of the New Testament Asserted and Explained* (1694) and William Lowth, *A Vindication of the Divine Authority and Inspiration of the writings of the Old and New Testament* (Oxford, 1692, an enlarged edition was published in 1699). John Williams delivered two series of Boyle Lectures in 1695 and 1696 defending the theme of Scriptural inspiration against Leclerc, Spinoza, and Simon. For a very brief account see S. A. Golden, *Jean Le Clerc* (New York: Twayne, 1972), pp. 133-7. See also M. C. Pitassi, *Entre Croire et Savoir. Le Probleme de la Methode Critique chez Jean Le Clerc* (Leiden: Brill, 1987) and M. I. Klauber, "Between Protestant Orthodoxy and Rationalism: Fundamental Articles and the Early Career of Jean Le Clerc," *Journal of the History of Ideas* 54 (1993), pp. 611-36.

[9] Locke, *Correspondence*, Letter 834, 2:748.

in a clear light."[10] Leclerc suffered in his ambitions of finding secure employment in England because of the reputation of his critical writings. In 1694, Gilbert Burnet, Bishop of Salisbury wrote to Leclerc explaining why he had been unsuccessful in finding a position for him: bad men took advantage of his writing while "weak minds are offended by it." Even though Leclerc had nothing but pious intentions in conducting his critical researches, Burnet warned him: "I pray God to direct you in this matter that your work may tend to raise and not sink the credit of the Scriptures." Two years later Burnet commented to Leclerc that he should expect a hostile reception upon his visit to Oxford because of his "discourses of inspiration."[11] It is little wonder, then, that Newton, faced with the example of the sort of antagonistic reception Leclerc's own work had received, was unwilling to risk his own public reputation suffering such a casualty.[12] Biblical criticism was an ambivalent enterprise, easily degenerating from piety to desecration.

Examining the content of the letters sent to Locke, and prepared for publication by Leclerc, has rarely been undertaken in historical accounts of Newton's thought. The "historical account," dealt as its title intimated, with "two notable corruptions of scripture": Newton's letters were contrived and presented as studies in textual criticism not as theological polemic, "tis no article of faith, no point of discipline, nothing but a criticism concerning a text of scripture wch I am going to write about." In the first two letters, Newton dealt at length with the status and textual authenticity of two scriptural passages (I John 5:7 and I Tim. 3:16). In the third letter, which has received even less scholarly attention than the first two, Newton, as he put it, mentioned "some others more breifly." As he continued, "For the attempts to corrupt ye scriptures have been very many & and amongst many attempts tis no wonder if some have succeeded. I shall mention those that have not succeeded as well as those that have, because the first will be more easily allowed to be corruptions, & by being convinced of those you will cease to be averse from believing the last."[13] All in all, he deals at some length with over a dozen passages before adding nearly as many again in the final pages of the dissertation, concluding controversially that "By these

[10] For Locke's reception of Leclerc's *Sentimens*, see Locke, *Correspondence*, 2: Letters 832, 834, 836. See, especially, Locke *Correspondence*, Letter 836, 2:755.

[11] See A. Barnes, *Jean le Clerc (1657-1736) et la Rèpublique des Lettres* (Paris: E. Droz, 1938), Appendix IV, Burnet to Leclerc, 17 November 1694, p. 252, and 24 January 1696, p. 255. Burnet was writing as late as November 1703 about the "odium theologicum" suffered both by himself and Leclerc, p. 258.

[12] On Newton's concern for his public reputation see R. Iliffe, "'Is he like other men?' The meaning of the *Principia Mathematica*, and the author as idol," pp. 159-76, in G. Maclean, ed., *Culture and Society in the Stuart Restoration: Literature, Drama, History* (Cambridge: Cambridge University Press, 1995).

[13] Newton, *Correspondence*, Letter 359, 3:129.

instances it is manifest that ye scriptures have been very much corrupted in ye first ages & cheifly in the fourth century in the times of the Arian Controversy."[14] The thrust of Newton's textual criticism is seen at its boldest in the third letter: scripture has been wrested and wried by the interests of the Churchmen: it was not just "hereticks" but "Catholicks" also. Newton exposed such "pious frauds" with the hope that Christians might be shamed out of such practices.[15] In the letter Newton examined the diverse translations, editions, and manuscripts of the New Testament that had survived: Greek, Latin, Syriack, and Ethiopick were all cited. By a process of comparing, collating, and contextualising the disputed passages, Newton attempted to illustrate how certain passages had been interpolated, added, and revised by human agency. The method was simple and historical although driven by the theological intention of exposing the workings of antichrist that informed much of Newton's other "Arian" essays in Biblical criticism.[16]

If the third short letter was implicitly challenging to orthodox understandings of the authenticity and inspiration of Scripture, the first and second letters were rooted in the soil of critical subversion. Gilbert Burnet had warned Jean Leclerc against the dangers of treating key Trinitarian texts such as I John 5:7 with any sort of textual asperity: denying the authority of any of St John's writings "is upon the matter to deny the whole New Testament for if some books (are) rejected for which we have as good authority as for the rest then all may be as well rejected."[17] Newton, for all his natural caution, was blunt in his first (and most important) essay. His method was straightforward: "I have here sent you an account of what the reading has been in all ages, and by what steps it has been changed, so far as I can hitherto determin by records." While one might expect pious frauds from the papal church "we must acknowledge it a greater crime in us to savour such practices." There was an historical variation in the text: exposing it was a service to truth. Given Locke's "prudence & calmness of

[14] Newton, *Correspondence*, Letter 359, 3:138.

[15] Newton, *Correspondence*, Letter 3590, 3:139.

[16] For important collections of essays on the wider ambition of Newton's hermenuetics see James E. Force and Richard H. Popkin, eds., *Essays on the Context, Nature, and Influence of Isaac Newton's Theology* (Dordrecht: Kluwer, 1990) and James E. Force and Richard H. Popkin, eds., *The Books of Nature and Scripture: Recent Essays on Natural Philosophy, Theology and Biblical Criticism in the Netherlands of Spinoza's Time and the British Isles of Newton's Time* (Dordrecht: Kluwer, 1994). See also Richard H. Popkin "Newton et l'interpretation des prophetes," in Jean Robert Armogathe, ed., *Le Grand Siècle et la Bible* (Paris: Beauchesne, 1989), pp 745-56; S. Mandelbrote, "'A duty of greatest moment': Isaac Newton and the writing of Biblical criticism," *Bulletin for the History of Science* 26 (1993), pp.281-302. For further discussion see Frank E. Manuel, *The Religion of Isaac Newton* (Oxford: Oxford University Press, 1974); idem, *A Portrait of Isaac Newton* (Cambridge Mass: Harvard University Press, 1968).

[17] See A. Barnes, *Jean Leclerc*, p. 252.

temper," Newton had no hesitation in "telling...my mind plainly."[18] Newton then succinctly outlined his theory of textual corruption: first "some of ye Latines interpreted the spirit water & blood of the Father, Son & Holy Ghost to prove them one." This false interpretation was then adopted by Jerome in the "express words" of the Trinity and inserted in his version of the New Testament. By transcriptional errors of incompetent scribes the addition "began at length to creep into the text." This mistranscription of the twelfth and thirteenth centuries was compounded by the advent of print technology: now "it crept out of ye Latine into the printed Greek against ye authority of all the greek MSS & ancient Versions." The core of Newton's case lay in the assertions of this short introductory paragraph: the rest of the manuscript, by examining the ancient codices, by contextualising the disputes and interpretations of patristic authorities such as Cyprian, Athanasius and Jerome, and by conducting a forensic analysis of post-reformation printed editions of the scriptures simply added the "truth of history" to his arguments.

As suggested above, Newton's method was simple if painstakingly established: put plainly, he conducted an historical examination of the literary history of the scriptural text. As he summarised, "by all the best inquiry yt I have been able to make it [I John 5.7] is wanting in the manuscripts of all languages but the Latine." Indeed it only occurred in Latin versions after the time of Jerome.[19] The pages of Newton's piece are littered with marginal references to biblical manuscripts and printed editions of all varieties: Slavonic, Ethiopick, Syraic, Arabic, Armenian, Turkish, French, Latin and Greek. Specific and important codices were cited by name, library, and shelf mark: encouragement to ocular examination accompanied many of these citations. Codices such as Vaticanus, Claromontanus, Bezae, and Alexandrinus were the staple of orthodox bibical critics; Newton supplemented these with the codices Lobiensis, Tornacensis, Buslidianus, and Rhodiensis among many others that were merely given their library call-marks. The "authority of ancient manuscripts" was crucial to Newton's argument. It is clear that he had not examined many of the codices he cited himself but relied upon authentic accounts given by trustworthy critics like Gilbert Burnet:[20] this did not mean Newton felt himself disabled from criticising other biblical scholars' mistakes, follies, and corruptions. Erasmus was reprimanded for including I John 5:7 in the third edition of his *Annotationes*, duped by the papacy. In the second letter, Newton went to great length to expose the faulty readings and transcription made by Theodore Beza: "for tho Beza tell us that all ye Greek

[18] Newton, *Correspondence*, Letter 358, 3:83.

[19] *Ibid.*, 3:94.

[20] *Ibid.*, 3:94.

MSS read θεοτ yet I must tell Beza's Readers that all his MSS read ο."[21] Newton, trusting the authenticity of the monuments of antiquity, uttered a challenge to any potential critics to produce any original testimony of the disputed passages: "Let manuscripts at length be produced, & freely exposed to the sight of the learned world, but let such MSS be produced as are of authority."[22] Newton in effect, although concentrating upon just two case-studies, offered a revision of the authenticity of the received text of Scripture. It was disingenuous of him to claim to Locke that his reflections on the textual status of I John 5:7 had no implication for doctrine or discipline: the authenticity of the text had been at the heart of Trinitarian debates against Socinianism since the 1640s.[23] Newton claimed, as noted above, that the "historical account" was simply "but a criticism concerning a text of scripture": in order to understand what Newton meant with this phrase, and in order to emplot his place in the texture of hermeneutical dispute of the period, it is necessary to reconstruct some context for exploring his contributions.

Undoubtedly the connection with Locke is an important place to start to try and reconstruct the critical discourse with which Newton was familiar and expert. John Marshall's recent study of Locke has gone a long way to expose the theological dimensions of Locke's interests: at the centre of this portrait is Locke's over-riding concern with the message and value of revelation. A "careful and intent" reading of Scripture was key to Locke's means for the construction of true and rational religious conviction.[24] The exchange of manuscripts between Newton and Locke is testimony to their mutual trust and ambitions: indeed it is in the nexus surrounding the aborted publication of Newton's "historical account" that lies one of the keys to unlocking one of the ways of construing the identity of their biblical criticism. It is a slender piece of evidence but once pursued it exposes the innovation of the "historical account" in a far more brilliant light. As noted in the introduction, Locke had approached Jean Leclerc to effect the translation of Newton's work, secure in the knowledge that the Genevan was both competent and experienced in matters of Biblical criticism. Not only had Leclerc disputed at length with the colossus of continental Catholic biblical criticism—Richard Simon—he was also engaged in his own critical editions of the Scriptures.[25] He was to publish a harmony of the evangelists

[21] *Ibid.*, 3:121.

[22] *Ibid.*, 3:109.

[23] See for example popular discussion of the falsity of the text in the 1640s reported in Thomas Edwards, *Gangraena* (1646), Part III, p. 90.

[24] See my review of Marshall's *John Locke* in *The Locke Newsletter* 25 (1994) pp. 110-19.

[25] Leclerc had started work on his Latin edition of the Old Testament in 1684. It was published in 1693. Sections had been circulated in England and Holland in 1690 for which he received encouragement from Gilbert Burnet. See Golden, *Jean Leclerc*, pp.148-150. See also R. Voeltzel , "Jean le Clerc (1657-

that attempted to justify and reconcile the variations in the gospel chronologies (English edition, 1701.) On receipt of the anonymous dissertations, Leclerc's response was to approve of their merit but also to recommend some further reading in the form of Richard Simon's recently published (in French and English) *Critical History of the New Testament* [CHNT](1689.) That Newton took note of these suggestions is instanced both in Leclerc's comments to that effect and also by the evidence of the original manuscript.

There are at least three sections of material derived from Simon's CHNT that were added to the "historical account" all to support Newton's argument against the authenticity of I John 5:7.[26] There is nothing controversial in the citations themselves (all from CHNT Book I, Part 2, chapter 18): Newton deploys Simon's account of various codices in French archives to support his case. But even in this moment of citation there is surely a great connundrum. Simon was a Roman Catholic: much of his writing was perceived by orthodox clerics in England as being contrived to attack the Protestant rule of Faith. Both Newton and Locke were virulently anti-popish. Leclerc, himself a radical protestant, had engaged in a bitter dispute with Simon since 1685 and so had no obvious incentive to recommend Simon's researches. So we have the peculiar situation of a Protestant Genevan recommending to his earnest English confreres the work of a French Roman Catholic polemicist whom he had reviled successfully in print. Yet Newton, perhaps obsessively sensitive about the wiles of the antichrist, cited Simon approvingly and without a warning. Understanding why and how Newton, the anti-papist, could appropriate and refashion the intellectual and scholarly resources of the erudite Roman Catholic Frenchman is an intriguing story.

2

There is no doubt however that both Locke and Newton were familiar with Simon's work. Locke, through his connection with Nicholas Toinard and the refugee Protestant Henri Justel, was familar both with Simon's work and the controversies surrounding it from the early 1680s. Locke had pestered Justel for a copy of the *Histoire Critique du Vieux Testament* [HCVT] (Paris, 1680) between 1679 and 1680, just as he was to keep on at Benjamin Furley in the late 1680s and early 1690s to supply him with a copy of Simon's later

1736) et la critique biblique," in *Religion, Erudition et Critique à la fin du XVIIe siecle et au debut du XVIIIe* (Paris: Presse Universitaire de France, 1968).

[26] Newton, *Correspondence*, Letter 358, 3:87, 90 and 100. Footnotes 21, 27, and 46 point out that each of these sections and paragraphs were added into the form of the original composition, either by marginal indication, or on the facing page.

work on the New Testament.[27] That Locke was close to the circle involved in the publication of the first volume (by Renier Leers in Rotterdam, 1689) is indicated by his correspondence with Toinard 1688-89. In late October, 1688, Toinard had written to Locke, "N'oriez vous point la critique de N.T. du P. Simon. Un de vos amis peut craind[re] que araverit in vintula ejus ca il a vu son ouvrage, comme trop d'otres gens."[28] In June, 1688, Locke had written to Limborch indicating that Simon's work was in press "and will come out in a month or two."[29] In reply to Toinard, Locke commented "I have seen and handled the Histoire Critique of the New Testament, though I have not read it, as it is not on sale yet; but expect to buy it this week."[30] Locke was also intimately connected with the radical Whig circle that included John Hampden, friend and patron of Simon's researches in the mid-1680s.[31] The evidence of both Locke's and Newton's library catalogues shows they owned and read the works of Simon.[32] Locke included a marginal citation from Simon in his work against Stillingfleet dated 1680-1.[33] Locke also used Simon's HCVT extensively in his notebook on the Old Testament: this provides a useful way of exploring how Simon's ideas might be absorbed or appropriated by Locke and Newton.[34] Locke's notes are detailed and systematic. The first pages summarised the key points and are followed by precise examples and references. Locke wrote: "Pere Simon brings proofs & instances. 1° That the several books of ~~Scripture~~ ye OT are collections & epitomes of other writings more ancient. 2° That in them there

[27] Locke's attempts to get hold of the second and third volumes of Simon's criticism of the New Testament can be followed in Locke, *Correspondence*, 4:Letters 1356, 1364, 1371, 1387, 1392, 1400, 1407, 1408, 1420, and 1464. For Locke's ownership of Simon's works and replies see, J. Harrison, P. Laslett *The Library of John Locke* (Oxford: Clarendon Press, 1971) esp. pp. 233-4.

[28] See Locke, *Correspondence*, Letter 1087, 3:514-5.

[29] See Locke, *Correspondence*, Letter 1058, 3:475.

[30] See Locke, *Correspondence*, Letter 1088, 3: 516-7. Interestingly, Locke commented that "the writer has ploughed with the heifer of a certain friend of mine": was this heifer Hampden or Furley's son? Locke organised a copy to be sent to Toinard. See Locke, *Correspondence* 3:Letters 1097 and 1109.

[31] For the role Hampden played in the transmission and translation of Simon's work in England in the 1680s and 1690s, see J. A. I. Champion "Pere Richard Simon and English Biblical Criticism, 1680-1696," in James E. Force and David S. Katz, eds., *"Everything Connects": In Conference with Richard H. Popkin* (Leiden: Brill, forthcoming).

[32] Both studies by Laslett and Harrison show copies marked or dog-eared.

[33] Many thanks to John Marshall for this information. He is currently working upon a critical edition of Locke's work against Stillingfleet.

[34] John Marshall in a personal communication has made the valuable point that, because Newton initialled passages in one of Locke's interleaved Bibles, Newton may have had access to Locke's other notebooks. It may be as well to think of Locke and Newton as having a close hermeneutical relationship. Bodleian MS Locke f. 32, which contains notes from HCVT scattered throughout, also has a reference to Newton at fol. 145. G. Reedy, *The Bible and Reason: Anglicans and Scripture in Late Seventeenth Century England* (Philadelphia: University of Pennsylvania Press, 1985), pp.113-4, has a short summary of this document.

are several things omitted. 3° Several things added. 4° Severall repetitions. 5° Severall varieties."[35] The first twenty-odd pages of the notebook follows this pattern of headings of general points substantiated by citations and proofs: "Compilators," "Stylus Veteris Testamenti," "Books lost," "Omissions," "that there were publique registers amongst the jews: proofs," "compilatio et epitome," "additions," "repetitions," "transposition," "style," "Varieties." After these more general headings, Locke included notes from Simon with those from other sources (Mede, Boyle, le Moyne, Spencer) under the title of the relevant Old Testament book. Locke's notes, with one exception, come from the first book of HCVT.[36] Locke's notes are considered and careful being a combination of direct citation (sometimes in French), translation, and summary. Under some headings, rather than transcribing quotations from HCVT, Locke instead marshalled Simon's Biblical citations and other scholarly sources used by his authorities. For example, under the heading "that there were publique registers amongst the Jews: proofs," Locke lists "Exod XVII.14 (p.47),[37] Deut XXXI.19 (p.46)," "testimonies cited L.I c 1. Joseph. Cont Appion p1. Euseb. de praep: evan. Theodoret in X Joh: 9.14. praef in Reg, in 2 Reg IV, in 3 Reg XI, Masius in Jos. X, Abrav: praef. sur Joh. Procop. in Reg iii.16, Sanctius."[38] This list was followed by more detailed transcription of the texts of some of the references. Here is not the place to examine every detail of a complicated document: it is enough to say that Locke had derived convincing points from Simon on the textual varieties of the Old Testament. At one point, Locke betrays his own convictions in reaction to reading Simon. Under the heading "Pentateuch," he had noted Simon's comments [HCVT pages 18, 24, 31, 32 (four times) and 45 (twice)] that Moses had not composed parts of the book: the telling line read "~~Deut: 11-12 could no be writ by Moses. JL~~."[39] This remark, evidence of Locke's own opinion, is sandwiched between similar extracts from Simon: the proximity in intent is significant. The attempted erasure is also indicative of Locke's doubts about the sensitivity of the remark.

[35] Bodleian MS Locke f 32 fol. 1.

[36] The references in MS Locke f 32 are to HCVT (1685). The one reference (fol. 5) not to the first book is marked by Locke as p. 639 (out of a book 667 pages long) which suggests that although the notes are confined to Book I he did at least glance through the rest of the work.

[37] References in parenthesis represent Locke's note of the page number in Simon.

[38] Bodleian MS Locke f 32 fol. 11-12.

[39] Bodleian MS Locke f 32 fol. 21.

3

At this point, in order to see how Locke and Newton were able to read Simon profitably, and in their sense piously, it is worth turning to the work of the Frenchman itself. The confines of this piece do not allow a comprehensive account of Simon's writings upon the Old and New Testaments, so discussion will be restricted to the substantive arguments. Simon came to the point in the preface to *The Critical History of the Old Testament* [CHOT] (London, 1682): the work was premissed upon the assertion that "it is impossible to understand thoroughly the Holy Scriptures unless we first know the different states of the Text of these books according to the different times and places, and be instructed of all the several changes that have happened to it."[40] Such a bold and un-polemical assertion of the historicity and textuality of Scripture challenged orthodox popular understandings of the universal, inerrant, and, ahistorical quality of revelation. Portraying himself as rebutting the errors of Spinoza, Simon stressed the critical, rather than theological, content of his work. Accurate critical understanding of the text was more fundamental than theologically correct understanding of doctrine or ceremony. As he wrote, "I am persuaded one cannot read the Bible with profit, if one be not first of all instructed in that which regards the Criticisme of the Text."[41] In Book One, Simon set out to deal with the codicological and archival remains of revelation starting from the conceptual position that "their first originals have been lost."[42] Since all surviving manuscripts were translations, abridgements, or incomplete, a careful process of reconstruction had to be made. The text of Scripture used by Christians was not original but the work of the "compiler" Esdras. All Scripture was the product of human time and thus subject to a variety of errors: transcriptional, lexiographical, grammatical, and translational. Books, chapters, sentences, words, and vowels had been misordered, lost, added, mutilated, and corrected. Given that the Old Testament was compiled at different times by different scribes it was a "collection" of diverse styles and voices. Simon went into complex detail over the problems of Hebrew linguistics: difficulties he insisted which many of the scribes had not understood. Some transcribers, for example, used their own Chaldaic orthography when working on the Hebrew text to deleterious effect.[43] The existence of various readings and contradictions did not throw the text into uncertainty but implied the need for critical

[40] R. Simon, *Critical History of the Old Testament* (London, 1682) Preface §1. (Cited hereafter as CHOT).

[41] CHOT I, pp. 2-3.

[42] CHOT I, p. 2.

[43] CHOT I, pp. 107-10.

restoration.[44] Many of the examples of "manifest errors" could be traced to the existence of two different traditions of manuscripts--one public and one private.[45] Simon hoped to supply a critical method for distinguishing good from bad manuscripts.[46] Much of the latter parts of Book One were spent explaining the complexities of the Masoretic system of pointing and the consequent problems and defects that had been introduced into later manuscripts because of ignorance over the purpose and motives of the grammatical system.[47] Much of the thrust of the first book was to show the complexity of criticism, especially the problems of translating and understanding Hebrew.

The second book of CHOT dealt with the "chief translations of the Bible." Here again Simon was bold: he was to survey all translations, from early Chaldean paraphrases and the Greek Septuagint to the achievements of Catholic and Protestant humanism, "but," he ruefully admitted, "we must agree that not one of them is exact, and that there are an abundance of faults in them all."[48] The Hebrew, the LXX, the Vulgate, the contributions of Jerome and Origen, and the more recent vernaculars were all problematic but could be restored by the application of the "Rules of Criticism."[49] What Simon recommended ran counter to orthodox practice which commonly decided that the Hebrew Bible, the LXX, or the Vulgate was the only "original" text: he suggested all sources should be consulted. So both the LXX and the Hebrew should be examined, although both might be defective: "One may luckily repair a Text by ancient Translations, and at the same time the translation by the text."[50] Importantly, though Simon was often represented as attacking the principle of vernacular translation, what he actually argued was that all translation was problematic: bad, grammatically ill informed, or textually narrow translations merely compounded the inherant problems of translation. So what he opposed was not the principle of vernacular translation (indeed he gave many examples of such), but bad translations. In his opinion many of the Protestant translations, because they had been led by theological points rather than the "rules of criticism," were poor and therefore useless. In the course of Book Two, Simon worked his way through every version of the Old Testament: ancient and medieval, Catholic and Protestant, Eastern and Western, giving a precise and acute analysis of their defects and benefits. Importantly, the

[44] See, especially, CHOT I, p. 113.

[45] CHOT I, pp. 137-46.

[46] CHOT I, pp. 147-51.

[47] See CHOT I, Chaps. 24-8, pp. 152-84.

[48] CHOT II, p. 15.

[49] *Ibid.*

[50] CHOT II, p. 24.

Catholic Vulgate was not excluded from this process. Here Simon's understanding of the difference between "originals" and "authentick copies" was central and radical. As he had already asserted no "first and true original" existed: only copies survived. Copies might be "authentick" or corrupt. They were "faithful" either when constituted authority so established them, or when the "rules of criticism" so established them. Simon noted that the Council of Trent had drawn a careful distinction between the language of authenticity and that of infallibility: simply because the Council of Trent had established the Vulgate as authentick did not mean that scholars might not revise and correct it. Simon saw the value of public authorisation in a similar way to Hobbes: this did not imply that private criticism was in any way restrained.[51] One can conceive why Locke and Newton, with their emphasis upon private study, might have approved of such arguments.

The third part of the CHOT has traditionally been regarded as the most anti-Protestant element of Simon's argument and thus perhaps the most difficult part for convinced reformers like Locke and Newton to appropriate.[52] Typically, Simon led off with an abrupt and radical statement: "that we have at present no exact Translation of the Holy Scripture: if we consider the difficulties which have been observed, it seems impossible for us to succeed."[53] The imperfection of the Hebrew manuscripts, combined with the inherant obscurity of the language, meant that all translation was bound to be compromised. Hebrew words were uncertain and equivocal: all translation was interpretation.[54] In one profound sense, this was calculated to undercut Protestant accounts of the theological transparency of Scripture. Protestants were mistaken in thinking that they receive "the pure word of God" when in fact they read merely "the Traditions of the Bible, which in many places contain things purely humane."[55] This was precisely what Newton undertook in his "historical account": a revision of *Protestant* textual certainties. The difficulties of style were manifest in the Old Testament: Books such as Job, Ecclesiastes, and Canticles were profoundly difficult. Even some "historical" books posed their own obscurities.[56] For Protestants, and especially Socinians, to insist that the meaning of Scripture was clear and easy was nonsense. Its inherant "intricacies" showed in the

[51] CHOT II, pp.75-76 (on Trent); pp.91-7 (on authenticity). For Hobbes see *Leviathan* (1651), Chap. 33.

[52] An excellently clear exegesis is given by W. McKane, *Selected Christian Hebraists* (Cambridge: Cambridge University Press, 1989), pp. 111-51.

[53] CHOT III, p. 1.

[54] For a contemporary assertion of Simon's point see W. Barnstone, *The Poetics of Translation. History, Theory, Practice* (New Haven: Yale University Press, 1993).

[55] CHOT III, pp. 7-8.

[56] CHOT III, pp. 12-20.

very fact that most Protestants could not agree amongst themselves.[57] What really irritated Simon, whether practiced by Christian or Jew, Catholic or Protestant, was mixing "theological argument" with critical examination.[58] Any person who failed to acknowledge the role of interpretation in any understanding of Scripture, who interlarded opinion with grammar, who tried to smuggle in mere belief under the disguise of pure linguistic meaning, was trying to assume the privileges of the politics of interpretation under the cover of poetics. Ultimately Simon wrote against interpretation.[59]

Simon did not press the concept of both the theological and linguistic obscurity to a completely negative conclusion. Because he opposed the injunctions of grammer against and above those of theology his remedy lay in advancing the techniques of criticism.[60] Surveying a history of different methods of translation and hermeneutics, Simon conveniently digested Aben Ezra's fivefold classification. There were digressive, allegorical, cabbalistic, and literalist methods none of which were suitable: the best recommendation was to follow the pattern of the Karaite Jews who followed the "plain" way. This was the best "Christian" hermeneutic.[61] Those later critics who came closest to this plain method were complimented. The art of true method lay in combining an understanding of tongues with "correct copies." The key theme was to make "Critical observations, which regard only the proper signification of the words, and the literal sense of the text." As a general lament, Simon commented that "there are few who have kept close to the literal sense."[62] That Simon was strict in this position can be seen in his critique of Walton's work. He had a high regard for much of Walton's scholarship[63] but could not accept the latter's claim that an understanding of true faith was not prior to the critical appraisal of the text of Scripture. For Simon, matters of theology were completely distinct from the text of Scripture: if Protestants claimed that true belief was only derived from the text of scripture, then they must only rely on "reason and criticism" as their rule of faith since the original "Word of God" was not to be found in the abridgement of scripture.[64]

[57] CHOT III, pp. 22, 90-1, and Chap. 16, "of Socinians," pp. 114-21.

[58] CHOT III, p. 88.

[59] It is no anachronism to suggest that in exploring the textuality of the Bible and the relationship between critic and text Simon opened the covers of modern critical theory. At least one modern critic, Michel de Certeau, notes this point in his essay "L'idee de traduction de la bible au XVIIeme siecle: Sacy et Simon," *Recherches de Science Religieuse* 66 (1978), pp. 73-92.

[60] CHOT III, p. 15.

[61] CHOT III, pp. 26-33.

[62] CHOT III, pp. 44, 105, 69.

[63] He commented that Walton et al. were rather "no Catholicks, than hereticks." CHOT III, p. 153.

[64] CHOT III, p. 165.

Simon was to re-iterate these points in his criticism of the New Testament published in both French and English in 1689. Although Simon opened his *Critical History of the Text of the New Testament* [CHNT] (1689) with the comforting assertion that the business of the critic "is not to destroy, but to establish,"[65] contemporaries were not convinced. In this first volume, and its sequel *A Critical History of the Versions of the New Testament* [CHVNT] (French, 1690, English 1692),[66] Simon brought his "art of the critic" to perfection.[67] The New Testament was shown to be in as poor a textual state as the Old: no original survived, apocrypha was rife,[68] the Gospels showed variations and contradictions. Importantly, Simon contextualised much of the New Testament (just as Newton was to do when exploring the origins of his two corruptions): for example the Gospel of Luke was composed to rebut those Marcionite heretics who were attempting to foist works on Paul.[69] In Chapter 18, Simon collected together all the critical evidence for the authenticity of I John 5:7--in all the ancient manuscripts he had examined it was either missing or clearly interpolated. Scholars such as Erasmus were berated for their ignorance in including the verse in their editions.[70] It is clear from a comparison of this chapter with Newton's 'historical account' that the two men agreed in their conclusions: it is also apparent that Newton's dissertation is the more detailed. Much of the rest of the CHVNT was spent giving an account of Simon's understanding of the inspired nature of the New Testament authors, against the rival arguments of Spinoza and, Simon's own critic, Leclerc. Importantly, having devoted chapters 23 to 25 to reviewing contemporary debates about inspiration, Simon then continued to undercut the relevance of any such discussion by re-focusing on the codicological and textual problems of transmission and translation: what would it matter if the original authors were inspired if their manuscripts had been lost, corrupted, and misordered? In the second volume, Simon examined the bibliographical history of the various versions of the New Testament from the patristic

65 R. Simon, *Critical History of the New Testament* (London, 1689), I, Sig. A4. (Cited hereafter as CHNT).

66 There is an interesting history of this publication to be researched. While it was actually published in 1692, it seems to have been licensed in November 1689 and registered in the Term Catalogues in 1690. The licenser James Frazer, a Whig, was later dismissed from his position (April, 1692) for allowing a book by Anthony Walker throwing doubt on the authorship of *Eikon Basilike* to be published. See R. Astbury, "The Renewal of the Licensing Act in 1693 and its lapse in 1695," *The Library* 33 (1978), pp. 298-322.

67 CHNT, p. 42.

68 Importantly, CHNT I, Chap. 3, pp.19-30, was the basis of much of the biblical criticism of John Toland. This unnoticed relationship will be explored in my forthcoming edition of *John Toland. Nazarenus* (Oxford: The Voltaire Foundation, 1998).

69 See CHNT I, Chap. 12, pp. 101-12.

70 CHNT II, Chap. 18, pp. 1-13.

times to his contemporaries again exposing the different manuscript traditions, inadequacies and indeterminacies of readings, grammatical infelicities, and errors. Again, in surveying the manifold editions and translations, Simon underscored his point that human interpretation and theological imperatives invariably lay at the root of all editions. "Critical science" was most often sacrificed to theological controversy.

It is difficult to summarise the content and argument of Simon's biblical studies: the detail and erudition is immense and, indeed, to the twentieth-century historian untrained in classical, patristic, and scholastic scholarship, almost incomprehensible. The profundity of detail, the mastery of manuscript and printed sources is bewilderingly dextrous. It must have appeared so to many of Simon's contemporaries. This monumental erudition did not however mean that the point of Simon's work was lost on his audience: the criticism is as important and radical in its form as it is in its content. Although a reader might not be competent to assess Simon's account of this or that linguistic obscurity, or would almost certainly be unable to consult the old manuscripts he cited, it would be clear, simply from the bulk of his discussion, that the assumption that Scripture was singular, inerrant, and universal was no longer tenable. After Simon, the Bible was a book with a historical bibliography, not simply the "word of God": Simon contributed to the erosion of "la lisibilité et puissance du texte." The Bible became a book with a material history: the "Word" became "language" to be examined.[71]

4

In his sermon against Simon, delivered in early 1683, Stillingfleet countered the Oratorian: the Old and New Testaments were not to be treated "as confused heaps of Collections, but as Authentick declarations of the will of God to mankind."[72] Such a reaction of devout horror at Simon's work was commonplace and reiterated in key public statements such as the Boyle lectures of John Williams and Ofsping Blackhall in the 1690s. The reactions of Locke and Newton were different. Both men were able to read Simon with profit. From Simon they drew encouragement, technique, and evidence. This is not to argue that somehow Simon "influenced," or prompted, either Locke or Newton to hold certain opinions, beliefs, or attitudes but that, on the contrary, in Simon's work the two Englishmen found a resource that was comfortable and resonated with their own ambitions. Both Newton and Locke were desperately interested in the material textuality of the Bible: the

[71] de Certeau, "L'idée de Traduction," pp. 74-5.

[72] G. Reedy, *The Bible and Reason*, p. 150.

evidence of their library catalogues shows that they both owned multiple copies and editions in Greek, Hebrew, Latin, French, English, and polyglot. Owning such a variety of versions would have enabled the men, perhaps, to check Simon's references and also would have reinforced any conclusions they might have drawn from his arguments: the Bibles, ranged on the shelves, would have been silent witnesses to Simon's point. Newton amended and corrected his Bibles with critical marginalia. John Locke had copies of the Bible interleaved so that he might include variations, commentaries, and explicatory detail more conveniently: he shared his thoughts and scholarship with Newton.[73] Simon provided both method and technique. Throughout his work, Simon had described the rules of good criticism and given advice on the proper grammatical and linguistic methods. Consistently, he had advanced the merits of the "plain" reading, coupled with a forensic survey of the correct text. As he wrote, "according to the laws of criticism, the reading which is the most plain, and is withal confirmed by the plurality of copies, ought to be accounted the best."[74] This language of the "plain" reading meshed well with Newton's thoughts as articulated in "Rules for interpreting the words and language in Scripture": as he explained, "he that without better grounds then his private opinion or the opinion of any human authority whatsoever shall turn scripture from the plain meaning to an allegory or to any other less naturall sense declares thereby that he reposes more trust in his own imaginations or in that human authority then in the Scripture."[75] Similarly, the injunction of a plain reading lay behind Locke's understanding as advanced in his *Reasonableness of Christianity* (1695.) Again, this is not to argue that Simon influenced or caused such ideas in Locke and Newton, but that they received intellectual support from his consonance. As will be discussed below, there were important divergances between Simon's position and that of the two Englishmen.

One of the ways in which men like Locke and Newton received help from Simon's work was through the example of its form, as well as the content, of its prescriptions. It gave them confidence to be critical. It perhaps needs to be reinforced precisely how intellectually dangerous and adventurous the sorts of Biblical criticism that Locke and Newton undertook exactly was. Newton's "historical account" was radical: he clearly realised that any public hint of his authorship would bring disrepute and ruin.

[73] See A. W. Wainwright, ed., *John Locke: A Paraphrase and Notes on the Epistle of St Paul*, 2 vols. (Oxford: Clarendon Press, 1987), Appendix X, 2:691-94, where Wainwright gives details of Locke's interleaved Bibles. Locke's Library No. 309, a 1648 octavo compiled in 1693, includes citations from Newton.

[74] CHNT II, p. 161.

[75] See Manuel, *The Religion of Isaac Newton*, pp. 118-9.

Similarly, his manuscript of the Apocalypse (Yahuda MS 4) where he had, verse by verse, annotated the various readings from every manuscript source he could find, was not simply an academic excercise, but a dangerous and potentially ruinous excursion into Simonian territory. John Locke, similarly, in his own notebooks, or his constant use of Toinard's *Harmonia*, probed the textual infrastructure of Scripture. Locke's correspondence, certainly from the early 1680s, is littered with inquiries about this manuscript or that, about new editions and recent finds: he was integrated into an international republic of biblical criticism.[76] Locke grappled with these problems throughout the 1690s, right to the end of his life. His *Paraphrases* was one attempt to resolve his criticism to an understanding.[77] Insight into his position can be gained from the exchange of letters between him and Richard King in September, 1703. King had written to Locke to ask about the "best method" for understanding scripture: how "to attain the true sense and meaning of the inspired writers." Locke's reply gave detailed advice correcting King's language: interpretation of Scripture was not the objective, "understanding" was. King must read the Bible "assiduously and diligently: and if it can be in the Original: I do not mean, to read every day some certain number of chapters, as is usual; but to read it so, as to study and consider, and not leave till you are satisfyed, that you have got the true meaning." Locke continued to advise King that he should seek the assistance of interpreters and commentators "such as are those call'd the Criticks." But he could not be assured of finding an expositor that he could rely on "blind-fold" so "you must make use of your own judgement." He continued to explain at length the precise method of reading and order: gospel harmonies were very useful (especially the English editions of Leclerc and Whiston) and would "save a great deal of time and trouble in turning the Bible."[78] This last remark evokes not only the material process of study and concentrated reading Locke envisaged, but also the way in which scripture could be disordered and reconstructed. Simon had shown how the seamless texture of Scripture could be unpicked, re-woven, and re-sewn: how the activity of criticism might be performed effectively and accurately.

Simon had not only supplied a massive historical account which might be a useful starting point for any individual starting out in the business of

[76] See the important discussion in M.C. Pitassi, "Le notion de communication dans l'exégèse biblique de la fin du XVIIe siècle," in H. Botts and F. Waquet, eds., *Commercium Litterarium. Forms of Communication in the Republic of Letters 1600-1750* (Amsterdam: Amsterdam Univerity Press, 1994), pp. 35-50.

[77] See M. C. Pitassi, "Le Philosophe et L'écriture. John Locke exégète de sainte Paul," *Cahiers de la Revue de Théologie et de Philosophie* 14 (1990).

[78] Locke, *Correspondence*, Letter 3333, 8:64; Letter 3339, 8:69-71. For later advice to the same man on the importance of interpreters as a method of avoiding the dangers of "willful ignorance," see Letter 3346, 8:76-7.

criticism, but he had also published accessible catalogues of Bible editions and critical learning.[79] These two slim pamphlets gave critical advice on the best and most convenient editions: which sources they used, whether they were of a manageable size and readable design.[80] Dictionaries, grammars, and lexicons were also recommended. Indeed, it is not an exaggeration to suggest that Simon was deliberately calculating advice to cultivate a practice of private reading and criticism: this is a model that Locke and Newton would have recognised and embraced as supporting and validating their own activity. Although Simon attacked the extremes of public disorder engendered by interpretative individual enthusiasm or institutional dissent (Protestantism), he actively encouraged private study and criticism as a legitimate and just activity. He made it clear that although key authorities like the Council of Trent had authenticated one public text in the form of the Vulgate, it did not forbid, "private persons to consult the originals for their own particular instruction." He explicitly denied that Roman Catholic prohibitions against seditious and innovative practice outlawed private criticism: vernacular scripture was acceptable and legitimate for private study.[81] Simon went beyond intellectual justification in making plans for his own polyglot edition of the Bible which "may be abridg'd and usefull for private persons." The project, loosely based on the London polyglot, which he had presumably discussed with its sponsor, the radical Whig friend of Locke and Newton, John Hampden, would consist of three texts—"the original Hebrew, the Septuagint translation and the Vulgar edition—marked with a clear marginal apparatus for variants and differences. This edition "convenient and useful" would provide "by this means the Scripture in its original."[82] Later in the same text, Simon suggested that individuals should own at least two texts: one public "the pure text" with all critical apparatus, concordances, summaries, and variants removed, the other private and complete.[83]

This model of plain and private critical understanding is one that was comfortable for Locke and Newton: not only did it give them critical hints, it reinforced their own sense of how to conduct themselves hermeneutically. Of course, there were differences of content. Although Locke opposed "systems of divinity" with as much virulence as Simon objected to theology

[79] Published with the 1682 issues of CHVT. Simon commented "we may find in several books the names of most printed Bibles, but those who publish these sort of Catalogues seldom give their own observations to make us understand the best editions." *A Catalogue of the Chief editions of the Bible*, p. 1.

[80] Simon, *A Catalogue of the Chief editions of the Bible*, pp. 3, 10-1.

[81] CHVNT, pp. 68, 377-8.

[82] Simon, *A Catalogue of the Chief editions of the Bible*, pp. 10, 12-3.

[83] Simon, *A Catalogue of the Chief editions of the Bible*, pp. 20-1.

being admixt with criticism, he was still committed to searching for some ultimate meaning for "what St Paul taught."[84] With indifferency, Locke searched for an understanding of the truth "and to the best of my understanding embrace it with equal readiness whether I find it among the orthodox or the heterodox."[85] This commitment to the priority of a correct text and a plain understanding was shared by Newton and combined with a millennial belief in the possibility of determining the meaning of prophecy. The rules of criticism for Newton could clear the way for understanding: if scripture was corrupt there would only follow heresy and impiety. The difference between this convinced and passionate search for real meaning and Simon's (almost) disdain for theological interpretation is acute. As a public Catholic it was acceptable and perhaps convenient for Simon to marginalise the need for individual interpretation and bow the knee to Papal authority in matters of doctrine. It is also possible to suggest that Simon's private intellectual disposition was also predisposed to underscore the impossibility of escaping the corruptions of interpretation. For Simon the task of criticism was endless, ultimately "meaning" was elusive. For Locke and Newton, on the other hand, the cycle of criticism could be closed because there was a theological truth that could be perceived: ultimately, the biblical text had meaning. What Simon had constructed, and what Locke and Newton enacted, was a *religio laici*.

The public reception of Simon's works, the furore that surrounded Leclerc's attempt to clarify the nature of Biblical inspiration in the early 1690s, all persuaded Newton that the dissemination of his private researches was fraught with danger. A failure of commitment to public debate did not mean Newton (or Locke for that matter) was any less convinced of the importance of the duty of criticism. In these moments of doubt and conviction lies an example of how the rival claims of religion and human knowledge were negotiating new networks of authority and power: the focus of this renegotiation was the material text of Scripture. This battle was literally fought out in the margins of orthodoxy. If this paper has achieved anything in thinking about the conundrum of how Locke and Newton could appropriate the "Catholic" arguments of Richard Simon, it should be that there is a need for a re-assessment of the confessional quality of Simon's work, or at the very least that historians need to pay attention not just to what texts say, but how they might be read.[86]

[84] Locke, *Correspondence*, Letter 3647, 8:414.

[85] Locke, *Correspondence*, Letter 834, 2:749.

[86] The warning comes, of course, from, M. de Certeau, *The Practice of Everyday Life* (California: University of California Press, 1988), pp. 131-77.

ESSAY 5

Those "Whose Business It Is To Cavill": Newton's Anti-Catholicism

ROB ILIFFE
Imperial College of Science, Medicine and Technology

> Their being in affliction increased their devotion. They invoked the saints before, but now they worshipped them with more devotion & ceremony. Now they began to invoke them in their publick supplications or Litanies for averting God's wrath. Now they began to make solemn processions to the churches where the supplications were to be made: Now they began to carry the Pictures & Images of <Christ &> the Saints in Solemn procession & for promoting their veneration to pretend that they were ancient reliques <& to attribute miracles to them> Now they began to be more offended at those men who opposed the placing of Pictures & Images in Churches of demolished those wch were there placed them those men that worshipped them. And now also they began to say Masses for souls in Purgatory. And these things were set on foot by the devotion of <Pope> Gregory the Great, a man who being a monk was much inclined to superstition; & being learned & very diligent & well skilled in business he now modelled the Roman service & ceremonies & by his preaching writing visible humility & sanctity of life & pretence of miracles gained such a reputation among the ~~people~~ clergy & people as gave credit & authority to his visitations, so that they were received & followed in after ages & still continue in use tho not without additions.[1]

In November 1690, Isaac Newton completed a study of two texts of Scripture which he thought proved them to be corruptions introduced at a later date. Sending it as "an historical account" to an anonymous recipient (actually John Locke), he began by examining the case against the

[1] King's College, Cambridge, Keynes MS 5, fol. 115-6. Angled brackets indicate additions; struckthrough words indicate deletions.

J.E. Force and R.H. Popkin (eds.), Newton and Religion, 97–119.
© 1999 *Kluwer Academic Publishers.*

authenticity of 1 John 5:7 and told Locke: "I have done it the more freely because to you who understand the many abuses wch they of ye Roman Church have put upon ye world, it will scarce be ungratefull to be convinced of one more yn is commonly beleived." Having discussed many issues of mutual interest at various meetings over the previous year, Newton was aware that he was writing to someone who shared many of his own theological views, and more generally, he and Locke were part of a Protestant intellectual culture which took it for granted that papists had altered Scripture and indeed the whole of the Christian religion to suit their own ends. Such abuses had to be challenged, and Newton noted that many commentators such as Luther and Grotius had queried the passage mentioned above, but:

> ye generality are fond of ye place for its making against heresy. But whilst we exclaim against the pious frauds of ye Roman Church, & make it a part of our religion to detect and renounce all things of that kind: we must acknowledge it a greater crime in us to favour such practises, then in the Papists we so much blame on that account. For they act according to their religion but we contrary to our's.

Against these perversions and what he believed were attempts to stifle enquiry into the truth, Newton justified his own right as an individual both to examine old manuscripts where more authentic readings might be extant, and to determine the best sense of dubious passages. "In disputable places," he told Locke, "I love to take up wth what I can best understand":

> Tis the temper of the hot and superstitious part of mankind in matters of religion ever to be fond of mysteries, & for that reason to like best what they understand least. Such men may use the Apostle John as they please: but I have that honour for him as to beleive he wrote good sense, and therefore take that sense to be his wch is the best ...[2]

In a further letter sent not long after the first, he added a further list of passages where Catholics had corrupted Scripture in order to lessen the import of passages which either implied that Christ suffered as a human, or proved that he was a deity. Most of these changes had been made in the fourth century in the times of the Arian Controversy, and Newton told Locke that, to the shame of Christians, "be it spoken ye Catholicks have been found much more guilty of these corruptions then the hereticks." In the

[2] Newton to Locke, 14th November 1690; *The Correspondence of Isaac Newton*, ed. H.W. Turnbull et al., 7 vols. (Cambridge: Cambridge University Press 1959-81), 3: 83-129, esp. pp. 83 and 108. 1 John 5:7-8 reads: "For there are three that bear record in heaven, the Father, the Word, and the Holy Ghost: and these three are one. And there are three that bear witness in earth, the spirit and the water, and the blood: and these three agree in one." For Locke's theological interests at this time, see John Marshall, *John Locke: Resistance, Religion and Responsiblity*, (Cambridge: Cambridge University Press 1994), Chaps. 8 and 9.

following two centuries, supposed heretics were always blamed for making additions to and deletions from the New Testament, but in fact it was the Catholics who "ever made ye corruptions (so far as I can yet find) & then to justify and propagate them exclaimed against the Hereticks & old Interpreters, as if the ancient genuine readings and translations had been corrupted." Catholics routinely changed the texts of older non-Scriptural writings so that "learned men blushed not in translating Authors to correct them at their pleasure & confess openly yt they did so as if it were a crime to translate them faithfully. All wch I mention," he continued, "out of the great hatred I have to pious frauds, & to shame Christians out of these practises." Needless to say, he had a special interest in examining such material beyond the disinterested scholarly goal of determining which was the most authentic text.[3]

These attempts to give an objective account of the history of textual perversions by Catholics and others scarcely hid the deep seated animus against the Roman Catholic Church which he had held since his adolescence, and a furious passion for his task underlies these letters to Locke. While few can have displayed such intense emotional commitment to the cause, his abhorrence of Catholic practices and doctrines did not of course, by itself distinguish him from the majority of his Protestant countrymen. Although he inherited his step-father's large collection of theological works when the latter died in 1653, there is no evidence that Newton immersed himself in theological concerns at such an early age. However, as a young man newly arrived at Trinity College, Cambridge, he wrote out a list of sins which he had hitherto committed and it is possible that this marked a new religiosity on his behalf. Within a few years, he had embarked on a major study of theology and the history of the Church, and arguably this became his major intellectual concern until his death. His central interest was in the history of the corruption of true worship; and his research concentrated on all the examples he could find of this both from the classical world and from the period since Christ. He devoted himself to examining the history of idolatry and wrote a major study on the history of this problem in pagan cultures, while his work on the Old Testament invariably centred on cases where the Jews and their enemies had engaged in depraved forms of worship. He developed what was possibly an unrivalled expertise in the early history of Christianity, and particularly in the troubles which erupted in the fourth century.[4]

[3] *Newton Correspondence*, 3: 129-44, 138-9.

[4] For the significance of the list of sins, see R.S. Westfall, "Short-writing and the State of Newton's Conscience, 1662," *Notes and Records of the Royal Society of London* 18, (1963) and F.E. Manuel, *A Portrait of Isaac Newton*, (Cambridge: Cambridge University Press 1968). For anti-Catholicism in the early modern period, see M. Aston, *England's Iconoclasts, Vol. 1, Laws against Images*, (Oxford 1988); Clifton, R., "Fear of Popery," in C. Russell, ed., *The Origins of the English Civil War* (London 1973),

According to Newton, that "The Church of Rome began to decay very much in vertue before the end of the Second Century, & to lay the stress of religion upon ceremonies and outward forms," and within two hundred years evil men had changed Christianity out of all recognition. Athanasius—the founder of monkery—had transformed the doctrines and practices of the Church into the very opposite of Christianity, introducing perversities such as the idolatry of the Trinity. Although general features of idolatry and the incomprehensible doctrine of the triune Godhead provided religious targets much wider than the Catholic Church, Newton saved his most venomous invective for Roman Catholicism. Nevertheless, his views were highly unorthodox and while most Protestant interpreters understood that the Reformation had been an overwhelmingly important phase in the history of the downfall of Antichrist, he did not. In the Book of Revelation, the images of the seals, trumpets, and vials pointed to the growth of the Roman Church and then its demise, and anti-Catholic rhetoric drew heavily from the rich language of the 1611 King James Bible. Within the tradition whose most able exponent was Joseph Mede, it was held that the sounding of the seven trumpets referred to the general period between the fourth century and the Millennium whose onset had either arrived or was close at hand, while the pouring of the seven vials depicted in more detail the increasing depravity and then crushing defeat of popery after the Reformation. Newton however believed that each identically numbered trumpet and vial referred to the same event, and the Reformation, with its retention of many perversions, did not mark a radically new beginning in the downfall of popery. In the one case where we do have evidence of contemporary reaction to his interpretation of prophecy, the Cambridge Platonist Henry More thought Newton's approach was idiosyncratic and liable to downgrade the significance of the Providentially supported advent of Protestantism.[5]

---

pp. 144-67; C. M. N. Eire, *War Against the Idols: The Reformation of Worship from Erasmus to Calvin* (Cambridge 1986); C. C. Hibbard, *Charles I and the Popish Plot* (Chapel Hill 1983); and P. Lake, "Anti-popery: The Structure of a Prejudice," in R. Cust and A. Hughes, eds., *Conflicts in Early Stuart England: Studies in Religion and Politics, 1603-1642* (London 1989), pp. 72-106.

[5] The Jewish National Library, Jerusalem, Yahuda MS 15.3, fol. 55r. The remark is on the same page as a heading for a chapter entitled "The Original of Popery" and the whole of Yah. Ms. 15 is a work on the history of the Church. For Newton's anti-trinitarian and anti-Catholic interpretation of Revelation, see R. Iliffe, "'Making a shew': apocalyptic hermeneutics and the sociology of Christian idolatry in the work of Isaac Newton and Henry More," in James E. Force and Richard H. Popkin, eds., *The Books of Nature and Scripture. Recent Essays on Natural Philosophy, Theology, and Biblical Criticism in the Netherlands of Spinoza's Time and the British Isles of Newton's Time* (Dordrecht: Kluwer Academic Publishers, 1994), pp. 55-88. For the English tradition of apocalyptic exegesis, see K. Firth, *The Apocalyptic Tradition in Reformation Britain, 1530-1645* (Oxford: Oxford University Press, 1978); P. Christianson, *Reformers and Babylon: English Apocalyptic Visions from the Reformation to the Eve of the Civil War* (Toronto: University of Toronto Press 1975); and B. W. Ball, *A Great Expectation: Eschatological Thought in English Protestantism to 1660* (Leiden: Brill. 1975). See also C. Hill, *Antichrist in Seventeenth Century England*, (Oxford: Oxford University Press, 1971).

In this paper, I examine the different strands of Newton's concern with the history of, and threat posed by, Roman Catholicism, and I show how—for Newton—it encapsulated the connections between idolatry and the deliberate misinterpretation and perversion of texts. I look at how he viewed Catholicism as an extension and intensification of older forms of idolatry, and at the ways in which these interests impinged upon his intellectual and political careers in the public sphere. His problem with the Jesuit critics of his work on light and colours derived from deeper suspicions about the theological basis of Catholicism, while his worry that the Catholic King James II might swamp the universities with papists galvanized him into action in the spring of 1687, when the monarch threatened to impose a Benedictine M.A. into Sidney Sussex College, Cambridge. His relatively public action on this occasion earned him the respect of his colleagues and friends, and it was this as much as the publication of the *Principia Mathematica* in the summer of that year that facilitated his move into public life after 1689. Much of the heterodox manuscript material presented in this paper was never seen by anyone in his own lifetime, and only a bare handful of people were ever deemed suitable to be made privy to the real nature of his theological interests. Someone who believed himself to be one of God's chosen few could not be too careful about his friends.

1. IDOLATRY

In its broadest sense, idolatry was Newton's chief dislike, and it was references to this which constituted the public's only awareness of his theological interests in his lifetime. In these glimpses of the vast projects which he carried out in private, corruption in religion and natural philosophy were intimately linked. In the "General Scholium" to the 1713 edition of the *Principia*, for example, he attacked both atheism and idolatry, while at the end of the final "Query" to his 1717 *Opticks*, he connected idolatry to the deterioration of the ancient philosophy. If the "inquisitive" scholar proceeded correctly according to the Newtonian modus, he argued, he might improve natural philosophy and

> if natural Philosophy in all its Parts, by pursuing this Method, shall at length be perfected, the Bounds of Moral Philosophy will also be enlarged. For so far as we can know by natural Philosophy what is the first Cause, what Power he has over us, and what Benefits we receive from him, so far our Duty towards him, as well as that towards one another, will appear to us by the Light of Nature. And no doubt, if the Worship of false Gods had not blinded the Heathen, their moral Philosophy would have gone farther than to the four Cardinal Virtues; and instead of teaching the Transmigration of Souls, and to worship the Sun and Moon, and dead Heroes, they would have taught us to

> worship our true Author and Benefactor, as their Ancestors did under the Government of Noah and his Sons before they corrupted themselves.[6]

Newton's theological notebook, containing entries which are mostly from the 1680s and 90s, is replete with both Old and New Testament examples of idolatry. Under the heading "Idololatria," for example, he noted how Moses and Aaron stood firm before the Pharaoh while magicians attempted to perform magic by means of frogs and inchantments and to lead them into idolatrous practices. He took down numerous passages where the Israelites were warned about the dangers of worshipping idols and graven images, and finished the section with a long note on Solomon. Solomon's fate was a warning about how idolatry could be avoided by having a right understanding about the natural world, but also about how a false view of the latter could lead to a corruption of the former:

> It is not probable that Solomon to whome God appeared twice 1 Kings 11.9 & who was the wisest of men & understood the nature of all trees, beasts, foules, creeping things & fishes 1 Kings 4.33 could be so little a Philosopher as to think that Images made of Gold or wood or Stone were Deities. We may rather suppose that ~~when~~ he fell into Idolatry, through a perswasion that there <was> some reality in the supposed Gods Chemosh, Molech, Asterosh &c to wch the Heathen Idols were dedicated. & so worshipped their Idols with respect to these. ffor if we consider yt in ye books of Moses & ye Judges there is frequent mention of Angels imployed on severall occasions to execute the will of God in the government of the world why might not his heathen wives & concubines by telling him stories of ye works & power of their Gods perswade him to think that the supreme God had committed the government of the world to Angels & such like intermediate beings & that of this sort were the Heathen Deities & theirfore might be supplicated for their assistance & worshipped as ye superintendents of mankind. And thus notwithstanding his wisdom he might be perswaded to go a whoring after them, as he did 1 Kings 11. 7, 33.[7]

[6] Newton, *Opticks, or A Treatise of the Reflections, Refractions, Inflections & Colours of Light*, 4th ed. (London 1730; repr. New York: Dover 1979), pp. 405-6. The last sixteen words were added to the third English edition of 1721. In Newton's own copy of the 1717 edition, on page 382, he adds by hand these extra words and then the following: "For the seven Precepts of the Noachides were originally the moral Law of all nations; & the first of them was to <have> but one supreme Lord God & not to alienate his worship; the second was not to profane his name; & the rest were to abstein from <blood or homicide & from> fornication, (that is from incest adultery & all unlawfull lusts,) & from homicide theft & all injuries, & to be merciful even to bruit beasts, & to set up magistrates for putting these laws in execution. Whence came the moral Philosophy of the <ancient> Greeks." A facsimile of this page is reproduced in F.E. Manuel, *Isaac Newton, Historian* (Cambridge: Cambridge University Press 1962), opposite p. 117.

[7] Keynes MS 2. On pp. 5ff of this notebook there are a series of notes on the "Idolatrous Egyptians & Israelites in Egypt [who] acknowledged the true God." These notes are probably from the early to mid-1680s.

At some point he was asked to deliver some sermons on a religious subject, and naturally he turned to idolatry. His subject was 2 Kings 17: 15-16, relating the story of those who became vain through following vanities: "they left all ye commandments of ye Lord their God & made them molten images even two calves & made a grove & worshipped all ye host of heaven & served Baal." Newton broke down his account of the idolatry of the captive Israelites into a number of different particulars. Firstly, they and those that succeeded them worshipped the true God at the same time as those gods of the gentiles; secondly, this was idolatry nevertheless, "& by parity of reason ... ye worship of ye true God by any Image wch he has not commanded, (as indeed he has commanded none) is Idolatry"; and, thirdly, the ancient idolaters who did revere the true God generally believed that the other gods were inferior and were the souls of great men who had died. Lastly, he concluded that the

> glory of ye true God wch Idolaters gave unto their Idols, was not any of his attributes as some are ready to suppose, but only this: To love, to fear, to trust in, & to to [sic] seek unto those Gods, for ye temporal blessings of this life, for wch ye true God has appointed us to love fear trust in & seek unto none but himself immediately. And consequently to do any of these things in a religious way to any other then the true god, is so far to depart from him & give his glory to another: the great crime of wch Idolaters stand accused.

The audience at Newton's lecture were then treated to a long list of other passages where Jews were punished or even destroyed for their idolatry, and few present could have been in any doubt that the sermon-deliverer was an expert in his subject.[8]

In a project which dated back at least as far as the 1680s, and which was to be the main thesis of an early version of the *Principia*, he ransacked—or reinterpreted—various classical works to show that the ancients had indeed possessed the correct knowledge of the natural world, but had then lost it through the adoption of increasingly idolatrous practices. Towards the end of his stay with Newton in May 1694, the Savilian Professor of Astronomy, David Gregory, observed that his host had "composed a tract on the origin of nations. Religion is the same at all times, but the religion which they received pure from Noah and the first men, they debased with their own inventions." According to Gregory, Newton held that "Moses began a reformation but retained the indifferent elements of the Egyptians (it was the Egyptians who most of all debased religion with superstition and from them it spread to other peoples). Christ reformed the religion of Moses."[9]

[8] Harry Ransom Research Center, the University of Texas. See also Babson MS 437 and Yah. MS 21.

[9] *Newton Correspondence*, 3: 338.

In a manuscript entitled *The Original of Religions*, possibly that seen by Gregory, Newton claimed that the vestal Noachid religion, in which worship had been organized around a central fire, was the most "rational of all": "twas one designe of y^{e} ~~true systeme of y^{e}~~ first institution of ye true religion to propose to mankind by ye frame of y^{e} <ancient> Temples, the study of the frame of the world as the true Temple of y^{e} ~~living~~ <great> God they worshipped." Corrupt worship began through the perverted adoration of heavenly bodies and men had been "led by degrees" to worship the sensible objects of the heavenly bodies represented in the vestal temple as "the <visible> seals of divinity." This worship "being more plausible than that of dead men & statues, mankind was more prone to it & hence it spread further." After this, there had been a general sort of corruption of the vestal religion in which men had come to worship "fals Gods & Goddesses in ye Prytanea." Next came a still 'grosser' form of idolatry, namely "introducing ye worship of dead men & statues."[10] This was constituted by the attempts of the first men to honour their ancestors by putting their names to the heavenly bodies, just as Galileo had done in the seventeenth century. Since the "nature of Mankind is prone to superstition they began to ~~honour~~ <pay respect to> these Gods in the Prytanea or Temples together wth ye true God," and finally they started to "honour ym wth ye worship of ye true God by sacrifices & to cover over this worship by applying the name of the true God Jehova" to them.[11]

As Gregory noted, Newton held that the original corruption of this kind happened in Egypt, and the Egyptians were the first to worship false gods and to set up their images. As a result of the increase in commerce they propagated this religion to the Greeks and Assyrians and it was not long before the offspring of heroes like Chus were worshipping false gods in their prytanea alongside the true God in their sacrificial Bacchanalia.[12] Rulers introduced "the veneration of their ancestors among ye people," which caused there "to be <soon created in the people a superstition towards them as Gods & by consequence a veneration of ye whole race of kings as>

[10] Yah. MS 41 ff.. 8r and 9r-v. Cf. also Yah. MSS 17.2 fol. 21v and 16.2 fol. 3r. See also R.S. Westfall, "Isaac Newton's *Theologiae Gentilis Origines Philosophicae*" in W. W. Wagar, ed., *The Secular Mind: Essays Presented to Franklin L. Baumer* (New York: Holmes & Meier Publishing, Inc., 1982), pp. 15-34.

[11] Yah. MS 41 fol. 9r-v. The reference to the Roman Catholic Galileo depicts him by implication as one of the latest in a series of idolaters.

[12] *Ibid.*, ff. 9r and 10r-v. There is a detailed account of the Egyptian religion (with an explanation of the relevance of characters like the crocodile) and worship at Yah. MS 16.2 ff. 44r, 67v-68v, and 73r (the latter is almost identical to Yah. MS 41, ff. 16r-17r). For the Egyptians' transference of their heroes' names to the stars, see Yah. MS 16.2, fol. 57r; Yah. MS 17.2, ff. 14r (on the introduction of "astrological superstitions" to astronomy) and 20v (on the duplicitous introduction by the Egyptian Priesthood of a "mythological chronology" and of stars "animated with souls which understood human affairs"); and Yah. MS17.3, ff. 1r-15r, esp. ff. 5v, 8r, 9r, and 11r-15r.

descended from these Deities." This naturally "tended to ye establishment & enlargement of ye Regal authority & dominion"; soon all kings were named after the greatest Gods and "after death to augment & perpetuate their honour, it grew in fashion to consecrate them (as was usually done to ye Roman Emperors) & from that time they were accounted Divi." "Thus," he concluded, "was ye world soon filled with Gods." In an extended work on the apocalypse, one of the few projects he ever came near to completing, Newton explicitly linked this to the same sin carried on by Catholics: "This worshipping of dead men was the crime for which the Roman Catholicks were punished ... & yet the more they were punished the more they worshipped them; Or, as is exprest in the two next Vials of wrath, They blasphemed God because of their pains & repented not."[13]

In an earlier Latin version of the *Original of Religions*, dating from the mid-1680s, Newton's specific target was the priesthood. The precise form of idolatry which they introduced after the worship of stars had begun and the cult of astrology had set in was that of the transmigration of the souls of the dead.[14] Priests began to use this fiction to state that the stars understood human affairs and were able to predict what would happen the terrestrial sphere, and finally they instituted "the veneration of sacred animals and plants and the cult of Bethels and columns and statues, as well as various prophetic and magic and necromantic incantations."[15]

Christ had reformed this abuse but the devil, Newton believed, had lain in wait and "played a cunning game." Like the older philosophical and religious corruption, modern Catholicism rested on a perverse misinterpretation. The ancients began to read *literally* the descriptions of the world-systems, understanding that the central symbol of the ancient world system referred to the Earth and believing incorrectly that the circular orbits of the sun and the other planets described real physical entities. Embodying an identical form of mis-understanding, the central rite of the Catholic faith was a theatrical sham, masterminded by "men of corrupt minds" who

> not attending to the relation wch the names of Christ have to the prophesies concerning him, & wch the severall parts of Scripture have to one another; but taking things in a litteral & ~~philosophical~~ natural sense wch were spoken allegorically & morally with relation to piety & virtue, & wresting the expressions of Scripture to the opinions of philosophers, have brought the Christian religion many philosophical opinions to wch the first Christians were strangers. So where Christ saith this is my body, meaning a type of his

[13] Yah. MS 41, ff. 10r-11r; Keynes MS 5, fol. 108r.

[14] Yah. MS 17.3, ff. 14r-v and 15r, is a detailed account of the early astral idolatry and its transformation into the worship of transmigrated souls.

[15] Yah. MS 17.3, ff. 5v and 8r.

body, the Roman Catholicks understand it litterally as if the bread was ~~transubstantiated~~ <changed> into Christ's body in a litteral sense.[16]

Catholicism shared the general features of ancient idolatry but was, in its details, far worse than anything ever perpetrated before the birth of Christ. It was the modern variant of the worst form of religious abuse Newton could imagine, and he devoted even more effort to pointing out the nature of these more recent corruptions than he did to the older perversions.

2. Catholicism as Sorcery

When writing on the uniquely pernicious religion of Roman Catholicism, Newton took down information from modern works as well as from older sources, both Scriptural and non-scriptural. He maintained a keen interest in the doctrinal issues which separated Protestants from Catholics, and in his theological notebook, he extracted pro-Arian passages from Falkland's discourse against infallibility. According to the writers from which Newton drew, both Catholics and the church fathers were inconsistent with one another and amongst themselves. In notes from Lord George Digby's letter to Kenelm Digby, along with the latter's replies, Newton recorded that Lord Digby appealed to the Rule of Faith and argued that Catholic adherence to the authority of the fathers was perilous:

> If you lay ye rule absolutely generall that what article soever is delivered directly or by intimation from ye fathers to have been a received doctrine of ye Church ought to be swallowed for an infallible verity, it will easily be made appear that this method must betray you not only into some Protestant tenents [sic], but also into beleifs on both sides confessed to be erroneous. It must draw you to be a Millenary, to hold a necessity of Children's partaking ye Eucharist, to abhor ye use of Images as Idolatrous, & to reject out of ye Canon those books wch wee esteem Apocryphal: for all these do ye fathers deliver wth somewhat more then intimations that they were taught to them as

[16] Yah. MS 15.3, fol. 47v. Leibniz is the main target here. Compare with Yah. MS 15.5, fol. 97r: "The grand occasion of errors in the faith has been the turning of the scriptures from a moral <& monarchical to a physical &> metaphysical sense & this has been done chiefly by men bred up in the <metaphysical> theology of the heathen Philosophers ~~(such as were Pythagoras, Zeno, Plato & Aristotle)~~. For men tainted with the metaphysical principles of Philosophers <Cabbalists & Schoolmen> upon reading the scriptures have been apt to strain everything to their own opinions. If it be said *This is my body*, meaning a symbol of my body, they take it in a metaphysical sense for transsubstantiation ... If God be called ο παντοκρατωρ the omnipotent, they take it in a metaphysical sense for God's power of creating all things <out of nothing> whereas it is meant <principally> of his <universal> irresistible monarchical power to teach us obedience." For a short passage on idolatry from the late seventeenth century which holds that transubstantiation was the worst form of idolatry, see Yah. MS 14.1, ff. 9r-10v.

derived from ye Apostles, & from generall receptions of ye Catholick Church.[17]

He drew from a number of different sources for his information on Catholicism, and one section ended with a seemingly unconnected remark on Marian atrocities. From a work entitled *No Protestant Plot*, he noted that in a period of five years, "according to ye truest account," there were no fewer than 284 Protestants burnt at the stake for professing the Gospel besides those who were exiled or died in prison. On the other hand, the author of the preface to Ridley's *De Cena Domini*, reckoned to be Grindal and hence, Newton remarked, of extraordinary credit, had noted that the figure was closer to over eight hundred in the first two years of her reign. Newton went on to point out that even future incumbents noted the viciousness of the behaviour of popes, and he further remarked that Aeneus Sylvius (later Pope Pius II) believed that "there was never any great Slaughter in Christendome, nor any great calamity happen'd either of Church or State, whereof ye Bishops of Rome were not the Authors." For good measure, Newton added, Machiavelli agreed with this view in his history of Florence.[18]

Although he was a meticulous note taker, his penchant for producing extended essays and treatises on various themes allowed him to marshal this evidence into a coherent format, and in most of these texts, attention soon turned to the perversions of Christianity introduced by Catholics. In a narrative from the mid or late 1680s—the period of the production and publication of the *Principia*—he launched into both ancient and modern idolaters although the language clearly betrayed popery as the target of his ire. Those described in Revelation chapter 9 verse 21 who "neither repented...of their murders nor of their sorceries, nor of their fornication, nor of their theft," were "sorcerers, yt is addicted to pious frauds, fals miracles & jugling tricks in matters of religion, using all kinds of fraudulent Arts & delusions for deceiving men into their idolatrous religion & carrying on ye interest of their Church." Newton went on to describe the heathen oracles

> who by the impostures of their Priests spake as it were miraculously by ye spirit of ye Gods, to ye seduction of ye world & advancement of idolatry. ~~The third miracle is~~ <'To these miracles may perhaps be referred' added by Newton> yt ye fals Prophet caused all men to receive a mark in their right hand or in their foreheads. Of this kind is a cross or crucifex adored in ye right hand for <its> miraculous powers & effects, & ye signe of ye cross when made wth ye right hand in the forehead ~~in nomine + Patris & + Filij & +~~

17 Keynes Ms 2, ff 3-5 and 8r.

18 Keynes MS 2, fol. 18v. The most significant of these texts was [R. Ferguson], *No Protestant-plot, or, The present pretended Conspiracy of Protestants against the King and Government; discovered to be a Conspiracy of the Papists against the King and his Protestant-subjects*, (London, 1681).

> <'or any where els in nomine Pa+tris et Fi+lij &' added by Newton> spiritus <'+' added by Newton> sancti to scare away ye Devil or produce any other supernatural effects.[19]

Any pretence at being a historical description of ancient idolatry soon disappeared as he tore into the subject of Christian sorcery and magic, again taking over from his amanuensis and rewriting what the latter had just copied. Humphrey Newton, the amanuensis concerned, took down the following:

> ffor sorcery & magick is nothing else then ye making & using certain figures & characters wth ceremonies & invocations to make spirits come & go, appear & disappear & to produce any other supernatural effects. To make ye figures is innocent, but to place a supernatural power in them is sorcery be it in any religion whatsoever, & Christians who do so are sorcerers, thô they do not know it.

In case the reader had not got the point, Newton himself put pen to paper and made the target of his writing clear:

> So the consecrating <Images Pictures> Holy water, Agnus Dei's, Psalters, wings, <Beads> wooden crosses & ye like by kissing them or touching ym wth reliques or performing any other ceremonies wth invocations & prayers for endowing ym with any kind of supernatural powers & vertues is a superstition of ye same kind wth ye Charmes & spells of ye old Heathens, & even wthout a figure may be truly called enchantment & sorcery: as may also ye turning a wafer into ~~Christ's <natural> body~~ <a man> by ye words Hoc est corpus meum.[20]

Such analysis was explicitly applied to the origins of the Catholic Church in one of his most protracted efforts at writing a history of the machinations of Athanasius, the fourth century Christian, which probably dated from the mid-1690s. Athanasius and his friends corrupted the records of the Councils of Nice and Sardica in order to impose their own trinitarian reading of the basic Creed agreed at Nice, and then he "set on foot the invocation of saints." Soon this new religion was spread into Syria and beyond:

> Nor did they only invoke the Saints as intercessors & mediators but also as ~~Guardians~~ Protectors, Rulers & Coadjutors. They accounted them ye bulwarks <towers> & guards of their Cities, & the Patrons & Guardians of their persons; ~~& pray~~ they ascribed deliverances, victories, cures of diseases & other miraculous operations to them & to their reliques; they prayed that they would assist, defend, govern <& prosper> them & their Churches, &

[19] Yah. MS 9.2, ff. 99r-100r. Humphrey Newton (no relation) was Newton's amanuensis between approximately 1684 and 1689; most of this text is in his hand.

[20] Yah. MS 9.2, ff. 99r-100r, 99v.

> after death receive them into heaven: <all> which is the highest degree of worship. <I will not give it a hard name but leave the reader to consider it.>.

Allied to this was the doctrine of miracles, and this age far surpassed any previous one in its superstition, "for their devotion towards the saints being enflamed by the cry of miracles ~~done by them~~, seems to have been in this age at so much <a> greater pitch than in those wch followed, as ye cry of miracles was <now> greater."[21]

When Athanasius found his ambitions thwarted by increasingly anti-trinitarian Councils, he and his colleagues attempted to persuade heathens to become Christians by pointing to the similarities between pagan worship and the religion which he was trying to introduce. To Newton this showed how he "contrived his religion for ye easy conversion of the heathens ~~as far as ye~~ by bringing into it as much of ye heathen superstitions as the ~~temper of those times would bear~~ name of Christianity would then bear." It was natural that the heathens would choose a religion which resembled their own, and he remarked: "this I take to be ye true original of heathen cermonies [sic] & superstitions wch continue to this day in the Greek & Latine Churches. ffor who sees not that those superstitions came into Christianity while heathenism was yet in being?" Emboldened by his success, Athanasius himself grew in stature:

> Twas this authority & none but this wch could bring into ~~the Christianity~~ ye religion of his party, ~~stupendious Legends~~ monstrous Legends fals miracles, veneration of reliques, ~~Invocation of Saints, ye doct~~ <~~the doctrine of Ghosts and Dæmons~~> charmes, ye doctrine of Ghosts or Dæmons, ~~ye invocation & adoration of Saints~~ & their <intercession> invocation & worship & such other heathen superstitions as were then brought in. ffor ~~had~~ as we have shewed that he did allow these things so had he disallowed them his party would not have received them.

At the same time he instituted the practice of false miracles, amongst the earliest examples of which were to be found in Athanasius's *Life of Anthony*.[22]

Newton marshalled a great deal of evidence to show that such miracle-working was pioneered by Athanasius in his position as Bishop of Alexandria. The Athanasians organized the exchange of relics, although they were not "content wth <dissecting &> distributing the bodies of ~~the~~ real martyrs, but daily pretended to the discovery of new bodies by dreams & revelations." So many martyrs and relics were discovered in this period that Emperors had to issue edicts against such practices, and in tandem with

[21] Clark Library MS. ff. 38r-39r, 54r-55r, 45v and 46v. I am currently preparing an edition of this text and 'The Original of Religions' for publication.

[22] Clark Library MS, ff. 48r and 61r.

these tactics, Athanasius inaugurated monkery and the writing of false legends, such as that of his *Life of Anthony*. However, the outraged Protestant don railed against the audacity of such a crude piece of deception:

> this life is full of prodigious stories such as are the devil's appearing <frequently> to Antony in several shapes & <bignesses &> talking with him & afflicting him & struggling with him & sometimes multitudes of devils appearing in various shapes. Christ's appearing in ye form of light to Antony & speaking to him. Antony's remaining as fresh & plump after long fasting as if he had not fasted, his curing diseases, casting out devils, escaping Crocodiles by prayer, frighting away devils <& curing demoniacks> by the signe of the crosse, seing the soul of Ammon ascend up to heaven, ~~having revelations, foretelling things~~ being <himself> lifted up into ye air, & having revelations & by a spirit of prophesy foretelling things.

Some had doubted that a saint like Athanasius could write such gibberish, and even Protestants thought that the incredible life now extant differed from a now lost, authentic one composed by Athanasius, but Newton knew otherwise. Such people objected "that <the life now extant contains> many fabulous ~~things~~ & incredible things, wch so pious & judicious a man as Athanasius would not write." "But," he continued, "I take it to be <rather> an argument that Athanasius did write them because they are of ye same kind wth other things wch Athanasius then wrote."[23]

The effects of this life were momentous. First, "it <did> set all the Monks upon an humour of of pretending to miracles: so that ye whole world presently rang wth stories of this kind." From this "it came to passe that ye lives of <~~almost~~> all ye first ~~monk~~ & most eminent Monks were filled wth apparitions: of Devils, miraculous cures of diseases <prophesies> & other prodigious relations." This was "ye original of those ecclesiastical Legends ~~& Romances~~ wch are still use in ye Church of Rome ~~& from wch ye name of Romances seems to have been taken.~~" Secondly,

> the superstitious or to speak more truly the magical use of the signe of ye Cross was hereby set on foot & propagated all over ye East. In the western Churches it was in use before. ffor they began ~~in~~ <~~before~~> <in> ye time of Tertullian <or before> to be tainted wth ye superstitions of the Montanists. Athanasius in his exile learnt it there & ~~makes Antony teach how by this signe th~~ <now turning it into an inchantment makes Antony> in a large discourse to ye Monks teach them how by this signe they may drive away the Devil & dissolve all kinds of ~~witch inchant~~ inchantments & witchcraft <And this I like to be the original of ye Greek & Latine Church using it for this end.>.

The life of Anthony brought the monks within Athanasius's sphere of influence, and caused monkery to "overflow ye Roman world like a torrent.

[23] Clark Library MS, ff. 64v, 80r, 81r, and 77r.

ffor this life going into all ye Empire struck the homousians wth admiration & made many of them desire to imitate Antony; ~~the~~ especially when they heard that ye guift of miracles continued ~~still~~ amongst ~~ye~~ his successors."[24]

The clergy became monks and Athanasius gained his Bishopric by dint of the standard techniques of sedition, lying, and violence. He and his successors began to triumph over their enemies, including the valiant anti-trinitarian Arians, while simultaneously promoting the view that their own lives were paragons of Christian virtue. Newton was not impressed. To him they were the "most superstitious part of mankind" that the world had ever seen or would see. Instead of avoiding temptation, they tried to wrestle with it, an activity in which they invariably and unsurprisingly failed. By trying to restrain their lust they fell right into it, and every losing battle with the temptations of the flesh left such indelible impressions that the thoughts returned with increasing vengeance. Other bodily practices, such as fasting, were taken to such extremes that the early monks were prone to hallucinations and ultimately madness. Newton had a ready diagnosis: "<The want of sleep & due refreshment ~~brings men~~ disorders the imagination & <at length> brings> men to a sort of distraction <& madnesse> so as to make them ~~think they~~ have visions of weomen conversing with 'em ~~& sitting upon their knees~~ & think they really see <& touch> them & heare them talk." He then regaled the reader with other deviant practices and attitudes, such as the lusting after passive sodomy, "their digging up the bodie[s] of dead weomen wth w[ch] they burned in lust," and above all, the relating of such tales "wthout blushing." The superstitious nature of their religion was proved by their placing so much of religion in "bodily exercises," such as fasting, ascetic auto-mutilation, and the adoption of unnatural corporeal postures, as well as "in using ye signe of ye cross, ~~holy water &~~ <dead mens bones, holy ~~oyle &~~ water, consecrated oyle &> other consecrated things to do supernatural operations; ~~in saying a certain number of prayers every day~~ & in praying by number." From the last practice, Newton concluded, "you have ye original of the Popish custome of praying by beads: & if you desire the original of praying ~~without [?]~~ wthout understanding or (wch is all one) in an unknown tongue..."[25]

The narrative of Athanasius's wrongdoings went to the heart of the origins of all Newton despised. This was a carnal and metaphysical religion, cynically modelled on the heathen religion to which it was intimately related. It was a religion whose authority rested upon ignorance and the

[24] Clark Library MS., ff. 66r, 67r, and 67Ar.

[25] Clark Library MS, ff. 67Ar, 75r, 76v, and 76r. For an extended treatment of Newton's ascetic care of the self, and of his views on the depraved bodily practices of Catholics, see R. Iliffe, "Isaac Newton: Lucatello Professor of Mathematics," in C. Lawrence and S. Shapin, eds., *Science Incarnate: The Bodily Presentation of Scholarly Self* (Chicago: The University of Chicago Press 1998), pp. 121-55.

misuse of power rather than upon understanding and the proper dissemination of truth, and its crafty proponents had an arsenal of techniques to blind and dupe ordinary people. Although his theological work was a wholly private affair, he got the chance to publicly confront the living spectre of Catholicism at an early juncture. While such textual researches informed his dealings with Catholics in the 1670s and 80s, these same experiences no doubt heightened the intensity of his writings on Catholicism.

3. Disputing Jesuits

Recent work has shown that, for a number of natural philosophers connected with the early Royal Society, the epistemological status of the knowledge they produced was grounded on the freely given assent of relevantly qualified witnesses. Promoted in particular by Robert Boyle, this empiricist approach to the validation of knowledge was part of a process in which an historically specific event such as an experiment or 'trial' could become more credible by being witnessed many times and in many different places by a large group of reliable people. This was also probabilistic, in that although one piece of 'probable' testimony could not guarantee the absolute certainty of a matter of fact, a concurrence of such probabilities increased its likelihood of being true and made it 'morally' certain. Derived explicitly from a legalistic framework, it enjoined anyone working according to its tenets to publish clear and accurate accounts of real events that actually took place. The wider the readership, the broader the group of people who could in principle attest to the matter of fact depicted in an experimental report.[26]

However, this *modus operandi* of natural philosophy clearly grated with someone like Newton, who wrote in a mathematical way and was used to the mathematical styles of proof and exposition he had adopted in the optical lectures which he had given on assuming the Lucasian Chair of Mathematics in 1669. In this world, the demonstrated experiment, in particular the 'crucial instance' described by Francis Bacon, shouldered the burden of proving the certainty of a matter of fact. Incapable of giving the absolute certainty which was obtainable in the mathematical disciplines, and prone to the vagaries of human error, reliance on witnesses was a perverse confusion of the worlds of the natural and the social. Disdaining the authority of the vast number of witnesses who might vouch for the authenticity of a matter of fact, Newton told Hooke in February 1675/6: "What's done before many

[26] See S. Shapin and S. Schaffer, *Leviathan and the Air-Pump: Hobbes, Boyle and the Experimental Life*, (Princeton, 1985), pp. 55-60; R-M. Sargent, *Diffident Naturalist: Robert Boyle and the Philosophy of Experiment*, (Chicago and London: The University of Chicago Press, 1995).

witnesses is seldome wthout some further concern then that for truth: but what passes between friends in private usually deserves ye name of consultation rather then contest." In this section I examine how Newton believed that the Jesuit critics of his theory of light and colors also turned the world of natural philosophy into a law-court.[27]

In early February 1671/2, he published a paper on light and colors in the *Philosophical Transactions* in which he demonstrated that white light was composed of a number of different rays, each of which was associated with a particular color and each of which had its own degree of refrangibility when passed through a prism. Initially he claimed that he had first noticed something strange when a beam of sunlight passed through such a prism had resulted in an extended oblong shape, a phenomenon not accounted for by the older 'modification' theory of light. Newton's historically specific account of an event (which actually never took place) was couched in the conventional Boylean language of reporting until Newton described a 'crucial experiment' and then suddenly announced that to continue in this historical vein would be 'tedious.' After this he laid down the 'doctrine' in a manner redolent of mathematicans, and provided another experiment to illustrate the doctrine.[28]

After the theory's lukewarm reception from a number of critics such as the Jesuiit Francis Linus, Newton tried on a number of occasions to 'withdraw' from the Royal Society. However, having been lured back into the public sphere to respond to these issues at the end of 1675, he was suddenly faced with new criticisms from another Jesuit, Anthony Lucas, in May 1676. Lucas claimed that he had performed a number of relevant experiments contradicting Newton's views on the constant refrangibility and colour of individual rays, and that according to standard practice, Newton needed to provide a large number of experiments before his readers could be convinced. In response, Newton pointed him to the authority of the crucial experiment mentioned in his original article and argued that Lucas should have determined the truth of Newton's *printed* experiments before he questioned whether Newton had, as he had claimed, performed a number of other unpublished experiments. He confirmed that he did have such a store and had "written a Tractate on the subject wherein I had set down at large ye principal of those experiments wch Mr Lucas has now sent me," but it was wholly unnecessary to release these if the trials proposed in his February 1672 letter to Oldenburg were "demonstrative."[29]

[27] Newton to Hooke, 5 February 1675/6, in *Newton Correspondence*, 1: 416.

[28] Newton to Oldenburg, 6th February 1671/2 in *Newton Correspondence*, 1: 92-107, esp. 94-97. Oldenburg removed the remark about the tedium of the Boylean style in the published version of the paper.

[29] Newton to Oldenburg, 11th May and 18th August 1676 and Lucas to Oldenburg, 17th May 1676 in *Newton Correspondence*, 2:6-7, 8-12, 76-81, esp. 79. See S. Schaffer, "Glass works: Newton's prisms

Seeing his word being called into question by a member of a religion he despised, Newton insisted on the validity of accurate measurement, mathematical rigour, and his own veracity as a credible relator of his own experiment. He repeatedly referred Lucas back to the *experimentum crucis* and warned of the consequences for truth if their controversy degenerated into the disputacious argot of law:

> should I be drawn from [this], it might create us both ye trouble of a long dispute, & by ye multitude of words cloude rather then clear up ye truth. For if it has already cost us so much trouble to agree upon ye matter of fact in ye first & plainest experiment, & yet we are not fully agreed: what an endless trouble might it create us, if we should give our selves up to dispute upon every argument that occurs, & what would become of truth in such a tedious dispute.

In a covering letter he mentioned that a series of new protocols given to Lucas and his allies in order to replicate successfully his crucial experiment were "perhaps more to ye purpose considering who I have to deal wth, whose buisiness it is to cavill."[30]

Lucas remained unsatisfied by Newton's response and wrote again to Oldenburg in October 1676. Drawing Newton's "new Theory into a close Syllogisticall methode," his letter confirmed Newton's belief that the Jesuits merely wanted to dispute in the barren legal mode of the monks. Such a style of correspondence triggered Newton's loathing and he informed Oldenburg that in responding to their criticisms, "I see I have made myself a slave to Philosophy." If he could finish the business, he would "resolutely bid adew to [Philosophy] eternally. For I see a man must either resolve to put out nothing new or to become a slave to defend it." As for Lucas, Newton once more returned him to the crucial experiment, asserting that anything else was an "attempted digression." What mattered was the "truth of my experiments," for this was the basis of his theory. "Which is of more consequence," he asked, "ye credit of my being wary, accurate and faithfull in ye reports I have made or shall make of experiments in any subject, seeing yt a trip in any one will bring all ye rest into suspicion [?]" The question being about the length of the image, Newton had added yet more directions for determining whether "ye Prisms of men abroad" were similar to his. Once this was resolved, he had hoped that this would "take off all suspicion of my misrepresenting matter of fact ... yet I cannot but say yt I

---

and the uses of experiment", in D. Gooding, T. Pinch and S. Schaffer, *The Uses of Experiment: Studies in the Natural Sciences*, (Cambridge: Cambridge University Press, 1989), pp. 67-104 and A. Shapiro, "The gradual acceptance of Newton's Theory of Light and Color 1672-1727", *Perspectives in Science* 3 (1996), pp. 59-139.

[30] Newton to Oldenburg, 18th and 22nd August 1676, 2nd September in *Newton Correspondence*, 2:80-1, 83, and 94.

had rather have my expt confirm'd by his assent than ye suspicion only taken of."[31]

Lucas himself was irritated by Newton's reply. He *had* examined the crucial experiment and it was *not* "a new attempted digression," as Newton had insinuated. Nor was Newton's professed intention to examine only one or two select experiments the proper way to decide the issue. Such a way of proceeding Lucas condemned as "illegal, unlesse the defect of these two experiments, considered by him, be shown to run throught the rest allso." Swiftly losing patience, in two drafts of a letter of March 5th 1677/8, Newton described Lucas as making a "stirr" about his objections. He finished one draft with the accusation that Lucas had "treat[ed] me wth ill language for my unwillingness to meddle wth" the experiments he had proposed, and remarked: "if you will not yet freely let us know your mistakes but hope to mend ye matter by new disputes, you are at your liberty." Above all, he wrote, he himself was not about to be "put upon disputes of this Nature ..."[32]

In a second draft of this letter he railed against Lucas's attempts to "reconcile" him with the truth, and denied that the language of the law court was apposite for this kind of affair:

> You endeavour to oblige me into a new dispute wth you by calling me *a Demurrer* & your new objections *your main Debate* &c. And my profer to answer one or two of your objections wch you should recommend for ye best you despise as if it was Illegal (as you term it) if I answer not at all. Do men use to press one another into Disputes? Or am I bound to satisfy you?

Newton did "not think this a fit Subject to dispute about", and told Lucas that on these grounds this and the accompanying letter were to be judged private: "I hope you will consider how little I desire to explain your proceedings in public & make this use of it to deal candidly wth me for ye future."[33]

In May, 1678, Robert Hooke told him that he had seen a copy of yet another missive from Lucas, and remarked "I much admire your patience that you will trouble your self wth such an extravagantly impertinent—, who will never yeald be the matter never soe plain." Soon after this, Newton told John Aubrey that he had heard rumours that "instead of becoming more moderate [Lucas] still betakes himself to sharper language than ever." Now Lucas had received an answer from Newton he was "for contending further," and he had begun to persuade Aubrey that Newton was "hot & that

[31] Lucas to Oldenburg, 13th October 1676, Newton to Oldenburg, 18th and 28th November 1676; *Newton Correpondence* in 2: 104-8, 182-3, 183-5.

[32] Newton to Lucas, 5th March, 1677/8 in *Newton Correspondence*, 2:259-60.

[33] Newton to Lucas, 5th March, 1677/8 in *Newton Correspondence*, 2:262-3.

he gives me all this time to grow cooler in." He continued: "I did not think that in a dispute wch ought to be grave & sober we should have come to flurting one another, much less could I have expected that so learned a man would ever have sent a letter so far from Liege to London on no more material an errand than this." Again Newton ranged back over what he took to be the bones of the Jesuits' position. Ignoring the quest for truth, the Jesuits had tried to find new observations to support some old ones of Linus "though in candor & aequity they ought to have sent an immediate acknowledgement to take of those suspicions of error & dishonesty in me..." A supposed "reconciliation" of Linus with Newton and truth was merely a "trick" and the addition of new objections an attempt "to divert ye readers eye as I presently apprehended & run of ye business into a new dispute." On the other hand, Newton had sent his last letter "not to prolong disputes but to be published wth his letter if he thought that more his interest rather then to keep quiet."[34]

Like other apologists for the new philosophy, Newton thought that the 'tricky' method of inquiry favoured by the Jesuits was one of fruitless dispute, an obfuscating procedure which clashed with his own views on the correct way to interpret both nature and Scripture. Above all, they tried to embroil a good Protestant in disputes which were verbal sorcery designed to trick the unwary. The truth was simple and susceptible only to the kind of experimental enquiry which Newton himself proposed. Disputes were also liable to mystify religious truths, and in a presentation of his own religious creed he remarked that it was

> short and easy to be remembered as the symbols of religion ought to be <all its Articles are in the Scriptures in express words & so liable to no disputes. It conteins not repetitions nor> Theories like some of the Articles wch we have omitted, but practical truths on wch the whole practice of religion depends.... All the Articles of faith necessary to baptism were put into the Creed by the Apostles & nothing unnecessary <nothing disputable or mystical> was to be added afterwards.[35]

For Newton, truth concerning nature or church history could never result from legal wrangling. The latter was Jesuitical and metaphysical, and it smacked of the kind of behaviour carried out by the corrupters of

[34] Hooke to Newton, 25th May 1678, Newton to Aubrey, ? June 1678 in *Newton Correspondence*, 2:265, 266-7, and 268.

[35] Yah. MS 15.3, ff. 45r; 15.4, and 67r. This creed was likened to milk for babes, while Newton went on to say that "we must believe that there is a holy prophetic spirit by which Moses & the Prophets & Apostles were influenced that we may study their writings as the Oracles of truth, & thereby grow in grace & in the knowledge of our Lord Jesus Christ to the end of our lives ... These [i.e. more difficult truths] the Apostle compares to strong meats for men of full age who by use have their senses exercised to discern both good & evil [but] the strong meats ... ought not to be mixed with the milk wch is to be given to babes." Yah. MS 15.3, ff. 46r, 15.4, and 67r.

Christianity from the fourth century. Indeed, when Jesuits failed with such tactics they fell back on accusing their opponents of their very own crimes:

> They found by experience y^t their opinions were not to be propagated by disputing & arguing, & therefore gave out that their adversaries were crafty people & cunning disputants & their own party <simple> well meaning men, & therefore imposed this law upon the Monks that they should not dispute about ye Trinity. Thus they left ye success of their cause to ye working of miracles & spreading of Monkery.[36]

Having seen off Jesuits, Newton soon confronted a much more perilous threat, this time posed by own monarch.

4. DEFENDING CAMBRIDGE

His brush with members of the Catholic Church resulted in his withdrawal into the private rooms of Trinity College, and only a major event would draw him back into the public sphere. The threat of intrusions of Catholics into Cambridge University in the late 1680s constituted such a situation, and not yet having completed the *Principia*, he burst into action in early 1687 to defend the University against the efforts of the Roman Catholic James II's attempts to impose the Benedictine monk Father Alban Francis into Sidney Sussex College, Cambridge. Newton soon gained election to a delegation from Cambridge which appeared before a newly set up Ecclesiastical Commission in April and May 1687 and in his preparation for the appearance before the Commission—headed by the notorious Judge Jeffreys—Newton composed an essay on the legality of the King's dispensing with the laws forbidding anyone to take an M.A. in Cambridge without swearing an Oath of Allegiance to the King.

While divine right theorists held that the King was the sole judge of necessity in cases where it was deemed "necessary" that certain laws should be suspended or dispensed with, Newton began his full-frontal assault on the actions of the present monarch by stating that the people had as much right as the king to dispense with and transgress laws in instances of necessity, and neither could dispense on the grounds of mere caprice. As significant as any Protestant view about the evil of Catholicism was the fear of the arbitrary tyranny of a popish King. James's actions conformed to this stereotype and Newton pointed to a number of instances in which he had already infringed the liberties and properties of Englishmen by means of unjustified suspensions of the law. Such illegal dispensations included the removal of the Test Act in favour of Catholic officers, since it was "only the

[36] Clark Library MS, fol. 63r.

Kings will & pleasure <first> to have an~~y~~ Army & then to have those officers rather then others."[37]

Dispensations based on the principle of necessity could be invoked only to safeguard the common good of the people "& not for ye interest of ye King alone against ye peoples," and Newton added that a dispensation "therefore is not valid against their consent." This would be to lessen the status of the very laws which were supposed to protect the people and then to do that "in order to a further injury." From all this

> its plain that ye King is not ye sole judge of cases of necessity. His prerogative is bounded by law & he is not ye sole Judge of those bounds. If ~~the~~ <any King> shall say there's a necessity of raising money wthout consent of Parliament at a time when ye nation sees not ye necessity or of condemning men without triall, or of altering ye legal course of justice and setting up arbitrary Courts in civil matters, to ye danger of property, or of setting up an High Commission Court against law whereby any man whatsoever may be excommunicated & so lose his property for not obeying all ~~his~~ <ye King's> commands: tis not ye King's pretending necessity yt will ~~justify~~ <make> these things legal & therefore ye King is not ye sole judge.

Not any case "wch they will call so" was one of necessity: "ffor on this principle <any people> may as well justify rebellion as ~~the~~ <any> King may oppression: which God forbid." The necessity had to be extraordinary and appear to both sides to their "common satisfaction."[38]

When the delegates appeared before the Commission they failed to make any headway with Jeffreys, and the Vice-Chancellor of the University, John Peachell, was stripped of his office. On the 7th of May, the delegates hastily produced five drafts of an "Answer" to the questions Jeffreys had asked Peachell, and to which the latter had failed to satisfy the Judge. In an addition which survives on only one draft, Newton defended the action of the messengers by indicating that they had felt themselves obliged by their religion, which had been "established & supported by ye laws they are commanded to infringe." Papists and Protestants could not "subsist happily nor long together" in the same university, while if these institutions, "ye fountains be once dryed up ye streams hitherto diffused thence throughout ye Nation ~~will~~ must soon fall of." It was their religion and Church and not their preferments which "men of conscience" wished to preserve and, on behalf of the delegates, Newton lamented that if that religion "must ~~sink finish~~ fall they implore this mercy that it may fall by ye hands of others." A tantalizing reference to "a great ... wickedness" which the delegates might be forced to commit was broken off in mid-sentence. As a result of

[37] Keynes MS 121, fol. 1r.

[38] Keynes MS 121, fol. 1r-v.

resistance such as this, James ultimately failed in his attempts to recatholicize Oxford and Cambridge.[39]

5. Conclusion

Contemporary bishops and archbishops such as Gilbert Burnet, John Tillotson, Edward Stillingfleet, and Thomas Tenison, spent a great deal of time in print attacking the supposed idolatry of the Catholic Church and Newton was as much an expert in these matters as any of them. Others, such as Henry More, drew more explicitly from the apocalyptic exegetical tradition developed in a stunning new way by Joseph Mede and continued with equal intensity by Newton. When More and his colleagues sought a mathematically sophisticated exegete in August, 1680, to interpret some of the more abstruse passages in Revelation and thus help them understand the historical significance of the Popish Plot, it was to the Lucasian Professor of Mathematics that they turned. Unfortunately, on this occasion, Newton's apocalyptic hermeneutics turned out to have consequences for the understanding of More's historical predicament that were far too radical for the task.

In this paper, I have tried to show how the hatred of Catholicism permeated nearly every aspect of Newton's life. An adept in the history of the Church, he spent the bulk of his time at Trinity with the bunker mentality proper for one of the Elect, and engaged in theological study and not in scientific pursuits. Nevertheless, even when engaged upon the latter, his religious views informed the way in which he responded to what was mostly intelligent criticism from Catholic natural philosophers, while no doubt these experiences in turn reinforced his views concerning the true, historical nature of Catholicism as it was described in the Book of Revelation. The Jesuits' barren mode of disputation infuriated someone who believed that his good word as a Protestant scholar was being called into question by a group of men whose religious attachments left them chained to a perverse and outmoded way of examining the natural world. A decade later, when the call came to fight on behalf of his religion, Newton answered it and took the most important stand against the religion of the Beast he had made in his life. His Protestant understanding of history told him that a select few could stand against the crude efforts of the antichristian religion to extirpate the righteous; having read and written about the growth and demise of popery, it was now time to play his part in its downfall.

[39] Keynes MS 116, fol. 2r.

ESSAY 6

Newton, Corruption, and the Tradition of Universal History

ROBERT MARKLEY
West Virginia University

1

Since the publication of Frank Manuel's *Isaac Newton, Historian* in 1963, most commentators have recognized that Newton's unpublished manuscripts on the history of religions fall into three broad, if occasionally overlapping, categories: those devoted to the interpretation of biblical prophecy; those concerned with the histories of ancient empires and the corruption of a primitive monotheistic religion; and those that challenge trinitarian histories of the Church in the fourth and fifth centuries.[1] Although several scholars have argued that these manuscripts reveal what Betty Jo Dobbs has called the "the ultimate unity" of Newton's thought, few have been willing to claim that this unity manifests itself as a methodological imperative in his historiography.[2] In this essay, I want to investigate—and contextualize—the

[1] See Frank Manuel, *Isaac Newton, Historian* (Cambridge: Harvard University Press, 1963), and *The Religion of Isaac Newton* (Oxford: Clarendon Press, 1974); Richard Westfall, *Never at Rest: A Biography of Isaac Newton* (Cambridge: Cambridge University Press, 1980), and "Isaac Newton's *Theologiae Gentilis Origines Philosophicae*," in *The Secular Mind: Transformations of Faith in Modern Europe*, ed. W. Warren Wagar (New York: Holmes & Meier, 1982), pp. 15-34; James E. Force and Richard H. Popkin, *Essays on the Context, Nature, and Influence of Isaac Newton's Theology* (Dordrecht: Kluwer Academic Publishers, 1990); Stuart Peterfreund, "Saving the Phenomenon or Saving the Hexameron? Mosaic Self-Presentation in Newtonian Optics," *The Eighteenth Century: Theory and Interpretation* 32 (1991), pp. 139-65; R. Iliffe, "The Idols of the Temple: Isaac Newton and the Private Life of Anti-Idolatry," diss., Cambridge, 1989; Kenneth Knoespel, "Newton in the School of Time: *The Chronology of Ancient Kingdoms Amended* and the Crisis of Seventeenth-Century Historiography," *The Eighteenth Century: Theory and Interpretation* 30 (1989), pp. 19-41; Betty Jo Teeter Dobbs, *The Janus Faces of Genius: The Role of Alchemy in Newton's Thought* Cambridge: Cambridge University Press, 1991); and Dobbs and Margaret C. Jacob, *Newton and the Culture of Newtonianism* (Atlantic Highlands: Humanities Press, 1995).

[2] Dobbs, *Janus Faces of Genius*, p. 253.

J.E. Force and R.H. Popkin (eds.), Newton and Religion, 121–143.
© 1999 *Kluwer Academic Publishers.*

dialectical tensions between the redemptive history of Newton's writings on biblical prophecy and the postlapsarian history of those manuscripts, such as *The Original of Religions* and the *Theologiae Gentilis Origines Philosophicae* [*The Philosophical Origins of Gentile Theology*], concerned with humankind's Fall from a pristine Noachian religion. If Newton's manuscripts on Daniel and the Apocalypse draw on a rich tradition of prophetic interpretation, and confirm their author's self-perception as one of "a remnant, a few scattered persons which God hath chosen...to search after truth," then his unfinished, obsessively rewritten, and often self-censored *Origines* represents the dark underside of such a faith—the anatomizing of human corruption and the psychological insecurity of a truth-seeker irrevocably implicated in the fallen world that he condemns.[3] I have argued previously that Newton's inquiries in a variety of disciplines—mathematics, biblical scholarship, optics, alchemy, and history—constitute heuristic, and always partial, efforts to understand the workings of a mysterious providence.[4] Precisely because these attempts to celebrate the "absent" guarantee of a divine order must employ the fallen languages of postlapsarian existence, they become, in Newton's mind, rearguard actions fought against the forces of metastasizing corruption, embodied in trinitarian theology by Athanasius, in a deterministic, or at least non-voluntarist, physics by Leibniz, and in politics by the entwined threats of tyranny and Catholicism. In what follows, I want to extend my argument to suggest that Newton's fascination with the origins, corruption, and cyclical renovations of a pristine monotheism locates his sprawling, if fragmentary, historical project in the context of the dominant genre of seventeenth-century historiography—universal history.

Universal histories, with their imaginative readings of pagan myths and fanciful chronologies, are either dismissed as "serious" history by modern scholars, relegated to the status of antiquarian curiosities, or consigned to the realm of poetic fables. Rather than undertaking the thankless task of developing a formal definition of this diffuse genre, I shall examine these self-styled histories in terms of their authors' stated intentions. Universal histories, to paraphrase Milton, are efforts to justify the ways of God to humankind by finding evidence of providential order in the millenia-long cycles of moral depravity, political disorder, economic hardship, and ecological devastation that constitute postlapsarian history. Such histories often swell to prodigious lengths in attempting to forestall real and imagined objections to their global narratives; they celebrate, often to the point of obsession, the authority of the Mosaic account of history, rejecting all

[3] Yahuda MS 1.1; cited in Westfall, *Never at Rest*, p. 325.

[4] See Robert Markley, *Fallen Languages: Crises of Representation in Newtonian England, 1660-1740* (Ithaca: Cornell University Press, 1993), pp. 131-77.

suggestions that historical events are random or inexplicable. By invoking a divine justice at work in the affairs of humankind, then, universal histories assume the burden of explaining the entire historical record—from Moses to Mantheo (the semi-mythical historian of ancient Egypt)—as parts of a coherent narrative whole. But in drawing on pagan as well as Judeo-Christian texts, these histories oscillate between providentialist and materialist explanations of humankind's Fall from virtue and material abundance to sin and scarcity; their authors' divided allegiances mark a difficult—and incomplete—transition from theological metanarratives to what we might now call cultural anthropology.

In this essay, I want first to set these histories within the materialist contexts of their production; quite simply, to understand Newton's relationship to this often vexed tradition we need to examine the history of the seventeenth century as well as the historiography of the ancient world. In lamenting the disorder and deprivation that they find endemic in postlapsarian existence, the authors of universal histories find themselves confronting what modern historians have termed the "general crisis" of the seventeenth-century.[5] The political, moral, ecological, and economic corruption of their own times encourages writers in the 1600s to read history as an ongoing fall from a golden age of prosperity and virtue and, consequently, to see their own tasks as the recovery of universal patterns of meaning in an otherwise meaningless world. Secondly, I examine briefly the Christian and pagan epistemologies that compete for narrative authority in works by Godfrey Goodman, Peter Heylyn, and Sir Walter Ralegh to suggest that universal histories—marked by significant anxieties about the physical deterioration of the earth as well as the spiritual degeneration of humankind—need to be read as experiments in political economy, as attempts to explain the diminishing returns on investment that characterize the postlapsarian world. Finally, I look at the ways in which Newton resists some of the values and assumptions of this tradition to offer an alternative to widespread visions of history as the irrevocable decline from a mythic golden age. In *The Original of Religions*, Newton sketches a fragmentary history of the Fall into idolatry and offers at least glimpses of his program for reinvigorating a corrupt world: religious purity, technoscientific knowledge, and the dedication of a handful of individuals committed to demonstrating scientifically, theologically, and historically the interlocking principles that can sustain an unsullied faith.

[5] On the "general crisis" of the seventeenth century, see particularly Jack Goldstone, *Revolution and Rebellion in the Early Modern World* (Berkeley: University of California Press, 1991).

2

The near-obsession with assessing the consequences of the Fall among seventeenth-century historians cuts across political allegiances and theological beliefs; it is a fascination conditioned, in part, by more than a century's experience of increasing demographic and ecological pressures in England and, after 1640, the turmoil of the Civil War. In their efforts to make sense of ecological, demographic, economic, and political crises which overwhelm the analytical vocabularies which they have inherited, universal histories are forced to contend with crises that, as Jack Goldstone argues, can be understood only by deploying a "demographic/structural model of history."[6] Consider what it must have been like to read, say, Polybius on the fall of empires during the middle of the seventeenth century. Between 1500 and 1650, the population of England more than doubled to over five million; the increase in "marginal" groups (younger sons, sisters, and daughters among the upper and middle classes; displaced agricultural laborers who gravitated toward London or retreated to the waste areas of fens and forests among the propertyless classes) exceeded the growth of the general population by severalfold.[7] Grain prices during this century and a half rose by six hundred percent, in part because landlords sought to increase rents to compensate for a general inflation provoked by the increased population of marginal groups. Wages during the period, however, rose by only two hundred percent. Fixed rents and a greater population of underemployed laborers and younger sons of the gentry clamoring for government positions created intensifying pressures on the Crown to extract more money from duties and taxes to provision larger military forces. The cost of sending a given percentage of the total population into the field as a fighting force in 1640 was approximately twenty-four times what it had been in 1500 (in fixed terms) and probably twelve times that figure when adjusted for inflation and the discrepancy between rising food prices and the much slower increase in land rents and other revenues. Charles I's revenues were only ten percent higher than Elizabeth's had been in the 1580s, but the population of England had grown by two-thirds in this half century, and the gentry, "always hungry for patronage," had tripled.[8] The costs of running government in France, the Ottoman Empire, and China as well as in England

[6] Goldstone, *Revolution and Rebellion in the Early Modern World*, pp. 24-6. The ensuing discussion is indebted to Goldstone's analysis. Some useful reservations about revisionist histories of the seventeenth century are voiced by Ellen Meiksins Wood, *The Pristine Culture of Capitalism* (London: Verso, 1991), pp. 117-60. Although a full-scale articulation of crucial areas of agreement between Goldstone's structural/developmental model of history and sophisticated Marxian analyses lies beyond the scope of this essay, I believe that these approaches are less antagonistic than complementary.

[7] Goldstone, *Revolution and Rebellion*, pp. 32-4 and 87-92.

[8] *Ibid.*, p. 100.

by the mid-seventeenth century, Goldstone argues, fragmented elite classes, provoked rebellion among workers whose living standards had been declining rapidly, and led to fiscal crises that traditional state institutions and tax structures, which favored the elite classes, could not solve. The breakdown of the old order in these states was halted only after warfare, civil unrest, starvation, and disease abruptly—and tragically—reversed decades of population growth. During the period between 1650 and 1730, as population levels slowly rose to where they had been in 1640, state income rose in real terms, land became less scarce, labor shortages led to higher wages, and ecological pressures eased.[9] To writers such as Ralegh and Newton who witnessed the ongoing disintegration of a supposedly "timeless" order, chaos indeed seemed to have come again.

Goldstone's demographic/structural argument brings the seemingly arcane debates of seventeenth-century universal history within the realm of a cultural anthropology that offers us the opportunity to consider the crises that beset ancient empires and seventeenth-century states in terms of *intensification*. As Marvin Harris describes it, intensification is "the investment of more soil, water, minerals, or energy per unit of time or area," which is humankind's "recurrent response to threats against living standards."[10] But even if it raises living standards temporarily, intensification invariably proves counter-productive over time because, in Harris's words, "the increased effort sooner or later must be applied to more remote, less reliable, and less bountiful animals, plants, soils, minerals, and sources of energy." As resources become scarcer, living standards decline, unrest spreads, until cultures collapse into anarchy or "invent new and more efficient means of production which sooner or later again lead to the depletion of the natural environment."[11] Harris's description calls attention to the complex interactions among technology, rates of resource extraction, environmental degradation, population pressures, fluctuations in living standards, distribution networks for food and water, communication, labor, political power, and the creation of symbolic systems of value that favor those in positions of authority. These factors, to a great extent, are interdependent; if their complex interactions define the "general crisis" of the seventeenth century, they also resist hard and fast analysis, whether in the twentieth century or the seventeenth.[12]

[9] *Ibid.*, pp. 83-92 (on England); pp. 356-91 (on China).

[10] Marvin Harris, *Cannibals and Kings: The Origins of Cultures* (New York: Vintage, 1977), p. 5.

[11] *Ibid.*

[12] On the problems of identifying reasons for the disintegration of complex societies, see Joseph A. Tainter, *The Collapse of Complex Societies* (Cambridge: Cambridge University Press, 1988). His description of "diminishing marginal returns," in many ways, parallels Harris's account of intensification.

In this context, seventeenth-century universal histories can be read, in part, as attempts by their authors to find coherent narrative forms to describe what seem to be overwhelming crises of intensification. By setting themselves within the framework of providentialist and often eschatological thought, such histories foster the view that contemporary crises are the effects of thousands of years of corruption and harbingers of an apocalyptic end to history. Among those writers from Ralegh to Locke who study the origins of corruption, Newton may be the most determined to describe an alternative to apocalyptic narratives of decline by emphasizing the crucial role that scientific knowledge can play in overcoming the grim logic of intensification.

In their efforts to make sense of the crises of intensification and diminishing marginal returns that beset ancient empires, seventeenth-century universal historians are caught between competing historiographic methodologies, narrative modes, and conceptual models: a Judeo-Christian perception of history as the mysterious unfolding of God's will and a pagan view that human experience resists teleological explanation.[13] History as ideal form is set against history as contingent experience. Although universal historians invoke Judeo-Christian narratives to assess, and frequently denigrate, pagan "myths" of the rise of civilization, they nonetheless rely on a conceptual framework indebted to Hesiod's and then Ovid's accounts of the degeneration of humankind from the Golden, to the Silver, to the Brass, and finally to the Iron Age as well as to Polybius's cyclical history of the rise and fall of empires.[14] In practice, most universal histories undertake euhemeristic projects to align the various chronologies of ancient civilizations in coherent timetables that either demonstrate the moral authority and historical accuracy of the Old Testament or prove that the accounts of Chaldeans or Chinese retell, with some distortions, the story

[13] See Arnaldo Momigliano, *On Pagans, Jews, and Christians* (Middletown, Connecticut: Wesleyan University Press, 1987), pp. 11-57.

[14] For representative works that deal with universal history, see John Racin, *Sir Walter Raleigh as Historian: An Analysis of The History of the World* (Salzburg: Institut fur Englische Sprache und Literatur, 1974); Anthony Grafton, "Joseph Scaliger and Historical Chronology: The Rise and Fall of a Discipline," *History and Theory* 14 (1975), 156-85; Arthur B. Ferguson, *Clio Unbound: Perception of the Social and Cultural Past in Renaissance England* (Durham: Duke University Press, 1979), pp. 347-79; Paolo Rossi, *The Dark Abyss of Time: The History of the Earth and the History of Nations from Hooke to Vico*, trans. Lydia G. Cochrane (Chicago: University of Chicago Press, 1984); Frank Lestringant, *Mapping the Renaissance World: The Geographical Imagination in the Age of Discovery*, trans. David Fausett (Cambridge: Polity Press, 1994); Kenneth Knoespel, "Milton and the Hermeneutics of Time: Seventeenth-Century Histories and the Science of History," *Studies in the Literary Imagination* 22 (1989), pp. 17-35; and David Katz, "Isaac Vossius and the English Biblical Critics," in Richard H. Popkin and Arjo van der Jagt, eds., *Scepticism and Irreligion in the Seventeenth and Eighteenth Centuries* (Leiden: Brill, 1993), pp. 142-84.

of Genesis.[15] But these efforts ironically tend to decenter the Bible's authority by implying that the Mosaic account either requires or is buttressed by the independent confirmation of outside sources. This relativizing of the Old Testament, in turn, calls into question its narrative of humankind's Fall and leads to often tortuous efforts to reconcile seemingly antithetical interpretations of the origins of corruption.

In Genesis, humankind is exiled from Paradise because Eve and Adam sin. Their expulsion from the Garden into a world of labor and scarcity makes the Fall of nature an effect of willful disobedience: we have, in other words, only ourselves to blame for our misery and for the niggardliness of a demonized nature. Pagan myths, in contrast, emphasize the material prosperity of the Golden Age and suggest that virtue is, in effect, a by-product of an idealized existence of ease and plenty; they offer no clearcut causal mechanism for the degeneration from virtuous abundance to sinful scarcity. In such accounts, the Fall of Nature is explained by organic analogies—things are tough because the earth is growing old and its productive capacity is decaying. Confronted by these rival traditions, historians tend to straddle both sides of the epistemological questions raised by Christian and pagan narratives; logical contradictions between these accounts, as Arthur Ferguson notes, are finessed rather than resolved.[16] The Golden Age, consequently, becomes both a state of edenic innocence and a postlapsarian era in which justice reigns and the efficient distribution of resources takes place peacefully. John Locke, for example, spends a good deal of time in the first *Treatise of Government* arguing against Robert Filmer's interpretation of Genesis and offering his own authoritative hermeneutics, clearly assuming that the Bible represents a true, incontrovertible record of the Fall from Paradise; yet, in the second *Treatise* he explicitly invokes a "*Golden Age*" of good government that takes place after the Fall but "before vain Ambition, and *amor sceleratus habendi*, evil Concupscience, had corrupted Mens minds into a mistake of true Power and

[15] See Gerardus Vossius, *De Theologia Gentili et Physiologia Christiana* (1641; rpt. New York: Garland, 1976); Isaac Vossius, *Dissertatio de vera aetate mundi, qua ostenditur Natale mundi tempus annis minimum 1400 vulgarem aeram anticipare* (The Hague, 1659); Georg Horn, *Arcae Noae sive historia imperiorum et regnorum a condito orbe ad nostra temporum* (Leiden, 1666); John Webb, *An Historical Essay Endeavoring a Probability that the Language of the Empire of China is the Primitive Language* (London, 1669); Matthew Hale, *The Primitive Origination of Mankind, Considered and Examined According to the Light of Nature* (London, 1677); Francis Tallents, *A View of Universal History* (London, 1695); and Jean LeClerc, *Compendium Historiae Universalis: A Compendium of Universal History from the Beginning of the World to the Reign of Charles the Great* (London, 1699). On seventeenth-century chronologies, see Knoespel, "Newton in the School of Time," pp. 19-41, and Donald Wilcox, *The Measure of Times Past: Pre-Newtonian Chronologies and the Rhetoric of Relative Time* (Chicago: University of Chicago Press, 1987).

[16] Arthur B. Ferguson, *Utter Antiquity: Perceptions of Prehistory in Renaissance England* (Durham: Duke University Press, 1993), p. 2. See also A. Bartlett Giamatti, *The Earthly Paradise and the Renaissance Epic* (1966; rpt. New York: Norton, 1989).

Honour."[17] Locke's invoking of the "*Golden Age*" does not encourage his readers to question the Old Testament account of the Fall so much as it invites them to suspend selectively their assumptions about what it means to live in a fallen world. In this respect, the golden age is a "real" historical epoch, but its claims to reality rest on the paradox that Locke can demonstrate its truth-value only by deploying it as a political and economic ideal to anchor his vision of a hierarchical, but non-authoritarian, society that both protects property rights and distributes resources equitably.

Universal histories, like Locke's *Two Treatises of Government*, internalize tensions between the Judeo-Christian view that history has a beginning, middle, and end and pagan assumptions that the world cycles through epochs of prosperity and corruption or is locked into a process of irreversible decline. These tensions are expressed characteristically as the faith that there is a meaning *to*, but not *in*, history, and thus universal historians describe a fundamental paradox: the study of history can offer only retrospective assurances of a divine intention at work in human affairs; it cannot predict much of anything except its own apocalyptic end. In his posthumously published *Observations upon the Prophecies of Daniel and the Apocalypse of St. John*, Newton attacks the "rashness" of those "interpreters" who try to read Revelation in order to "fortel times and things": God "gave [Revelation] and the Prophecies of the Old Testament," he asserts, "not to justify men's curiosities by ennabling them to foreknow things, but that after they were fulfilled they might be interpreted by the event, and his own Providence, not the Interpreters', be then manifested thereby to the world."[18] As Newton's comments suggest, in order to justify the ways of God to man, universal histories must discover among the detritus of corruption the traces that justify their authors' faith in prophetic truths that have yet to be revealed. In this regard, while their discussions of agriculture, politics, economics, property law, and religion reconfirm sin as the ultimate cause of both the Fall from Eden and the ensuing degeneration of ancient civilizations, universal histories paradoxically are structured by the *absence* in the historical record of an ordering principle. This absence is the imaginative space in which universal historians debate the age of the Chinese empire, the location of Mount Ararat, and the nature of the language which Adam and Eve spoke in the Garden of Eden. The historian, then, must establish his status as one of the chosen "remnant" by undertaking a heroic quest to discern the pure narrative of providential

[17] John Locke, *Two Treatises of Government*, ed. Peter Laslett (Cambridge: Cambridge University Press, 1989), 2; para. 111, p. 342. See Laslett's effort in the note on this page to make sense of the contradiction between Christian and pagan notions of the prelapsarian state.

[18] Isaac Newton, *Observations upon the Prophecies of Daniel and the Apocalypse of St. John* (London, 1733), p. 251.

meaning that can be salvaged from the fables, distortions, and misreadings of postlapsarian history. But this faith in his own role as an interpreter of Mosaic history embroils writers from Gerardus Vossius to Newton in controversies that erupt during the seventeenth century over the different chronologies offered by the competing Masoretic (Hebrew) and Septuagint (Greek) texts of the Old Testament, over the political uses to which the Bible could be put, and over the thousands of variant readings of the sacred text that biblical scholars offered.[19] These controversies testify both to the corrosive effects of textual and theological disputes on biblical authority and to the herculean task faced by any latter-day Mosaic historian intent on writing an unchallenged, truly universal history.

3

In his efforts to describe a means to recover the tenets of an ancient scientific knowledge and pristine Noachian religion, Newton must work against two interlocking traditions within Christian theology: one denigrates the natural world and adapts pagan models of mundane senescence to argue that the world was, according to Cyprian, "testifying to its own decline by giving manifold concrete evidences" of the process of "decay"; the other displaces crises of intensification into various sins which doomed the pagan empires of the ancient world.[20] In the seventeenth century, views of the decline of nature characteristically blur moral cause and environmental effect, creating a kind of hermeneutic feedback loop which has had significant implications for subsequent perceptions of humankind's relationship to the environment.[21] Theologians such as Godfrey Goodman invoke God's condemnation of Adam, "Cursed be the ground for thy sake" (Genesis 3.17), to suggest both that Adam's disobedience caused the fall of nature and that God cursed nature to punish Adam; the former view makes humankind responsible for ecological devastation, the latter renders scarcity

19 See Christopher Hill, *The English Bible and the Seventeenth Century Revolution* (New York: Penguin, 1994).

20 Quoted in Tainter, *Collapse of Complex Societies*, p. 44.

21 On perceptions of ecology during the period, see the contrasting views of Carolyn Merchant, *The Death of Nature: Women, Ecology, and the Scientific Revolution* (New York: Harper & Row, 1980), and Richard H. Grove, *Green Imperialism: Colonial Expansion, Tropical Island Edens, and the Origins of Environmentalism 1600-1860* (Cambridge: Cambridge University Press, 1995), esp. pp. 4-6, 57-8. On perceptions of the natural world in Bacon's work, see Rebecca Merrens, "Exchanging Cultural Capital: Troping Women in Sidney and Bacon," *Genre* 25 (1992), pp. 179-92; Alvin Snider, *Origin and Authority in Seventeenth-Century England: Bacon, Milton, Butler* (Toronto: University of Toronto Press, 1994); Robert Stillman, *Philosophy and Universal Languages in Seventeenth-Century England: Bacon, Hobbes, and Wilkins* (Lewisburg: Bucknell University Press, 1995); and Desiree Hellegers, *Handmaid to Divinity: Science, Politics, and the Gendered Poetics of Resistance from John Donne and Francis Bacon to Anne Finch* (Norman: University of Oklahoma Press, 1999).

as divine retribution. As his title, *The Fall of Man, or the Corruption of Nature* implies, Goodman argues this question both ways—if sin marks a fall into scarcity, then a Hobbesian competition for scarce resources is both the cause and effect of humankind's corruption. Going beyond laments such as Cyprian's, Goodman claims that nature "gives man an infinite desire, intimating that she hath an infinite treasure; but our desires are therefore infinite, because wee receive no contentment at all, and so still wanting, still we desire."[22] In Goodman's moral cosmology, Nature becomes an active agent, provoking humankind's "infinite desire" by "intimating" that her inexhaustible resources can be converted endlessly into "treasures." If nature's "treasures" were truly infinite, he implies, then scarcity and environmental degradation could forever be held at bay, and humankind could return to the prosperity of the Golden Age. But nature, Goodman complains, has no "infinite treasure." Alienated from God and man, "she" becomes the scapegoat for humankind's economic and ecological miseries:

> That a generall scarcitie and penury may appeare, consider the prices of things, how excessive they are in these our latter dayes, and how they are raised above measure beyond all expectations.... [N]ow we seem to sustain a continuall famine, notwithstanding our peaceable times, and our great labors (necessity so enforcing us); and therefore we can blame none but nature, nor can we thinke that she is casually distempered...for we see no tokens, signes, or appearance of any recovery.[23]

Goodman casts crises of intensification into a narrative of humankind's tragic fall from a lost Edenic state.[24] Nature assumes pagan characteristics as a feminized entity that can be attacked both for the temptations she offers and the promises of inexhaustible resources that she fails to keep. The degeneration that Goodman describes—more work for less output, greater divisions between the rich and poor—is the default condition of universal history, a condemnation of nature's false promises that marks, in moralistic terms, his and his culture's recognition of the consequences of intensification: rising prices, deprivation, and diminishing marginal returns. As the corruption of nature plays itself out in economic hardship, disease, and hunger, the vision of its "infinite" resources leaves humankind with frustrated desires that both gnaw inwardly as private vices and explode externally in violence, rebellion, and disorder.

22 Godfrey Goodman, *The Fall of Man, or the Corruption of Nature* (London, 1616), p. 49.

23 Goodman, *Fall of Man*, pp. 369-70.

24 On the popularity of the notion of the decay of nature in the seventeenth century, see Clarence Glacken, *Traces on the Rhodian Shore: Nature and Culture in Western Thought from Ancient Times to the End of the Eighteenth Century* (Berkeley: University of California Press, 1967), 379-89. Glacken admits that he has not read Goodman and bases his argument on the characterization of Goodman in George Hakewill, *An Apologie, or Declaration of the Power and Providence of God in the Government of the World* (London, 1627).

The grafting of Christian theology onto cyclical explanations for the rise and fall of empires allows historians, such as Peter Heylyn, to anatomize the sins that lead inexorably to the degeneration of each civilization. At the beginning of his *Cosmographie*, Heylyn asks rhetorically, "If now we look into the cause of that desolation which hath happened in the Civill State of those mighty Empires; to what can we impute it but their crying sins? the pride of the *Babylonians*, the effeminacy of the *Persians*, the luxury of the *Greeks*, and such an aggregate of vices amongst the Romans (or western *Christians*) before the breaking in of the barbarous Nations, that they were grown a scandal unto Christianity."[25] For Heylyn, humankind's propensity for sin explains why even virtuous and prosperous societies degenerate into vice, luxury, contention, and rebellion. His terse description of the fall of ancient empires relies on sets of binary oppositions—good versus evil, Christianity versus paganism, purity versus corruption, order versus disorder, civilization versus barbarism, and masculine virtues versus feminized vices—to offer a conceptual framework for his historical geography, an eleven-hundred page catalog of the religions, cultures, technologies, governments, and racial and national characteristics of ancient empires and modern peoples. This logic of explanation both foregrounds and represses crises of intensification; like Goodman, Heylyn casts the diminishing marginal returns on investments of labor, energy, and capital into a moralistic narrative. The "crying Sins" of ancient empires arise when empire-building has reached its limits, when colonization no longer provides necessary raw materials, and when the demands of the upper and middle classes for continuing distributions of material rewards for their loyalty to central administrations go unmet. In the body of his *Cosmographie*, the metanarrative of redemptive history paradoxically manifests itself in anatomizing carefully, even obsessively, the ways in which various cultures—from ancient Egypt to contemporary Britain—relive the fall into "crying sins."

Goodman's jeremiad and Heylyn's history represent only two instances of the seventeenth century's fascination with the fall from virtuous subsistence societies to profligate empires, from proto-constitutional (or at least benevolent) monarchies to tyrannies, from honor to iniquity. In an important sense, the fall from edenic grace is defined as the fall into history, and therefore the ultimate task of the universal historian becomes describing

[25] Peter Heylyn, *Cosmographie in Foure Books, Contayning the Chorographie & Historie of the Whole World, and All the Principal Kingdomes, Provinces, Seas, And Isles, Thereof*, second ed. (London, 1657), A2r. Heylyn's tome went through eight editions and several additional reissues by 1702. Despite his importance, no modern biography of Heylyn exists. See the rival biographies by George Vernon, *The Life of the Learned and Reverend Dr. Peter Heylyn* (London, 1682), and John Barnard, *Theologo-Historicus, or the True Life of the Most Reverand Divine, and Excellent Historian Peter Heylyn* (London, 1683).

the lost origins of virtue and prosperity that remain outside, and yet paradoxically anchor, the narratives of postlapsarian history. Despite the differences between the accounts they offer, Ralegh and Newton recognize the crucial importance of demonstrating the historical reality of an edenic or golden age that will allow their readers to find providential meaning in successive cycles of sin and deprivation.

In his *History*, Ralegh argues vehemently against the view that Eden is "altogether mysticall, and allegorical," and insists that "Paradise was a terrestriall garden, garnished with fruits, delighting both the eye and taste," a region south of Armenia that even in its fallen state retains vestiges of its originary fertility: "so great is the fertilitie of the ground," he asserts, "that the people are constrained twice to mow downe their cornfields, and a third time to eat them up with sheep.... Besides, those fields are altogether without weeds (saith Plinie) who addeth...That the second yeare the very stubble (or rather falling downe of the seedes againe) yeeldeth them a harvest of Corne without any further labour."[26] Eden, for Ralegh, must remain a historical reality because it confirms the Mosaic account of God's authority, the identification of virtue and abundance, of sin and scarcity. In postlapsarian times, the geographical location of Eden preserves some of the attributes of a vanished Golden Age, notably a fertility that offers fallen humankind a glimpse of what existence must have been like before the descent into scarcity, labor, tyranny, and warfare. Writers who deny the existence of Eden risk fostering the dangerous notion that nature is subject to endless transformations, a view that, in turn, undermines Mosaic history by buttressing the heretical belief in the eternity of the world. But, ironically, because he is writing from within the corruption of historical time, Ralegh must describe Eden as a back formation—it exists only as the *negation* of a postlapsarian nature defined by scarcity, economic competition, war, and ecological devastation. For Ralegh as for Newton, then, the Fall motivates and structures a theocentric narrative in which the authorizing presence behind the meaningfulness of history—God's plan for humankind—can be described only as the negation of a negation, as the absence of sin rather than the demonstrated presence of virtue. Therefore, as Ralegh argues, "times and places are approved witnesses of worldly actions" because an accurate biblical history is the only means available to conjure into being the symbolic guarantee of a Creator who remains irrevocably beyond human attempts at understanding or representation.[27]

Because universal history is, in effect, the history of corruption, the account of Adam's and Eve's sin seems less problematic for most seventeenth-century commentators than the story of Noah. To an extent that

[26] Sir Walter Ralegh, *The History of the World* (London, 1614), pp. 34-5; 37; 56.

[27] *Ibid.*, p. 39.

few historians or critics have recognized, the crucial moral and ecological debates among competing seventeenth-century histories center on the questions about the relationship between original sin and a pristine re-origination of humankind and nature after the Flood.[28] The fascination with the story of Noah and the second Fall into sin and scarcity provokes seemingly endless speculation about the size of the ark, numerous efforts to calculate and recalculate how long it took to repopulate the earth, and various theories about the extent of the Flood and the dispersion of the sons of Noah in the years leading up to the building of the Tower of Babel.[29] The second fall from a postlapsarian Golden Age leaves seventeenth-century interpreters searching for a mechanism to explain the reemergence of sin. Ralegh maintains that "all the posteritie of our first Parents were afterwards borne and bred in a world, suffering the perpetuall Eclipse of spirituall light. Hence it was that it produced plants of such imperfection and harmefull qualitie, as the waters of the generall floud could not so wash out or depure, but that the same defection hath had continuance in the very generation and nature of mankinde." Moral corruption, in other words, survives the Flood so that "the sonnes of Cham did possesse the vices of the sonnes of Cain, and of those wicked Giants of the first Age."[30] Antediluvian sin persists, according to Ralegh, because it inheres in the frame of nature as well as in the corrupted souls of men. The mutually reinforcing "defection[s]" of nature, sin, and idolatry that survive God's cleansing of the earth come close to describing an almost manichean view of creation; rather than the paradox of the fortunate Fall, Ralegh implies that humankind is cursed by an originary alienation from a remote deity. In a world given over to pervasive corruption, there is no way back to the Garden, and nothing to learn from history except the confirmation of our own corruption: "In a word, we may gather out of *History* a policy no lesse wise than eternall; by the comparison and application of other mens fore-passed miseries, with our owne like errours and ill deservings."[31] Consequently, Noah cannot become, for Ralegh, a symbol of the human capacity for spiritual regeneration or a

[28] The most frequently cited work on this subject remains Don Cameron Allen, *The Legend of Noah: Renaissance Rationalism in Art, Science, and Letters* (Urbana: University of Illinois Press, 1949), but this work must be used cautiously: its approach is dated, its research uneven, and its conclusions dubious.

[29] On the size of the ark see John Wilkins, *Essay Towards a Real Character and a Philosophical Language* (London, 1668), pp. 360-65; on calculating repopulation, see Hale, *Primitive Origination of Mankind*, esp. pp. 145, 180-88; and on China, Webb, *Language of the Empire of China*, pp. 3-79. Webb and others suggest that the Flood was restricted to Palestine and contiguous lands and that, therefore, China, founded by Noah, escaped destruction and preserved, more or less intact, a pristine morality as well as the Adamic language. See Edwin J. Van Kley, "Europe's 'Discovery' of China and the Writing of World History," *American Historical Review* 76 (1971), pp. 358-85.

[30] Ralegh, *History*, p. 84.

[31] *Ibid.*, p. A2r.

rejuvenation of a corrupt nature because history remains trapped in a downward spiral: "as the Deuill our most industrious was ever most diligent: so is he now more laborious than ever: the long day of mankinde drawing fast towards an evening, and the worlds Tragedie and time neare an end."[32] Because history has been given over to the Devil, historical change can be described only as degeneration, the fall from primitive idolatry to the false worship in modern times of "the high and shining Idoll of glorie, the all-commanding image of bright Gold."[33]

His vision of history as tragedy leaves Ralegh little choice but to spend hundreds of pages anatomizing the corruption that he finds everywhere in the historical record. Unlike Newton, he gives short shrift in his *History* to the development of new technologies, advances in agricultural productivity, and the New Science of the early seventeenth century. Instead, positive images of virtue, comfort, and sociability are described consistently as relics of a prelapsarian existence. His visions of the abundance of Nature, in this regard, are located so securely in the past that even Pliny can be invoked as a near-contemporary. To some extent, though, Ralegh seems as much a prisoner of his era as he does of the circumstances which landed him in the Tower. Confronted by the depressed economic conditions of the early part of the century, he, like Goodman, universalizes the downswings of economic and ecological cycles into general and seemingly incontrovertible principles of historical degeneration. Eighty years later, with new technologies in place and new strategies of intensification (large-scale coal production, for example) underway, it becomes possible for writers such as Locke to imagine, once again, nature as bounteous—that is, as infinitely exploitable—and begin to write progressivist narratives of history. At the risk of generalizing, we might say that Newton occupies a midpoint between the historical visions of Ralegh and Locke. In the *Origines*, Newton follows the elder Vossius, Ralegh, and others in describing at length the fall of humankind into idolatry and corruption, but, crucially, he also begins to develop a different kind of narrative, a fragmentary account of ways to recover the pristine religion of Noah. These ways are lit by a divinely-sanctioned technoscientific knowledge.

4

In an important sense, Newton's historical manuscripts represent efforts to move beyond what their author apparently sees as the conceptual dead end of relying on the Bible to secure both the meaningfulness of the history and

[32] *Ibid.*, p. 97.

[33] *Ibid.*

the coherence of the natural world.[34] The decentering of the Bible as an absolute guarantee of God's covenant with humankind leads Newton to seek principles of order in the natural world: "there is no way," he claims in the *Origines*, "(wthout revelation) to come to ye knowledge of a Deity but by the frame of nature."[35] In the writings of other seventeenth-century physico-theologians, the metaphor of the two books—the Bible and Nature—elevates the principle of a mutually implicative order to an article of faith: "the Book of Grace," Robert Boyle assures his readers, "doth resemble the Book of Nature; wherein the Stars...are not more Nicely nor Methodically plac'd than the Passages of Scripture."[36] In Boyle's voluntarist theology, the Bible guarantees the unity and coherence of the physical world; without the underpinnings of revelation, the natural world threatens to disintegrate into the demonic realm of Calvinist thought.[37] For Newton, however, the trinitarians' corruption of the biblical text (save for Daniel and Revelation) undermines the metaphor of the two books; the coherence of the "frame of nature" cannot be deduced directly from the mystical unity of a disputed text but can be made manifest only through a painstaking program of inductive investigation guided by a voluntarist faith in God's omnipotence. This intensive research is necessary for Newton precisely because the argument from design is not self-evident. In a fallen world, corruption—physical as well as textual—threatens to overwhelm reasoning by analogy. To a greater extent than Boyle or any of his other contemporaries, then, Newton must struggle to balance seemingly incommensurate views of nature: the natural world is *both* the fallen world of Protestant theology *and* the "frame" which provides incontrovertible evidence of the existence of a supreme being. His fragmentary universal history becomes the analog of his scientific efforts to uncover the means to reclaim humankind from corruption.

As Westfall suggests, at the heart of Newton's belief in a pristine Noachian religion lies the symbol of the Temple—the eternal flame on a sacrificial altar at the center of a geometrically precise representation of the heliocentric solar system.[38] Although the Temple of Solomon is the most complete symbolic embodiment of the mathematical laws of the universe, its formal integrity as the true representation of a divine order derives from

34 See Hill, *English Bible and the Seventeenth-Century Revolution*, pp. 413-35.

35 Yahuda MS 41, f. 7; also cited in Westfall, "Newton's *Origines*," p. 25. Westfall notes that Newton "inserted the parenthetical phrase 'wthout revelation' above the line as an afterthought."

36 Robert Boyle, *Some Considerations Touching the Style of the Holy Scriptures* (London, 1661), p. 53.

37 On Protestant attitudes toward the corruption of the natural world, see Norman O. Brown, *Life against Death: The Psychoanalytic Meaning of History* (Middletown, Conn.: Wesleyan University Press, 1959), pp. 202-34.

38 Westfall, "Newton's *Origines*," pp.. 22-4. See also David Castillejo, *The Expanding Force in Newton's Cosmos* (Madrid: Ediciones de Arte y Bibliofilia, 1981).

older places of worship that date to the time of Noah and the repopulation of the earth. Newton begins the second chapter of his English precis of the *Origines*, *The Original of Religions*, by asserting that "the religion most ancient & most generally received by the nations in the first ages [after the Flood] was that of the Prytanea or Vestal temples."[39] The symbolic integrity of the prytanea provides Newton with incontrovertible evidence that a universal Noachian religion antedates all forms of idolatry. The sacrifices of "clean...beasts & fowels" in the Temple recall the original act of Noah's faith in Providence and his sacrifices after the Ark landed on Mount Ararat; they testify to the continuity of religious observance from before the Flood: "so soon as Noah came out of ye Ark he built an altar & offered burnt offerings of every clean Beast & every clean fowel unto the Lord."[40] In contrast to Ralegh, Newton de-emphasizes the corruption that survives the Flood. Like the rainbow, the mark of God's covenant with humankind, the biblical record of "clean" sacrifices becomes evidence, for Newton, of the persistence of a true faith and true scientific knowledge that survive the ravages of sin, corruption, and the universal deluge. This knowledge is embodied in the prytanea:

> So then was one design of ye first institution of ye true religion to propose to mankind by ye frame of ye ancient Temples, the study of the frame of the world as the true Temple of ye great God they worshipped. And thence it was y^{t} ye Priests anciently were above other men well skilled in ye knowledge of ye true frame of Nature & accounted it a great part of their Theology.[41]

By "frame" Newton does not mean the material existence of celestial objects but the mathematical form that remains a timeless guarantee of a divine intention, a design unaffected by the successive cycles of decay and the corruption of the material world. The prytanea functions, then, as a kind of computational code by which true knowledge can be demonstrated and transmitted without interruption, interference, or corruption. In this respect, the descriptions of temples that Newton culls from ancient historians elevate the prytanea to the status of a universal form—the heliocentric solar system—which is represented by, yet transcends, its material incarnations: ancient temples may be destroyed but the knowledge which they encode endures. The divine wisdom that reveals itself in the form of the prytanea remains unrepresentable, but the ancient temples themselves serve as injunctions to study nature, to learn its mysteries. The mathematical structure of these temples depends, in other words, on the absence of graven images on their central altars; their form is a kind of mathematical proof that

[39] Yahuda MS 41, f. 1r.

[40] Yahuda MS 41, ff. 3r-4r.

[41] Yahuda MS 41, f. 7r.

God is universal, abstract, indivisible, and all-powerful. The lack of idolatry in the "first ages" thus confirms the Judeo-Christian belief that God remains irrevocably beyond representation.[42]

The sacred flame in Newton's account of the Noachian religion becomes a complex symbol of both moral renovation and technoscientific knowledge and, in both respects, it can be identified with the purifying flame of alchemical sublimation. As Dobbs has demonstrated, alchemy, for Newton, is not simply the search for the philosopher's stone but a complex inquiry into the structure of matter; in this sense, it is an idealized image of a scientific knowledge that promises the purification of the soul as well as a means to unlock the secrets of matter theory.[43] Although alchemical knowledge is typically envisioned as the province of a few adepts, Pamela Smith has argued persuasively that alchemy could also be used heuristically to describe the development of a credit economy in the early modern period by providing images of the production of value from investment, the manipulation of exchange rates, and increases in the money supply.[44] To extend her insight, we might recognize that the alchemical fire of sublimation does not consume but produces value, and therefore can be linked symbolically to the eternal flame of the prytanea. To the extent that Newton identifies the fire under his retort with the sacrificial fire of Noachian monotheism, he can assign it a complex of meanings: it is a spiritual light; a symbol of the divine wisdom manifest in the heliocentric solar system; the transforming flame of alchemy; and a testimony that technoscientific knowledge can be used to transcend the logic of scarcity. The sacred fire thus offers the promise of access to the continuing bounty of Nature as a sign of God's grace.

The prytanea, as the image of the true temple of God, becomes a means for Newton to explore the ways in which faith and knowledge operate within history. The course of the purifying flame traces a lineage at once geographical, theological, and technological:

> Noah & his sons carried wth them ye sacred fire from ye Tower of Babylon into ye Land of Shinaar, ...& Abraham carried it wth him to offer Isaac & Aeneas carried it wth him from Troy & the ancient Kings of Greece & Persia carried it along wth them when into the field they went to make war: so the sons of Noah when they went from him into their several countries took this

[42] See Jean-Joseph Goux, *Symbolic Economies after Marx and Freud*, trans. Jennifer Curtiss Gage (Ithaca: Cornell University Press, 1990), pp. 221-34.

[43] See Dobbs, *Janus Faces of Genius*, passim.

[44] Pamela Smith, *The Business of Alchemy: Science and Culture in the Holy Roman Empire* (Princeton: Princeton University Press, 1994).

> fire along wth their several families & the like was done by their sons & grandsons as oft as they went from one another in a distinct polity.[45]

The structure of this sentence is revealing: clauses are strung together by "&s" which occlude the complicated chains of circumstance by which the sacred fire was transmitted from one family or culture to another. In this regard, the flame itself, as evidence of God's commandment to sacrifice to Him, gives theological form to what otherwise might seem the random distribution of sacrificial practices across the ancient world. It comes to mark the political transition from familial affiliations to "distinct polit[ies]." True worship thus defines the advance of civilization and becomes the only means, Newton implies, to construct a progressive narrative of technological, scientific, and political development. The flame consumes material goods—clean beasts and first fruits—and offers in return the continuation of a pristine knowledge, an uncorrupted theology, and, significantly, the redemption of the natural world through the promise of an unending productivity.

Despite Newton's emphasis on the dynamics of renovation, however, the *Origines*, like other universal histories, devotes much of its attention to the metastasizing corruption which defigured the ancient world. His account of the fall of humankind from monotheism, like those of the elder Vossius and Ralegh, follows a familiar trajectory: men discard an absolute faith in God for a "veneration" of the secondary effects by which his wisdom can be apprehended, thereby confusing form and content, the ideal and the material, the timeless and the corrupt. Worshiping the mere representations of divine order—the Sun, stars, and planets—turns men and women away from technoscientific knowledge and true faith and makes them subject to self-willed delusions. The "corruption of this [pristine] religion," for Newton, occurs when

> the frame of the heavens consisting of Sun Moon & Stars being represented in the Prytanea as ye real temple of the Deity men were led by degrees to pay a veneration to these sensible objects & began at length to worship them as the visible seals of divinity. And because ye sacred fire was a type of ye Sun & all the elements are part of that universe wch is ye temple of God they soon began to have these also in veneration. For tis agreed that Idolatry began in ye worship of ye heavenly bodies & elements.[46]

The worship of "sensible objects" marks humankind's self-corruption from the mathematical principles that demonstrate the divine wisdom manifest in creation. In the last sentence of the *Opticks*, Newton makes explicit the stakes that are involved in this rejection of a theocentric mathematics: "if the

[45] Yahuda MS 41, f. 4r.

[46] Yahuda MS 41, f. 8r.

Worship of false Gods had not blinded the Heathen, their moral Philosophy would have gone farther than to the four Cardinal Virtues; and instead of teaching the Transmigration of Souls, and to worship the Sun and the Moon, and dead Heroes, they would have taught us to worship our true Author and Benefactor, as their Ancestors did under the Government of Noah and his Sons before they corrupted themselves."[47] Reading this sentence in the context of Newton's descriptions in *The Original of Religions* suggests the extent to which his historiographic and scientific narratives interpenetrate. Idolatry—the worship of planetary objects and "dead Heroes"—is the inability to discern the ideal form, the only true evidence of the incorruptible structuring principle of the "frame of the world." It deifies the natural world and thus makes humankind subject to this false goddess's capricious elements. By rejecting the technoscientific faith that alone offers a vision of transcending the ecological constraints that, as we have seen, are identified insistently with a fallen nature, humankind condemns itself to seemingly endless cycles of scarcity, intensification, and collapse.

5

It is difficult to read through the mass of manuscripts that seem intended for the *Origines* without sensing that the image of the sacred fire at the center of the prytanea ultimately becomes, for Newton, a symbolic legitimation of his unfinished universal history. To justify the ways of God to humankind, he must envision his project as a mode of resistance to the threats of trinitarianism, tyranny, and idolatry. The *Original of Religions* must argue, then, that the "truths" which he seeks to recover are not irrevocably lost; even as the practices that Newton associates with Noachian monotheism are fouled by idolatry, the mathematical form that can redeem humankind remains uncorrupted, even by those peoples who mistake "sensible objects" for eternal truths. If "ye old religion of the Egyptians was ye true religion tho corrupted before the age of Moses by the mixture of the worship of fals Gods wth that of ye true one: [then] by consequence ye religion of ye Jews was no other then that of Noah propagated down in Egypt till ye age of Moses."[48] Moses's recovery of the true religion becomes a model, as Stuart Peterfreund suggests, for Newton's "self-fashioning" as the heir to a tradition of prophetic authority that underwrites his scientific as well as theological endeavors.[49] The project of reclaiming an ancient scientific wisdom and a pristine religion, then, is reserved to those few, the "remnant,"

[47] Isaac Newton, *Opticks*, fourth edition (1730; New York, Dover, 1952), pp. 405-6.

[48] Yahuda MS 41, f. 5r.

[49] Peterfreund, "Mosaic Self-Presentation in Newtonian Optics," pp. 139-65.

capable both of interpreting correctly the corruption of Noachian theology and of comprehending the underlying forms which persist as a mark of God's covenant: the mathematical truths of physics, the optics of the rainbow. Newton's emphasis on the form of the prytanea and, in other manuscripts, the form of the Temple effectively restricts the understanding of history itself to the adepts of higher mathematics. Because universal history, Newton contends, can be verified only by an objective mathematical knowledge, mathematics becomes akin to revelation, both analytical method and moral shield against the corruption of the physical world.

Not surprisingly, then, the *Origines* reproduces the dynamics of what Bernard Cohen has called the "Newtonian style," the demonstration of "purely mathematical counterparts of simplified and idealized physical situations" which can then be calibrated carefully against observed empirical phenomena.[50] But there are no "physical situations" in postlapsarian history, only the fragmentary records of ancient civilizations which, in Newton's mind, reconfirm an originary alienation of corrupt human beings from an ideal knowledge. I term this alienation "originary" because the method and metaphysics of mathematics is itself irrevocably implicated in the regime of scarcity—that is, in the need to count; measure; allocate; calculate rates of use, expenditure, and interest; estimate future stores and requirements; and develop means of exchange whereby jugs of water, bushels of wheat, and herds of cattle can be converted symbolically into correlative values. As Goodman's repetition of "infinite" desires and "infinite" treasures may suggest, visions of a Golden Age place humankind beyond counting or the need to count—the abundance of the earth registers not as excess, luxury, or surplus value but as what we might call the infinite elasticity of use-value, the harmony between humankind's "infinite" desires and the means that Nature offers to satisfy them. In one respect, the development of mathematics marks the fall from an edenic abundance: if resources are inexhaustible and equally available or equitably distributed, then counting becomes superfluous. What Brian Rotman has termed an "infinitarian mathematics" depends ultimately on the fiction of infinite or inexhaustible resources. If such a mathematics, Rotman contends, is irrevocably idealized, decontextualized, and disembodied because it is predicated on "the fantasy of the absolute: absolute sameness, identity, equality, and permanence of mathematical objects for all time and all conceptual regimes and all historical materializations," then this idealization depends on the mutually reinforcing idealization of a nature that is endlessly productive.[51] Nobody counts, tabulates, or rechecks columns of figures in

50 I. Bernard Cohen, *The Newtonian Revolution* (Cambridge: Cambridge University Press, 1980), p. 37.

51 Brian Rotman, *Ad Infinitum: The Ghost in Turing's Machine* (Stanford: Stanford University Press, 1993), p. 156. For an extended discussion of the mutually reinforcing logics of infinite resources and

the Golden Age; no one has to be concerned with the incremental rates of change in infinitesimals because the Golden Age exists effectively outside of time, before history. In an odd way, then, the mathematical knowledge embodied by the Temple can guarantee the order of the universe and the possibility of infinite productivity precisely because it remains "absolute." In the context of western culture's reflexive beliefs in an infinitarian mathematics, the infinite exploitability of nature, and the possibility of unending profits, it is worth keeping in mind that Newton's most hubristic statement of his alchemical process, in the manuscript *Praxis*, offers a formula for "multiplication" that will "encreas [gold's] vertue ten times,...perhaps a thousand times." "Thus," Newton concludes, "you may multiply to infinity."[52] Although Westfall attributes this claim to psychological stress, we can glimpse in Newton's alchemy the germ of a political economy that is predicated on trying to satisfy infinite desires by holding out the promise of a mystical and scientific knowledge to lead humankind from scarcity to abundance.

Ultimately, however, Newton's approach to universal history rests on the unacknowledged irony that the motivation for both idolatry and for humankind's developing mathematical knowledge (the only means to appreciate the theocentric design of the universe) stems from the same set of initial conditions—crises of intensification, the depletion of a once bounteous environment. Value can be displaced, according to Newton's own logic, from abstract truths to physical objects only when the resources identified with those objects—first fruits, clean beasts—can no longer be taken for granted, when they become worth counting precisely because they have become scarce. The worship of "dead Heroes," in this regard, is a form of misplaced nostalgia, a harkening back to a mythic time when resources were more bountiful and the earth less degraded. The original sin is the alienation from nature, an alienation that gives rise not only to violence, idolatry, and tyranny but to the technologies and forms of socioeconomic organization which define civilization. In this respect, technoscientific progress and religious corruption are not moral alternatives but dialectically-related aspects of the recognition of, and response to, the crises of diminishing marginal returns which afflict complex societies.

The fall from the Golden Age, then, is precisely the fall into the calculations of political economy, into rudimentary and not-so-rudimentary cost-benefit analyses which presuppose a scarcity of resources. As people corrupt themselves, self-interested rulers and ambitious individuals cynically weigh the costs of various kinds of labor, including warfare,

---

infinite productivity, see Robert Markley, "'So Inexhaustible a Treasure of Gold': Defoe, Capitalism, and the Romance of the South Seas," *Eighteenth-Century Life* 18 (1994), pp. 148-67.

[52] Babson MS 420, p. 8. Cited in Westfall, *Never at Rest*, p. 530.

against expected rewards. The fall from innocence and prosperity is measured by what Aphra Behn in her poem "The Golden Age" calls "Rapes, Invasions, Tyrannies," the advent of a militaristic aggression mystified by the language of nobility and heroism.[53] For Newton, idolatry provides a means to describe the rise of tyrannical forms of government: "whilst [idolatry] created in ye Egyptians & eastern people an honour for their kings as being descended from ye Gods & tended to ye establishment & enlargement of the regal authority & dominion it was not only promoted by those Kings but also greedily received by others."[54] In "The Golden Age" and elsewhere, Behn imagines the "enlargement of the regal authority & dominion" in graphic descriptions of violence and self-perpetuating dictates of honor that lead individuals to make themselves miserable in the quest for the false idols of abstract morality; in contrast, tied to the binary logic that structures the narrative tradition of universal history, Newton describes the advent of tyranny as an effect of theological fraud and apostasy. If the tradition of universal history depends on defending the authority of Genesis, then Newton's efforts to reveal the workings of God in the conservation of a true knowledge of the cosmos must end not only in a demonstration of that knowledge in the *Principia* but also in an all-out assault on the origins of corruption. In his writings on the history of the Church in the fourth and fifth centuries, Newton articulates what amounts to a conspiracy theory of history that takes as its targets Athanasius and his cohorts. His response to the mysterious workings of iniquity, therefore, differs from Behn's analytic; where she investigates ideological constructions of self and society, he scapegoats those latter-day idolaters, Catholics and trinitarians, for the ongoing corruption of the true faith. In his mind, universal history—the justification of the ways of God to humankind—never can be distinguished from its corollary, the demolition of false knowledge represented by the fraudulent systematizers of theology, from Athanasius to Leibniz. The *Origines* thus remains incomplete even as Newton embarks on its sequel: his antitrinitarian history of an apostate Church.[55]

Westfall's "solitary and dejected scholar of Cambridge" may have resented his status as a sizar and his mother's stinginess, but his dejection, in some measure, is characteristic of several generations of self-styled historians who looked back across the ages and saw in text after text

[53] Aphra Behn, *The Works of Aphra Behn*, ed. Janet Todd (Columbus: Ohio State University Press, 1992) vol. 1, p. 32. On "The Golden Age" see Robert Markley and Molly Rothenberg, "The Contestations of Nature: Aphra Behn's 'The Golden Age' and the Sexualizing of Politics," in *Rereading Aphra Behn: History, Theory, and Criticism*, ed. Heidi Hutner (Charlottesville: University of Virigina Press, 1993), pp. 301-21.

[54] Yahuda MS 41, f. 7r.

[55] See Markley, *Fallen Languages*, pp. 153-69.

unequivocal anticipations of the crises of the seventeenth century.[56] In one respect, universal histories can be read as efforts to explain the threats posed to civilized order as a whole in terms of their author's senses of alienation. The moral authority claimed by historians such as Ralegh and Newton is paradoxically conferred on them by their exclusion from centers of power, privilege, and corruption. Universal meaning inheres in personal tragedy, and universal history becomes the last refuge of the condemned courtier, the failed revolutionary, the unheeded prophet: Ralegh wrote his *History* in the Tower, Milton books eleven and twelve after the Restoration, and Newton during solitary hours as the isolated truth-seeker surrounded by the malevolent forces of trinitarianism. In the Introduction to his *Cosmographie*, Heylyn, Archbishop Laud's right-hand man before 1642 and an avid royalist until his death in 1662, states that he wrote his massive historical geography when "by the unhappiness of [his] Destiny, or the infelicity of the times, [he was] deprived of [his] preferments, and divested of [his] Ministeriall Function."[57] Although at the opposite end of the political spectrum from Milton, Heylyn, too, explains his destiny in universal terms—his ejection from the pulpit and the confiscation of his library allow him to claim for his work a moral authority and historical objectivity that he identifies explicitly with the workings of God's providence. In all of these cases, the universal historian arrogates to himself a privileged position from which to gaze backwards across history and reassemble theocentric narratives that justify his own efforts to God and man. The historian's strategies to resist the pessimism of a Goodman or Ralegh are varied: Milton's faith in the ideal of the Commonwealth, a faith that persisted after 1660; Webb's efforts to relocate the Golden Age not in remote history but in a pristine China; and Newton's quest to find in the natural world a mathematical order to recall humankind to true religion. Universal history, in the tumultuous seventeenth century, gives voice to the always unrealized hope that the prophet in the wilderness can transcend—or escape—the history of his times.[58]

[56] Westfall, *Never at Rest*, p. 75.

[57] Heylyn, *Cosmographie*, B2r.

[58] I would like to thank Kenneth Knoespel, Bernadette Andrea, Desiree Hellegers, Jean Marsden, and Rebecca Merrens for their valuable comments on various drafts of this essay.

ESSAY 7

Newton's *Of the Church*: Its Contents and Implications

MATT GOLDISH
University of Arizona

Sir Isaac Newton's histories of the primitive church and its corruption are perhaps the least studied documents in the body of his unpublished manuscripts yet they are among the most enlightening for those seeking insight into Newton's understanding of God, religion, and even science. It will be my goal here to point out how two much debated issues in Newton's relationship with the church are clarified by these histories: his often criticized decision to remain within the Anglican church; and his attitude toward Latitudinarianism and Deism. In addition, I will adduce an example of a scientific issue on which light may be shed by the church histories, viz., Newton's debate with Leibniz, which was raging at the time Newton was composing these writings.

1. NEWTON'S *OF THE CHURCH*

I will begin by reviewing the role of church history in Newton's theology briefly. Newton understood the nature of God's Providence to be expressed, among other ways, in His repeated attempts to offer mankind a return to right religion throughout history. After humanity failed to uphold the simplicity of Noachide natural religion, the tradition was carried on by Abraham and his posterity. Men were offered another opportunity to derive correct religious beliefs through the Pentateuch. When the descendants of Abraham erred by rejecting Jesus, who was sent to bring yet another chance to mankind, the tradition shifted to the Christians. This recrudescence of right religion, embodied in primitive Christianity, lasted for some time before it was corrupted. But corrupted it was, and the simple love of God and man which it taught was displaced by a heretical Christianity embracing emanation doctrines, foreign worship and other errors.

J.E. Force and R.H. Popkin (eds.), Newton and Religion, 145–164.
© 1999 *Kluwer Academic Publishers.*

The history of the fall of primitive Christianity is a partial fulfillment of the prophecies in Daniel (particularly Chap. 8) and the Apocalypse. The Pope and his Roman Church, the symbol of the corrupted ecclesiastical polity which incorporated heretical metaphysics into the Christian creed, is identical with the Antichrist and the "Man of Sin" described in those prophecies.[1] The true church and holy people are Daniel's "Host of Heaven," and Jesus is "Prince of the Host."

The major work in which Newton records his ecclesiastical annals is *Of the Church*, which is extant in two major manuscripts (formerly together as lot 249 at the Sotheby sale of Newton's MSS)—MS Yahuda 15, currently housed at the Jewish National and University Library in Jerusalem and MS Bodmer, at the Bibliotheca Bodmeriana in Geneva. While David Castillejo claims that neither of these is the truly complete fine original,[2] he did not see MS Bodmer, which contains many indications that it is the edition prepared for publication. MS Yahuda 15 is a series of rough drafts of the various chapters, while MS Bodmer is a much more polished and edited version of the whole work, though it is often rough as well. MS Bodmer has a series of theoretical tables of contents, none of which seems to be fully consistent with the chapters actually written in the preserved MSS. Chapter heads almost all have an editor's comments in a Baroque hand, though it is unclear when these were written.[3] In addition to these two more or less full copies, there are thousands of pages of manuscript notes in both the Yahuda collection and others which are preparatory materials toward the creation of this treatise. The most noteworthy of these drafts is that called *Irenicum*, found in King's College, Keynes MS 3. Newton worked hard on church history, and *Of the Church*, written circa 1710, was the culmination of these researches, possibly earmarked by the author for eventual publication.

Newton was never fully satisfied with *Of the Church*, and made myriad revisions and drafts; as noted above, even the table of contents exists in about a dozen forms. In order to present the content of the book, I will describe briefly the themes of the chapters which were actually written, as

[1] On the application of these figures to the Roman church beginning in the sixteenth century, see Christopher Hill, *Antichrist in Seventeenth-Century England* (London: Oxford University Press, 1971); LeRoy Edwin Froom, *The Prophetic Faith of Our Fathers*, 4 vols. (Washington D.C.: Review & Herald, 1948), 2:373-94.

[2] David Castillejo, *The Expanding Force in Newton's Cosmos* (Madrid: Ediciones de Arte y Bibliofilia, 1981), p. 57.

[3] Catherine Conduitt, Newton's niece and heiress, wrote in her will that Newton's "church history compleat" should be considered for publication. This was the cause of the fine copy being sent to Rev. Arthur Ashley Sykes, who was charged with examining the work. Sykes died, according to Castillejo, before returning it and it was reportedly lost. See: Castillejo, *Expanding Force*; Frank Manuel, *Isaac Newton: Historian* (Cambridge: Cambridge University Press, 1963), p. 57.

preserved in MS Bodmer.[4] In Chapter 1, "Of the Host of Heaven & the Prince of the Host," Newton discusses, in general terms, to whom these terms (found in Daniel 8:10-11) might refer. The Host of Heaven, in Newton's view, is the original pristine Church with its simple creed and unity of believers (Jews and gentiles) who resisted outside corruptions. Jesus is Prince of the Host. In Chapter 2, "Of the Holy Covenant," Newton deals with the covenant between God and His people to preserve the simple belief in love of God, Christ and fellowman. The differentiation is made between this basic doctrine for all believers, called "milk for babes," and optional further theological discourse and study, to be undertaken only by scholars, called "meat for men of full age." The important thing, Newton stresses, is that we must never fight about these more complicated issues, nor allow outside influences to affect us, but remain unified in our simple creed of faith. Unfortunately, the Church became infected, especially in the period of the early councils, and thus began to break the covenant.

In Chapter 3, "Of the Faith once delivered to the Saints," Newton discusses the original creed, written in Hebrew and transmitted orally. Newton again stresses the disaster which occurs when this creed becomes contaminated, when the "meat for men of full age" begins to contaminate the "milk for babes." The logical next historical step follows in Chapter 4, "Of the Theology of the Heathens, Cabbalists and ancient Hereticks." In this very important chapter Newton puts forth his opinion that the contamination of the pristine Church in the early Christian centuries occurred when converts from heathenism and Judaism insinuated their metaphysical ideas into Christianity. Platonism and Kabbalah were particularly insidious, and the theory of God's emanation which they taught can be detected as it begins to appear in the creed, thereby corrupting it and bringing disaster to the Christian religion. The Gnostics were essentially converted Jews, who brought Kabbalah with them and consequently overthrew the true pristine faith.[5]

Chapter 5 is given different titles in two draft title-pages: "Of the Host of Heaven and the corruptions which crept into it" and "Of the disputable opinions which crept into the religion of the primitive Church."[6] Here

[4] The MS opens with a discussion concerning the history of the Old Testament books, which seems more connected with Newton's work on prophecy than his church history—it is essentially identical with the beginning of the posthumously published *Observations upon the Prophecies of Daniel, and the Apocalypse of St. John* (London, 1733). The introduction comes before the tables of contents which leads me to believe it might have been haphazardly stuck on to the rest.

[5] As the editor has noted, all the drafts of Chap. 4 following the first (they are all entitled *Of the Workings of the Mystery of Iniquity*) seem to be the original basis for the first four chapters as they appear and, therefore, not independently relevant.

[6] Here MS Bodmer's organization and neatness begin seriously to deteriorate as the editor has noted on the title page.

Newton goes over the intrusion of foreign metaphysics again, adding a more detailed historical perspective of its progression during the reigns of the various Roman Emperors. He again stresses the unity and peace enjoyed among the earliest Christians, before they were spoiled by arguments about faith deriving from the impingement of metaphysics on the creed. In those halcyon days, Jewish Christians could continue living happily according to the law with their gentile Christian brethren who were not commanded to keep the law. Conditions began to deteriorate soon after the time of Justin Martyr. In Chapter 6, "A further Account of the Host of Heaven and the Corruption which crept into it," Newton continues these themes but also discusses how decisions of the various early Church councils (especially concerning the wording of the creed) institutionalized the errors of converted metaphysicians.

In Chapter 7, "Of the Christian Religion and its corruption in Morals,"[7] and Chapter 8, "Of the Corruption of the Church in language and Opinions," Newton further develops his history of Church corruption. The most dangerous corruption, of course, was the transmutation of metaphysical emanation theories into Trinitarianism, a concept which Newton stresses did not exist in the primitive creed and which, sadly, became a dogma with the rise of Athanasius and defeat of Arianism.

Chapters 9 and 10, "Of the Sanctuary of Strength, or Of the Temple & Synagogues of the Jews & Churches of the Christians," contain Newton's detailed description of the Holy Temple at Jerusalem and its management, and shows that the primitive Church was governed in imitation of the Temple and synagogues. It was this government of the original church, along with its creed and unity, which were swept away with the (unsanctioned) rise of the Bishop of Rome to supreme power in the Church. In Chapter 11, "Of the Roman Empire in relation to its Ecclesiastical Dominion and legislature in matters of Religion," and Chapters 12 and 13, "Of the Rise and Dominion of the (Roman) Catholic Church," Newton discusses how, by compromising with the metaphysical heresies and fomenting disunity in the Church during the fourth century, the rise of Rome was engineered. The Pope, in Newton's view (as in that of many of Newton's contemporaries), is identical with Daniel's "Mystery of Iniquity." Popes Damasus, Valens, and Valentinian jockied for control of the whole Church and, when they could not succeed in these political maneuvers, they caused the split of Western and Eastern empires. They pretended to be anti-heretical, but their true allegiance was to expedience and, in fact, they promoted heresy within the Church.

[7] Between Chap. 7 and Chap. 8 there are a few pages of a chapter entitled "Of the corruption of the Christian Religion in discipline & morality."

In Chapter 14, "Of the Abomination of Desolation," Newton examines the many specific abuses of the Roman Church. Some of these abuses could be traced to the erroneous metaphysics which had crept in to corrupt pristine Christianity, while many others were clearly heathen practices which Catholicism simply absorbed. This was the case with the idolatrous invocation of saints and adoration of relics, whose development Newton describes. The remaining chapters of *Of the Church* are chaotic drafts but these chapters clearly reflect the main body of the work Newton intended to publish.

In *Of the Church,* Newton indeed tells the explicit story of the early church and its corruption; the bulk of the work is taken up with long discussions about church councils, versions of the creed, and acts of various popes. There is, however, a clearly implicit message for contemporaries, as there was in each of the many studies of the primitive church written by Englishmen during this era.[8] In Newton's case, the message is that *schism* caused the downfall of primitive Christianity and it is to be avoided at almost any cost. A careful study of church annals reveals that all evils proceed from schism created by people who wished to add metaphysics to the simple faith; to mix "meat" with the "milk." Newton rails constantly against Catholics, the worst schismatics, who are to be despised. At the same time, however, he believes it imperative not to create new schisms among Christians. If one is not coerced to uphold an idolatrous position by one's church, one must not split from it. This deep fear of schism is undoubtedly a product of Newton's memories from the Civil War era, memories not shared by his more outspoken students, such as William Whiston.

2. ISAAC NEWTON: HONEST NEWTONIAN?

It is well known that Newton was deeply anti-Trinitarian and that the Church of England to which he belonged throughout his life was Trinitarian. Scholars have suggested various reasons for Newton's choice to remain within the English church, none of which redound to his credit in the view of many 20th-century interpreters. The most common explanation has been that Newton feared losing his stipend as Lucasian Professor at Cambridge which would have been one of the results of any personal rift with the Church of England (witness what happened to his student William Whiston, James Force's "honest Newtonian," when Whiston publicly declared his

[8] The earnest study of ancient church history began, in fact, in an attempt under James I to present the Church of England as the heir of primitive Christianity. See David Harris Willson and Stuart E. Prall, *A History of England*, 3rd ed. (New York: Holt, Rinehart and Winston, 1984), p. 276. John Jewel and Richard Hooker were early practitioners.

anti-Trinitarian position.) Others have suggested that Newton, feeling the winds of change blowing his way, chose not to rock the boat with any public declarations.[9] The evidence seems conclusive. James Force, having examined Newton's most specific statements on the English church, found in MS.Keynes 3, entitled *Irenicum*, and published by McLachlan,[10] concluded that Newton "throughout all the versions of the *Irenicum* ...argues implicitly that the Church of England is an apostate Church, a 'synagogue of Satan'."[11]

It is Force himself, however, who has laid the foundation for a different understanding of Newton's position, in his paper "Newton, the Lord God of Israel and Knowledge of Nature."[12] Force was the first scholar to point out Newton's careful division between dogma and adiaphora (docrtrines which are true but in which faith is not dogmatically required.)

The Apostle Paul differentiates between the simplest Christian truths and those requiring discernment of reason, called by him "milk" and "meat" respectively:

> ...But you are still in need of milk, and not strong meat. For every man whose food is milk is unfamiliar with the word of righteousness; for he is a babe. But strong meat belongs to those who are of full age, even those who by reason of use have their senses exercised to discern both good and evil. (Heb. 5:12-14.)

Newton takes this very general image and turns it into a paradigm for almost his entire theology. "Milk for babes," in Newton's view, refers to essential scriptural doctrines of faith of which there can be no doubt or dissent. One cannot be a communicate Christian without knowing and believing them and one is a schismatic for adding anything to them. These "milk" dogmas are the sole content of the primitive Christian creed which alone is to be taught to all Christians before baptism. Anyone adding to them breaks with the true original Christianity:

[9] See, for example, Richard S. Westfall, "Newton: Theologian," in Edna Ullmann-Margalit, ed., *The Scientific Enterprise: The Bar-Hillel Colloquium: Studies in History, Philosophy and Sociology of Science*, Vol. 4 (Dordrecht: Kluwer, 1992), pp. 228-30; Frank Manuel, *The Religion of Isaac Newton* (Oxford: Oxford University Press 1974), pp. 62-3 (including indications that Whiston and Haynes saw Newton in this light); Scott Mandelbrote, "A Duty of the Greatest Moment: Isaac Newton and the Writing of Biblical Criticism," *British Journal for the History of Science* 26 (1993), p. 289, where Mandelbrote calls Newton a hypocrite for his "outward conformity" to the Anglican Church.

[10] Herbert McLachlan, ed., *Sir Isaac Newton: Theological Manuscripts* (Liverpool: Liverpool University Press, 1950), pp. 36-8.

[11] James E. Force, "Newton, the Lord God of Israel, and Knowledge of Nature," in Richard H. Popkin and Gordon M. Weiner, eds., *Jewish Christians and Christian Jews* (Dordrecht: Kluwer, 1994), p. 146 and, in general, pp. 140-7.

[12] See the preceding note. This paper was delivered at the Clark Library conference on "Jewish Christians and Christian Jews," April, 1992. Its original title was "Milk for Babes and Strong Meats for Elders."

> [20r] For there were certain principles of religion preached by the Apostles from the beginning, & in which all men were to be instructed before baptism by catechizing them, & there were many other truths which they were to learn after baptism by studying the scriptures to the end of their lives. The Principles the Apostle compares to milk to be given to babes, the other truths to strong meat wch belongeth to men of full age...(Bodmer Newton MS, Ch. 2, fol. 20, at the Martin Bodmer Foundation, Geneva. Quoted by permission.)

Newton goes on to tell us what these "milk" doctrines are: "...repentance from dead works, faith in God, Baptism, laying on of hands, & the doctrine of the resurrection & day of judgment..."(*Ibid.*)[13]

The "meat" doctrines are truths which are to be sought out after catechism is complete and the mind more developed. Newton does call them truths but, on the other hand, the inherent impossibility of proving any adiaphoric position from straightforward, uninterpreted Scripture renders them impossible to demand as dogma. They are to be studied and taught, but not argued over. The particular doctrines which Newton places in the "meat" category are most instructive. Here are two passages:

> [40r] Besides the fundamental points of faith <requisite to baptism> wch all men were to learn before baptism, & wch the Apostel therefore compares to milk for babes there are in the scriptures many truths of great importance to be learnt afterward. And these the Apostle compares to strong meats for men of full age who by use have their senses exercised to discern both good & evil. With these truths the mind is to be fed continually as the body is with meats. And to these truths I referr what Christ did before his incarnation, & between his death & resurrection, what he doth [41r] now in heaven how the saints shall reign with him as Kings & Priests at his second coming & rule the nations with a rod of iron & what he or they shall do after the day of judgment, as also all disputable questions about Providence, Predestination, Free-will, Grace, the origin of evil, the nature of Gods justice & of the satisfaction made by Christ, the nature of Angels the state of the dead between death & the resurrection the nature of the bodies wth wch the dead shall arise, the power of the keys, forms of the Church government, the keeping of Easter & other holy days & the like. About such things we are to believe the scriptures, to learn all that we can out of them & to instruct one another, but not to fall out amongst our selves. We must not impose consequences of scripture but keep to the language of the scripture it self, & bear with one anothers interpretations thereof. For imposing of consequences tends to wrangling discord & schism, & puts the true Church upon a level in disputing wth wrangling hereticks. (Bodmer MS, Ch. 2, fol. 40r-41r.)

Elsewhere, Newton writes in a similar vein that:

[13] These are derived from Heb. 6:1-2, immediately following the "milk"/"meat" passage. One must be careful not to confuse the "milk" doctrines, which are conditions for communion, with natural religion whose sole tenets are love of God and man as expressed in the Noachide commandments.

> To make it an article of communion that there are or are not Antipodes, that the earth rests or moves about the sun, that there are or are not more <inhabited> worlds than one that matter is or is not out of nothing, that the souls of men are or are not praexistent, that they are or are not sparks of divine light, etc. would tend not to salvation but to strife faction & schism & therefore would be criminal... (Yahuda MS15.5, fol. 79r, at the Jewish National and University Library, Jerusalem. Quoted by permission.)[14]

Many of the "meat" doctrines in both these quotations are issues which were hotly debated in Newton's Europe. In the first text, the "meat" issues which Newton lists are solely theological. In the second text, both theological and scientific matters are represented as "meat."

In the context of church history, the differentiation between "milk" and "meat" doctrines was what allowed the Jewish-Christians (Nazarenes and Ebionites), who were obliged in the Law to live in peace with gentile Christians, who were not obliged in it. Schism came to the church when certain Ebionites began demanding that all Christians keep the Law as a requisite of communion, thus adding an adiaphoric doctrine to the minimal dogma of primitive Christianity. The church was ultimately corrupted by hastily converted pagan and Jewish metaphysicians (hereafter known as Gnostics), whose introduction of emanational metaphysics into simple Christian beliefs brought disaster.

Two categories qualify as schismatic in Newton's view. First and foremost is the church itself when it seeks to dogmatize adiaphora as the Ebionites did. "All narrowing of communion by humane impositions is schismatical," (Bodmer MS. Chap. 2, f. 42r) he tells us. The other type of schism is that of an individual Christian who fails to remain within the church when it forces no false dogma upon him. "Separation without a sufficient cause made them [Christians] factious seditious & schismatical members." (Bodmer MS, Chap. 5/5a [first full draft], fol. 3r) In order to restore the peace of the primitive church, then, the church itself must not demand an adiaphoric creed, and Christians must not leave the church if it has met this obligation.

In his passage on the Anglican Church, Newton very clearly and consistently goes through the principles of the English Church and separates them into "Milk for Babes" and "Meats for Men of Full Age." By an amazing coincidence, the Church of England's essential dogmatic program, as conceived by Newton, precisely coincides with his own "Milk for Babes": renouncing the devil and his works (carnal desires, idolatry, ambition, pride, and covetousness, called "repentance from dead works" by the Apostle); adherence to the Apostles' Creed; observance of the ten

[14] This passage is in fact referring to the damage done by Gnostics, one of the parties guilty of mixing meat for men of full age (the said doctrines) with milk for babes.

commandments, as explained in the Church catechism; baptism; and the laying on of hands. "These...are the fundamentalls requisite to communion in the Church of England." A man may not be excommunicated for transgression of any other principle.

The most obvious problem for Newton is that the Church of England includes the Trinitarian Nicene, and Athanasian Creeds among its doctrines. Newton simply claims that while these are accepted creeds, they fall in the category of adiaphora *according to the Church of England itself*: "The Constantinople Creed usually called the Nicene Creed & the Creed usually called the Creed of Athanasius are not therefore any part of the milk for babes in the Church of England but are to be referred to the strong meats for them that are of age." (Keynes MS 3, fol. 51, at King's College Library, Cambridge. Quoted by permission.) How does Newton ascertain this? He points out that the 21st article of the Anglican Church expressly admits that church councils may make human errors; and its 8th article states that the Nicene and Athanasian Creeds are acceptable because their principles are taken from Scripture. Its 6th article claims that all truths necessary to salvation are found in the Bible so that anything not found there cannot be an article of faith. Newton thus concludes that these creeds, which were established by particular church councils or convocations, may be compared with Scripture and rejected in any particulars which are not derived directly therefrom, *by permission of the Church of England*—they are "Meats." Newton adds an ingenious proof to show that the English church has itself rejected a dictate of the Council of Ephesus. Newton has, of course, found many aspects of those creeds (such as the "Filioque") to be incompatible with Scripture, and thus rejected them.[15]

An examination of the 1719 Oxford edition of the Book of Common Prayer reveals that in neither the "small" catechism nor the ceremonies of baptism and confirmation is there reference to a Trinity of divine persons or any other doctrine considered heretical by Newton. In fact, the Apostles' Creed recited by catechumens defines the same dogmas as his "Milk for Babes." The Athanasian Creed, Greater Catechism and other problematic declarations are reserved for either special occasions or use of clerics alone (sig. C_3-C_4.)

By this simple stratagem Newton concludes that the Church of England does not *really* mix "meat" with "milk," because the heretical creeds it upholds are not fundamentals of its faith! This explains why Newton was willing to remain within the church, why he did not ever decry it, and why

[15] Note also the following passage: "The fundamentals in all Churches are only those things w[ch] are required in order to baptism & Confirmation. And in the Church of England the fundamentals are only those things w[ch] are taught in the Church Catechism & in the Offices of Baptism & Confirmation." (Bodmer MS, Additional Chapters, "<Chap. Of the Church or Mystical body of the Messiah>", fol. 5r.)

he got angry at his student Whiston for openly declaring his Arianism and breaking with the church. If one could find a way to remain within the church without compromising his principles, he would himself be guilty of schism for breaking away. "Separation without a sufficient cause," as we saw above, "made them [Christians] factious seditious & schismatical members." (Bodmer MS, Chap. 5/5a [first full draft], fol. 3r)

None of this is to say that Newton was completely honest with the Church of England or with himself. While he did discover a way by which he could uphold his own principles and still rationalize remaining an Anglican, he was certainly aware that in practice the church did not regard the offending creeds as optional at all. It was, in fact, illegal to deny Trinitarianism. Furthermore, in order to abstain from saying the Nicene and Athanasian creeds, one had to willfully skip them within the service. What is interesting is that Newton went to such lengths to avoid schism in his own life; and while we can never really know his motives, his "honesty" extended at least to the point of inventing an elegant if factitious rationale.

3. Deism, Latitudinarianism, and Newton

In a famous article entitled "Isaac Newton's *Theologiae Gentilis Origines Philosophicae*,"[16] Richard Westfall attempted to prove that Newton was essentially a Deist by demonstrating that in writing his history of ancient religion, Newton did not grant ultimate authority to the Bible, but rather accepted other sources as being more reliable. While this attitude on Newton's part has not been denied, James Force has sought to prove from other aspects of Newton's thought, in which Newton demonstrates great respect for biblical truth, that he could not have been a real Deist.[17] While I agree with Force's conclusions, I would like to point out that no discussion of Newton's relationship with a stream of theological thought such as Deism can be complete without careful examination of the evidence from *Of the Church*. Indeed, there are various sorts of evidence in the church histories which bear on the issue of Deism, one of which I will point out here.

[16] Richard S. Westfall, "Isaac Newton's *Theologiae Gentilis Origines Philosophicae*," W. Warren Wagar, ed., *The Secular Mind: Transformations of Faith in Modern Europe—Essays Presented to Franklin L. Baumer, Randolph W. Townsend professor of History, Yale University* (New York: Holmes and Meier, 1982), pp. 15-34.

[17] See especially James E. Force, "Biblical Interpretation, Newton and English Deism," in Richard H. Popkin and Arjo van der Jagt, eds., *Skepticism and Irreligion in the Seventeenth and Eighteenth Centuries* (Leiden: E. J. Brill, 1993), pp. 282-305; Force, "Newton and Deism," in Änne Bäumer and Manfred Büttner, eds., *Science and Religion/Wissenschaft und Religion. Proceedings of the Symposium of the XVIIIth International Congress of History of Science at Hamburg-Munich, 1-9 August 1989* (Bochum: Brockmeyer, 1989), pp. 120-32.

Because my own research has concerned the role of Judaism in Newton's theology, my evidence will concern views on that topic.

I have spoken of the great interest in church history among Englishmen of Newton's day, all polemical in some degree and fashion. It is possible to demonstrate that in his thinking about the Jewish influence on the ancient church, Newton falls into a line of ecclesiastical historians which gave birth to Deism. This tradition begins with John Selden and Edward Stillingfleet in the Civil War period (neither of whom was, of course, a Deist); and ends with Henry Stubbe and John Toland. These authors were interested in the role of the Jewish synagogue as the source of primitive church institutions, a conception which is not unrelated to the goal which J. A. I. Champion has shown to be central in deist circles: the reduction of clerical power.[18]

While there are some major differences in stress and opinion (particularly Newton's conception of a peaceful coexistence between Nazarenes and gentile Christians), the subjects and ideas in Newton's work very closely follow the pattern of the beginning of Stubbe's *An Account of the Rise and Progress of Mahometanism* (ed. H.M.K. Shairani, London, 1911); and that of Toland's *Nazarenus: or, Jewish, Gentile, and Mahometan Christianity* (London, 1718.) All three works deal with the Arianism of the early church; the history of the Nazarenes and Ebionites (including definitions of the latter term); the observance of the law by Jewish Christians and its nonobservance by gentile Christians as an established and protracted difference; the Noachide laws as natural law devolving upon everyone always; the danger of forcing unnecessary laws or doctrines on Christians; and the continuity of church institutions from those of the synagogue.

This last point is perhaps the most important for the present purpose because the rule of synagogues, as these authors perceived them, was far less hierarchical and priest-centered than their predecessors, the biblical and Temple polities. Stillingfleet, Newton and Stubbe all specify that the church grew out of the synagogue, not out of the Jewish Commonwealth polity, as most Civil War era thinkers believed. "Christ and his Apostles therefore instituted no new form of Church government," says Newton, "but continued the ancient form of government which was in use among the Jews in their Synagogues before his coming: And the Christian Churches are nothing else than Synagogues." (Bodmer MS, Chap. 9/10, fol. 8r.) Stubbe asserts with confidence that:

> The whole Constitution of the primitive Church Government related to the Jewish Synagogue...But since all learned men do now agree that the Christian Church was govern'd according to the pattern of the Jewish Synagogue, there

[18] J. A. I. Champion, *The Pillars of Priestcraft Shaken: The Church of England and its Enemies, 1660-1730* (Cambridge: Cambridge University Press, 1992).

> can be little doubt but that every Officer of the Christian Synagogues resembled those of the others as wel in Office as Name, & that as they retain'd the Rites & Customs of the Jewish Synagogues in all other things, even to structure. (Stubbe, *An Account*, pp. 18-9.)

The ultimate source for many of the ideas of Stubbe, Toland, and Stillingfleet about both Judaism and polity was John Selden,[19] whom they read themselves and also through the filter of James Harrington,[20] whose works were republished by Toland. Selden was a radical thinker but until recently his religious thought has been little studied by scholars. W. K. Jordan recognized that behind Selden's Erastianism stood a belief that no consensus of religious truth is possible and the only way to guarantee the necessary freedom of inquiry and avoid religious coercion is to prevent the church from gaining sway in politics. He fought both "high" Episcopalianism and Presbyterianism, and generally found clerics distasteful. Only a secular state with control over the church was acceptable to Selden.[21] Stubbe took from Selden the view that the ancient Hebrews practiced toleration, and the minimal creed of the seven Noachide laws; these themes appear again in Newton and Toland.[22] While it is well known that Selden used his Hebraic learning to scotch Presbyterian Hebraist arguments at the Westminster Assembly,[23] it was Selden who put the New Testament context in terms of the Jewish "synagogue" rather than the Hebrew Commonwealth of the Bible.[24] Whereas the latter admitted of a

[19] For Selden, see Jonathan R. Ziskind, *John Selden on Jewish Marriage Law: The "Uxor Hebraica"* (Leiden, 1991), esp. Introduction and p. 1, n. 1, with bibliography; Isaac Herzog, "John Selden and Jewish Law," *Judaism: Law and Ethics* (London: Soncino, 1974); W. K. Jordan, *The Development of Religious Toleration in England*, 4 vols. (Cambridge, MA: Harvard University Press, 1936-40), 2:479-88; *Dictionary of National Biography*, s.v. "Selden, John". Martha Ziskind's unpublished Ph.D. dissertation on Selden (University of Chicago, 1972) has not been available to me. For Selden's influence on Stubbe, see James R. Jacob, *Henry Stubbe: Radical Protestantism and the Early Enlightenment* (Cambridge: Cambridge University Press, 1983). For his influence on Toland, see, e.g., *Nazarenus*, pp. 25, 30.

[20] Harrington was also an Erastian, and he made much use of Selden, particularly in those areas in which Judaism was concerned. He was convinced that the rabbinic tradition was correct in its claim that the laying on of hands to pass along authority was continuous from Moses until Jesus' time. See Jordan, *The Development of Religious Toleration* 4:281-91; S. B. Liljegren, "Harrington and the Jews," *Bulletin de la Société Royale des Lettres de Lund* 4 (1931-2), pp. 65-92.

[21] Jordan, *The Development of Religious Toleration*, 2:479-88. Note the affinity of Selden's Erastianism to that of Hobbes, *Leviathan*, Chap. XLII ("Of Power Ecclesiastical.")

[22] See: Jacob, *Henry Stubbe*, pp. 31-3, 158, 163; Stubbe, *Account*, pp. 13, 24-5; Toland, *Nazarenus*, p. 46, n.64, and Chap. XI-XIII passim.

[23] Jordan, *Religious Toleration*, 2:485; Haller, *Liberty and Reformation*, pp. 230-1.

[24] As my colleague Kalman Neuman has pointed out, it is important to note that the term "synagoga" is often used simply to indicate any Jewish government, as a companion to the Christian "ecclesia." But in Selden's case, the entire description of the "synagogue" discussed in *De Synedriis* and elsewhere, including specific institutions (e.g. the "hazan," etc.), refers specifically to the houses of worship found in each community which we still call "synagogues," and not to the Temple or Commonwealth. While

clerical hierarchy with political power, the former was a confederation of independent institutions, each with its own "bishop" *and* "presbyters." This idea had echoes in Harrington, Stubbe (trying to modify Harrington), and Newton.[25] Newton particularly stresses the continuity of the church from the synagogue and relies heavily on Selden (though he hardly credits him) for identifying the functions of its officers.[26]

Certain drafts of material which would show up in *Of the Church* are entitled *Irenicum* (King's College, Keynes MS 3) or, more specifically, in one draft, *Irenicum or Ecclesiastical Polyty tending to peace.*[27] Despite the plethora of tracts with similar names in this period (which my colleague Robert John Arias has pointed out), it seems most likely that the name and subject of the *Irenicum* are derived directly from Edward Stillingfleet, a leading Latitudinarian, who had written a work with an identical title.[28] Part of Newton's design was certainly the same as that of his friend Stillingfleet: to propose a set of suggestions on how to achieve a stable peace within the Anglican church by recourse (among other things) to the example of ancient Jewish Christianity.[29] Stillingfleet, like Selden (whom he often quotes), is interested specifically in the synagogue rather than the Temple or biblical government as a precursor to church polity.[30] Unlike Harrington or Hooker (who were in any case less interested in the synagogue than in biblical polity), Stillingfleet wishes to show that the ancient synagogue did have a specific form of government which indeed influenced that of the early church heavily but that that specific polity is not "necessary" as the eternal

---

the synagogue had been discussed and described previously by the Buxtorfs and others, its place in the history of the early church was not developed (to the best of my knowledge) before Selden.

25 The importance of the synagogue in Stubbe as an example of non-hierarchic, irenical, ecclesiastic government is brought out by Jacob, *Henry Stubbe*, p. 67; Stubbe, *Account*, p. 18.

26 This continuity of offices is also found in Stubbe (*Account*, p. 19), though he does not develop the parallels. Lightfoot, as has been noted, was also a major source for Newton in this area.

27 MS Keynes 3, fol. 21.

28 Edward Stillingfleet (Bishop of Worcester), *Irenicum. A Weapon Salve for the Churches wounds. Or the Divine Right of Particular Forms of Church-Government; Discussed and examined according to the Principles of the Law of Nature, the positive Laws of God, the practice of the Apostles and the Primitive Church, and the judgment of Reformed Divines. Whereby a foundation is laid for the Churches peace, and the accommodation of our present differences,* 2nd ed. (London, 1662). This is the edition which Newton had in his library. See John Harrison, *The Library of Isaac Newton* (Cambridge: Cambridge University Press, 1978), #1564. I have quoted the title "in extenso" to clarify the purpose of the tract and its similarity to parts of *Of the Church.*

29 Newton says as much, in fact. To John Conduitt he states that he felt such an analysis could solve the problems of religious dispute as the *Principia* solved those of natural philosophy. Mandelbrote, "Biblical Criticism," p. 283, citing from Keynes MS 130.6, Book 3.

30 It is noteworthy that in introducing the subject of the synagogue's precedent for ecclesiastical institutions, Newton cites two very different authorities: Hugo Grotius, the great Remonstrant theologian; and Salmasius (Claude de Saumaise, 1588-1653), the Huguenot successor of Scaliger at Leiden, who used Jewish learning to uphold the divine right of monarchs. See: Stillingfleet, *Irenicum*, p. 239.

form of ecclesiastical government. He devotes a long and learned discussion to the institutions of the synagogue and their development within the church.[31] He also deals with the role of the Nazarenes and Ebionites, the Sanhedrin, and many other subjects Newton later discusses in *Of the Church*.

While the works of Selden and Stillingfleet were in Newton's library,[32] there is no record of his owning either Stubbe or Toland. Stubbe's work, on which Toland's is partially based, was written between 1671 and 1676, but it remained only in manuscript until 1911.[33] Since Toland began writing his *Nazarenus* in 1710,[34] almost exactly the time Newton was working on *Of the Church*, it seems very likely that he had access to Stubbe's work. On the other hand, one might wonder whether Newton was actually in communication with Toland or others of his circle concerning issues of ecclesiastical history during this period.

The goal of Stubbe in his *Account*, and to a lesser degree of Toland in *Nazarenus*, was to prove that pristine Christians (heavily based on Judaism) did not believe in true incarnation or in a Trinity and that this pure tradition had been preserved only among the Muslims. The obvious implication is that the Church of England, and certainly the Roman and other churches, had embraced heresies which could be historically proven false. All religion ought to require the essential Noachide commandments, the law of nature.[35] Christianity is simply a Jewish sect, a "reformed Judaism" as Selden put it.[36] Newton, in the company of these thinkers, believes that the return to pristine Christianity means a reinstatement of non-hierarchical synagogue polity whose only doctrinal task is to avoid the introduction of false metaphysics, such as Trinitarianism. Newton, then, though he is far from the Deists' position in his attitude toward scriptural prophecy, shares their essential program (as understood by Champion), of trying to restore a non priest-centered, simple Christianity.

Awareness of the Stubbe/Toland context for *Of the Church*, of Newton's Erastianism, and of his stratagem for remaining within the Church of England without compromising his religious principles, necessitates a

[31] This is drawn from Selden, who expresses natural religion and law in terms of the seven Noachide commandments in several of his works, most explicitly in *De jure naturali* and *De synedriis*. Selden himself was inspired by a much briefer reference in Grotius' *De iure belli ac pacis*.

[32] Harrison, *Library of Newton*,, #1481-4 (Selden); #1561-1565 (Stillingfleet).

[33] See: Jacob, *Stubbe*, Ch. 4, pp. 64-77; Champion, *Pillars of Priestcraft*, pp. 120-8.

[34] Champion, *Pillars of Priestcraft*, p. 125.

[35] This is drawn from Selden, who expresses natural religion and law in terms of the seven Noachide commandments in several of his works (see above, Chap. 3), most explicitly in *De jure naturali* (Argentorati, 1665) and *De synedriis* (Amsterdam, 1679). Selden himself was inspired by a much briefer reference in Grotius' *De iure belli ac pacis*.

[36] Selden, *De Synedriis*, vol. I, Ch. 8, quoted in Toland, *Nazarenus*, p. 30 ,n. 49.

reevaluation of Newton's relationship with Anglicanism. Newton is clearly an advocate of *comprehension* rather than of *toleration*; that is, he wants the church to minimize and purify its creed so as to swallow the maximum range of Christian doctrines, rather than to remain dogmatic within itself and merely tolerate views outside its own.[37] This is one of the beliefs for which he argues so cogently in his conception of "milk for babes" and "strong meats for men of full age." These attitudes raise the question of Newton's place in the spectrum of opinions found in late Stuart England, particularly his relationship to the "low" churchmen who also supported comprehension.

While Newton's participation in a historical tradition linked with certain Deists has been discussed, it must be noted that the conclusions just stated put him in close alignment with late seventeenth- and early eighteenth-century Latitudinarians.[38] Several scholars have recently attempted to identify Newton as a Latitudinarian but James Force has asserted that this label is simplistic and incorrect, that there was rather "a complex mixing of various strains of thought in Newton's philosophy."[39] This interpretation is undoubtedly correct and Newton's anti-Trinitarianism is the most obvious reason to exclude him from the Latitudinarians.

[37] In 1688-9, two bills were before Parliament: a Toleration Act and a Comprehension Act. It was the former, less revolutionary bill which passed, yet many held out hopes for the church by itself to become more comprehensive. At the political level nobody was thinking of comprehending dissenters, much less anti-Trinitarians. Only the Presbyterians and a few of the less revolutionary sects were to be included. Yet Newton was one of a not insignificant number of Unitarians, Socinians, Arians, and Deists (all of whom were anti-Trinitarian), as well as some otherwise conservative low-churchmen, who wished Anglicanism to drop the Athanasian Creed and several of the Thirty-nine Articles pertaining to Jesus' person so as to achieve a wider "swallow." See, e.g., Richard Ashcraft, "Latitudinarianism and Toleration: Historical Myth versus Political History," in R. Kroll , R. Ashcraft and P. Zagorin, eds., *Philosophy, Science, and Religion in England, 1640-1700* (Cambridge: Cambridge University Press, 1992), pp. 151-77, esp. p. 152; George Every, *The High Church Party 1688-1718* (London: S.P.C.K., 1956), Chap. 2, "Conflict over Comprehension in 1688-9," pp. 19-42. On the party associations of these issues, see the next two notes.

[38] Roger L. Emerson, "Latitudinarianism and the English Deists," in J. A. L. Lemay, ed., *Deism, Masonry, and the Enlightenment: Essays Honoring Alfred Owen Aldridge* (Newark, Del.: University of Delaware Press, 1987), pp. 19-48, seeks to show how different the religious views of the Latitudinarians were from those of the Deists. While this may be true, and I perceive Newton as ultimately much closer to Latitudinarianism than to Deism, it is clear that he draws on the same historical traditions concerning ancient Jewish-Christianity as did Stubbe and Toland.

[39] Force, "Sir Isaac Newton, 'Gentleman of Wide Swallow'?: Newton and the Latitudinarians," in James E. Force and Richard H. Popkin, *Essays on the Context, Nature and Influence of Isaac Newton's Theology* (Dordrecht: Kluwer, 1990), p.122, and, in general, pp. 119-41. The most important of the recent literature on Latitudinarianism includes: R. Kroll et al., eds., *Philosophy, Science and Religion in England*; Martin I. J. Griffin, Jr., *Latitudinarianism in the Seventeenth-Century Church of England*, annotated by Richard H. Popkin, edited by L. Freedman (Leiden: E. J. Brill, 1992); Barbara J. Shapiro, "Latitudinarianism and Science in Seventeenth-Century England," in Charles Webster, ed., *The Intellectual Revolution of the Seventeenth Century* (London: Routledge and Kegan Paul 1974), pp. 286-316. Barbara Shapiro and Margaret Jacob describe Newton as essentially a Latitudinarian. See Shapiro, "Latitudinarianism and Science"; Jacob, *The Newtonians and the English Revolution, 1689-1720* (Ithaca: Cornell University Press, 1976), pp. 29-36, esp. p. 33.

In his study of Latitudinarian religion, Martin I. J. Griffin has laid out certain characteristics of the major Latitudinarian figures which mark them off from their contemporaries: orthodoxy in accepting the traditional Christian creeds; conformity to the Church of England, including its episcopal government, Thirty-nine Articles, and Book of Common Prayer;[40] advocacy of reason in religion; theological minimalism; belief in an "Arminian" scheme of justification; emphasis on practical morality over credal speculation and precision; a distinctive sermon style; and connections with science and the Royal Society.[41] Nevertheless, Griffin is not doctrinaire about these characteristics: "This norm and standard as a test of full-fledged seventeenth-century Latitudinarianism need not be applied so rigidly as to deny the classification 'Latitudinarian' to all who did not hold every characteristic Latitudinarian doctrine."[42] As long as one specifies where the individual under discussion differs, he may still fit.

Griffin claims further that Latitudinarianism underwent deterioration after about 1690 and thereafter began to "swallow" even figures such as John Locke, Samuel Clarke, and William Whiston, whose ideas about the Trinity were clearly heterodox. These "must be considered atypical seventeenth-century Latitudinarians, but were unexceptionable ones of the eighteenth century."[43] Because Clarke and Whiston reflected Newton's theology fairly accurately, one would imagine that Newton might accurately be categorized as a kind of latter-day Latitudinarian. As opposed to Griffin, Force conceives the parameters of Latitudinarianism as "narrowing" and becoming more rigid in its later forms rather than loosening up after 1690 to accept even heterodox thinkers.[44] Force sees the case of William Whiston's persecution by both high and low churchmen, some of whom are specifically identified as Latitudinarians, as evidence that Latitudinarianism could not countenance the Newtonian heresies which Whiston espoused. Richard Ashcraft stresses this point as well and insists that Latitudinarianism was *not* a set of doctrines, but a political movement. Latitudinarianism must be judged by its actions rather than its alleged liberal doctrine; and among the Latitudinarians' actions was the persecution of dissenters and heretics like Whiston and Clarke.[45]

Having seen how Newton deals with the Church of England, and in view of Griffin's more "latitudinarian" definition of later Latitudinarianism, a fairly strong argument can be made for aligning Newton closely with the

[40] I would suggest an addition to this item: Erastian support of royal prerogative within the church.

[41] Griffin, *Latitudinarianism*, pp. 1-48, esp. p. 43.

[42] *Ibid.*, p. 45.

[43] *Ibid.*, p. 47.

[44] *Ibid.*, pp. 124-5.

[45] See Ashcraft, "Latitudinarianism and Toleration."

Latitudinarians. While Force does not see Newton's continuation within the Church of England as particularly strong evidence for his being a Latitudinarian, in light of the new evidence of his views in *Of the Church,* it should be given considerable weight. Newton succeeded in justifying adherence to Anglicanism by shifting the objectionable creeds out of the realm of essential Anglican dogma, thereby neutralizing cause for schism by many sorts of anti-Trinitarians. In the only open reference I have found in Newton to the Latitudinarians, he in fact places the highest value on their refusal to brook schism. At the time he is writing (ca. 1710), the term "Latitudinarian" is already used in a derogatory manner and he objects to that implication:

> And tho the [unnecessary theological] imposition should not be followed with actual schism, yet it may make the imposer a schismatick in his heart, & therefore all narrowing of communion by humane impositions is schismatical & all railing at those men as latitudinarians who endeavour to throw off humane imposition whereby the Christian world is divided & disturbed is railing at Charity & committing schism. (Bodmer MS, Chap. 2, fol. 42r.)[46]

For Newton, then, the Latitudinarians are to be praised for promoting what Griffin calls "theological minimalism" within the church. Though Newton still does not meet Griffin's criterion of adherence to the traditional creeds and Thirty-nine Articles, he is nevertheless able to remain a faithful member of the Church of England. Meanwhile, Newton strongly supports the other Latitudinarian positions, with the exception of his vague predestinarianism and the moot issue of sermon style. In other words, he has managed to "swallow" Latitudinarian religion.

What of the Latitudinarians' ability to "swallow" Newton? Concerning certain Latitudinarians' participation in the persecution of anti-Trinitarians, I would suggest that we can assign the attitudes toward such cases not only to differing periods and hierarchical positions among the Latitudinarians, but also to differences in their respective philosophies and levels of toleration. On the one hand, Latitudinarianism was put in the position of defending orthodoxy, thus being the body responsible for carrying out laws against heresies such as Arianism and Socinianism.[47] On the other hand, there were serious breaks in the ranks of the lower church over the

[46] See also Yahuda MS 15.7, fol. 186r for an almost identical statement. These are, to the best of my knowledge, the only references to Latitudinarians in Newton's writings.

[47] It must be understood that anti-Trinitarianism was against the law in England during the entire period with which I am dealing. Unitarians were not protected by the Toleration Act, and were liable to imprisonment under the Blasphemy Act of 1698 for propagating their beliefs. Yet this was a very widespread heresy in England throughout the later seventeenth and early eighteenth centuries. See M. R. Watts, *The Dissenters, vol. I: From the Reformation to the French Revolution* (Oxford: Clarendon Press, 1978), pp. 372-5.

definition of the Trinity.[48] Already during the later seventeenth century there were accusations that the Latitudinarians were really Socinian and though this may have been false at that time there was undoubtedly a great deal of convergence in the early eighteenth century.[49] In the early eighteenth century, the label of "Latitudinarian" could apparently swallow everyone from Whiston and Clarke to those who persecuted Whiston and Clarke, depending on *which* Latitudinarians are under discussion. Force's most cogent argument against considering Newton a Latitudinarian, then, the problem of Anglican Trinitarianism, is not a clearcut deviation from these later forms. It would seem to me that by Griffin's standards we could reasonably say that Newton was a Latitudinarian who, however, rejected Trinitarian creeds and Arminian justification.

4. A METAPHYSICAL ISSUE: NEWTON AND LEIBNIZ ON EMANATION

It has been noted that Newton saw pagan and Jewish emanationist metaphysics as the source of corruption in the ancient church which eventually caused the spread of the heretical Trinitarian doctrine. Now, with the new understanding supplied us by Allison Coudert's research into Leibniz's use of emanational kabbalistic metaphysics, we have a clearer understanding of Newton's motive for writing so vehemently against pagan and Jewish emanational heresy in the early church. The church histories of ca. 1710, when Newton was deeply embroiled in the debate with Leibniz, could hardly be a strictly antiquarian pursuit divorced from this issue. Frank Manuel, even without the evidence Coudert has presented, offered the intuition that "After the turn of the century, when Newton was engaged in writing a critical history of ancient heresies and their propagators, the duel with Leibniz was never far from his mind. Leibnizians and Cartesians were modern exemplars of the Cabbalists, Gnostics, and Platonists."[50] In 1974,

[48] The difficulty of understanding the Trinity in the Anglican Church was so acute that not only anti-Trinitarianism, but Tritheism, also became a problematic heresy. In the 1690s, William Sherlock attempted a refutation of the Unitarians with a new way of explaining the Trinity. Robert South accused him of Tritheism, Stephen Nye began accusing everyone of various heresies, and mutual recriminations were bitter. In the end, the monarchs had to issue two orders (in 1695 and 1714) to stifle creative explanations of the Trinity. See G. Rupp, *Religion in England* (Oxford: Clarendon, 1986), pp. 248-9; Edward Cardwell, ed., *Documentary Annals of the Reformed Church of England*, Vol. II (Oxford, 1839), pp. 339-41, 365-7.

[49] Griffin's statement that Whiston and Clarke were "unexceptionable" Latitudinarians of the eighteenth century has already been noted. On the accusations of Unitarianism in the 1670s and Gilbert Burnet's response, see Martin Greig, "The Reasonableness of Christianity? Gilbert Burnet and the Trinitarian Controversy of the 1690's," *Journal of Ecclesiastical History* 44, No. 4 (October, 1993), pp. 631-51; R. E. Sullivan, *John Toland and the Deist Controversy* (Cambridge, MA: Harvard University Press, 1982), pp. 82-3. On the real convergence in the 1690s and after, see Sullivan, *Toland*, pp. 89, 95, 103.

[50] Manuel, *The Religion of Isaac Newton*, p. 75.

when he published these words, Manuel could hardly have known how true they were.

Leibniz describes God's creation of substances in the following manner: "Now it is first of all very evident that created substances depend upon God who preserves them and can *produce them continually by a kind of emanation* just as we produce our thoughts..."[51] And again: "God alone is the ultimate unity or the original simple substance, of which all created or derivative monads are the products, and arise, so to speak, through the continual outflashings of the divinity from moment to moment..."[52] While he posits other forms of causation as well, emanation is central to Leibniz's cosmogony. Newton was aware (albeit in a more general way) of what modern scholars are now rediscovering: Leibniz drew upon Gnosticism and Kabbalah to formulate many of these ideas.

A brief example of the implication of Gnostic/Kabbalistic emanation theory for Leibniz and for Newton will suffice to show the significance of the Kabbalah polemic in their debate. Coudert has demonstrated that Leibniz is very close to the position of Francis Mercurius van Helmont, his Kabbalist contact, when, in the 1690s, he moves "from an essentially dualistic position [on the nature of matter] to a monistic one," according to which "spirit and matter are simply extremes on a continuum."[53] One of the implications of this position is a rejection of true creation *ex nihilo* in favor of a theory of prime matter (monads) which, as I understand Leibniz, to mean, God emanates constantly. Newton strongly condemns the Kabbalistic and Gnostic notion of prime matter for the precise reason that it is part of an heretical emanational cosmogony which contradicts voluntarism; he instead upholds a strict belief in creation *ex nihilo*. Newton will not countenance an emanational cosmogony which in his opinion also denies the absolute unity of God.[54]

For Leibniz, creation *ex nihilo* lacks sufficient reason both ultimately and (because Newton believes in the existence of vacua and rejects a plenum, so that there is no intervening medium) proximately. These are the same

[51] Gottfried Wilhelm Leibniz, *Discourse on Metaphysics*, in *Discourse on Metaphysics; Correspondence with Arnauld; and Monadology*, trans. G. Montgomery (La Salle: Open Court, 1902), p. 23 (my emphases).

[52] Gottfried Wilhelm Leibniz, *Monadology*, trans. G. Montgomery, p. 261.

[53] Coudert, *Leibniz and the Kabbalah*, p. 89.

[54] The anonymous reader of this paper has correctly pointed out, however, that much earlier in his career Newton himself had upheld a belief in emanation, though one quite different than that of Leibniz. See A. Rupert Hall and Marie Boas Hall, *Unpublished Scientific Papers of Isaac Newton* (Cambridge: Cambridge University Press, 1962), pp. 32, 37.

grounds upon which Leibniz rejects Newton's theory of gravity.[55] While it is true, then, that Leibniz never saw Newton's church histories, with their fulminations on the issue of Gnostic/Kabbalistic emanation ideas (perhaps nobody saw them during his lifetime), these writings bear directly on such central issues in their debate as sufficient reason, vacua, and the existence of prime matter.

5. Conclusion

A careful study of Newton's church histories teaches us a great deal about his attitudes to crucial contemporary debates and controversies, especially those concerning membership in the Church of England, Deism, Latitudinarianism, and the metaphysics of Leibniz, In showing how the church histories illuminate these aspects of Newton's life and thought, I hope that I have sparked some interest in these vitally important, but too long ignored, documents.

[55] For the debate between Leibniz and Newton (represented by Clarke), concerning sufficient reason, gravity, space and time, and the existence of vacua, see H. G. Alexander, ed., *The Leibniz-Clarke Correspondence* (Edinburgh: Edinburgh University Press, 1956). On Leibniz's idea of causation, see Coudert, *Leibniz and the Kabbalah*, Chap. 7, pp. 137-54.

ESSAY 8

The Seven Trumpets and the Seven Vials: Apocalypticism and Christology in Newton's Theological Writings

SARAH HUTTON
University of Hertforshire

> And he said, Go thy way, Daniel: for the words are closed up and sealed till the time of the end ... and none of the wicked shall understand; but the wise shall understand.
>
> Dan. 12:9-10

> The testimony of Jesus is the spirit of prophecy.
>
> Rev. 19:10

Newton's study of the apocalypse is, arguably, a point of intersection for his apparently diverse interests, be these mathematics or alchemy, astronomy or ancient religions, chronology or bible-study. Indeed the techniques he uses in his investigation of prophecies—the sorting and collation of a vast accumulation of data—is not dissimilar to the way he organises other areas of enquiry, whether scientific, historical or theological. The claim that Newton's apocalypticism is central to his theology and religious beliefs is nothing new. Through his writings on prophecy, scholars have gauged his anti-Catholicism, his anti-Trinitarianism, his providential view of history.[1]

[1] James E. Force and Richard H. Popkin, *Essays on the Context, Nature, and Influence of Isaac Newton's Theology* (Dordrecht: Kluwer Academic Publishers, 1990); Frank E. Manuel, *The Religion of Isaac Newton* (Oxford: Oxford University Press, 1974), chap. 4; Richard H. Popkin, "Newton's Biblical Theology and his Theological Physics," in P. B. Scheurer and G. Debrock, eds., *Newton's Scientific and Philosophical Legacy* (Dordrecht: Kluwer Academic Publishers, 1988), pp. 81-97; Richard S. Westfall, *Never at Rest: A Biography of Isaac Newton* (Cambridge: Cambridge University Press, 1988; first published 1980), p. 13. See also S. Hutton, "More, Newton and the Language of Biblical Prophecy" and R.Iliffe, "'Making a Shew': Apocalyptic Hermeneutics and the Sociology of Christian Idolatry in the Work of Isaac Newton and Henry More", in James E. Force and Richard H. Popkin, *The Books of Nature and Scripture. Recent Essays on Natural Philosophy, Theology, and Biblical Criticism in the Netherlands of Spinoza's Time and the British Isles of Newton's Time* (Dordrecht: Kluwer Academic Publishers, 1994), pp. 39-53, and 54-88. For Newton's study of history, see Frank E. Manuel, *Isaac Newton, Historian* (Cambridge, Mass: Yale University Press, 1963).

J.E. Force and R.H. Popkin (eds.), Newton and Religion, 165–178.
© 1999 *Kluwer Academic Publishers.*

His discussion of the apocalypse confirms his view of God the creator, "παντοκρατωρ, the supreme governor of the Universe,"[2] controlling all by his omnipotent will. This accompanies a narrow view of election and grace, which throws into relief his strictures about Judaism as much as his anti-Catholicism. Newton's writings on prophecy are, indeed, a rich source for Newton's theology. And his interpretation of the Book of Revelation gives us more insight into his Arianism than simply a means of identifying Arianism as a central component of the apostasy of the Roman church. The apocalypse records not just the onset and course of the trinitarian apostasy. Newton's interpretation of it tells us a good deal about the temper of his Christianity. In this paper I propose to examine Newton's approach to the interpretation of biblical prophecy as an indicator of his religious formation, both in respect of his theological preferences and in respect of his personal spirituality. I shall argue that although Newton denies the divinity of Christ, we should not underestimate the importance of the son of man in his anti-trinitarian theology. The significance of Newton's Christology becomes clear when his writings on the Apocalypse are set in relation to the account of religious fundamentals sketched in the *Irenicum* (Keynes MS 3). But before we can make sense of Newton's personal theology, it is important to recognise the received framework of interpretation within which he was working. And, above all, it is important to recognise just how seriously Newton took his study of the bible.

1. The Importance Of The Study Of Biblical Prophecy For Newton

That fundamentals hinged on the interpretation of the bible is clear from Newton's preamble to his MS treatise on the interpretation of prophecy (Yahuda MS 1). There he repeats his contention that we have a duty to study prophecies and "the danger by neglecting them is very great." And just as "the obligation to study them"[3] is enjoined to all Christians, so Newton felt it was his special duty to communicate the findings of his own studies. In so saying, he was mindful of the punishment meted out to the over-cautious servant in the New Testament parable of the talents:

[2] Cambridge, King's College Library, Keynes Newton MS 3, fol. 43. Quotations from the Keynes MSS are made by kind permission of the Provost and Fellows of King's College, Cambridge.

[3] Quoted from I. Newton, *Trattato sull' Apocalisse*, ed. with Italian translation by M. Mamiani (Turin: Bollati Boringhieri, 1994), p. 4. This work is translated from a manuscript in the possession of the Jewish National Library, Jerusalem: Yahuda Newton MS. 1.1. This and other quotations from the Yahuda MSS are quoted by kind permission of the Jewish National and University Library, Jerusalem.

> Having searched & by the grace of God obteined knowledg in ye prophetique scriptures, I have thought my self bound to communicate it for the benefit of others, remembring ye judgment of him who hid his talent in a napkin.[4]

For Newton, prophecies hold the key to understanding God's word: in both the *Irenicum* and his writings on the Apocalypse (Yahuda MS 1) he recommends study of prophecy for that purpose. Above all, the prophecies of Daniel and St. John are fundamental for confirming the truth of the Christian message. In the Book of Revelation the enigmatic prophecies of Daniel about the advent of Christ the Messiah are explained. The book with seven seals in St. John's prophecy is identified as the book sealed by Daniel, and its opening manifests its Christian import. This is why, as Newton says, "it is called the Apocalypse or Revelation of Jesus Christ."[5] Prophecy works by retrospective confirmation. The advent of Christ proved the truth of the "few and obscure Prophecies concerning Christ's first coming."[6] (The Gospels make comprehensible and confirm the prophecies of the Old Testament). The fulfillment of prophecies of the second coming of Christ (that is the prophecies contained in the Book of Revelation) will have the same effect (i.e. of revealing their truth content.) Furthermore, confirmation of the prophecies of the first coming was important for establishing Christianity, and also for showing "that the world is governed by providence."[7] The Prophecies of the Second Coming are even more far reaching in their importance:

> the many and clear Prophecies concerning the things to be done at Christ's second coming, are not only for predicting but also for effecting a recovery and re-establishment of the long-lost truth, and setting up a kingdom wherein dwells righteousness. The event will prove the Apocalypse; and this Prophecy, thus proved and understood, will open the old Prophets, and all together will make known the true religion and establish it.[8]

Just as the Old Testament prophecies of the first coming served in apostolic times for ye conversion of ye men of that age to the truth & for the establishment of faith," so:

[4] Yahuda Newton MS 1.1, fol.1r., in Mamiani, p. 2.

[5] Isaac Newton, *Observations upon the Prophecies of Daniel and St. John* (London, 1733), quoted from *Sir Isaac Newton's Daniel and Apocalypse with an Introductory Study of the Nature and the Cause of Unbelief, of Miracles and Prophecy* (London, 1922), p. 308.

[6] *Ibid.*, p. 306.

[7] *Ibid.*

[8] *Ibid.*

> the prophecies wch concern these latter times into wch we are fallen were in like manner intended for our use that in the midst of Apostacies we might be able to discern the truth & be established in the faith thereof.[9]

In Yahuda MS 1, Newton underlines the importance of "the designe of the Apocalyps" for sustaining true faith:

> Was it [the Apocalypse] not given for ye use of the Church to guide & direct her in the right way, And is not this the end of all prophetick Scripture.[10]

2. The Development Of Newton's Scheme

At what point Newton had worked out his own scheme definitively, it is not possible to say. Biblical prophecy was a subject he studied all his life, not just in his dotage, as used to be believed. We know of his discussions with Henry More in 1680 when differences between his scheme and More's are mentioned. He was still refining the scheme in 1691 when he discussed details with Locke and Lady Masham at Oates. On 7 February of that year Newton wrote as follows to John Locke:

> I should be glad to have your judgment upon some of my mystical fansies. The son of man Dan. 7 I take to be the same with the Word of God upon the white hors in heaven Apoc 19 and him to be the same with the Man child Apoc. 12 for both are to rule the nations with a rod of iron: but whence are you certain that the Ancient of days is Christ? Does Christ any where sit upon the Throne.[11]

This passage in the letter is marked by Locke for attention. On 30 June 1691, Newton wrote to Locke, "I thank you heartily for your being so mindfull of me and ready to assist me with your interest." And he notes that "My Lady Masham and you have done me much honour in looking into my Book and I am very glad to have the approbation of such judicious

[9] Yahuda Newton MS 1, fol. 2v-3r, in Mamiani, pp. 4-6

[10] Yahuda Newton MS 1, fol. 4r, in Mamiani, p.10.

[11] Newton to Locke, 7 February, 1691, in *The Correspondence of John Locke*, ed. E. S. de Beer, 8 vols. (Oxford: Clarendon Press, 1979), 4: 197. The references are to Daniel 7:9—"And I beheld till the thrones were cast down, and the Ancient of days did sit, whose garment was white as snow and the hair of his head like the pure wool: his throne was like the fiery flame and his wheels as burning fire"; Rev. 19:11-15—"And I saw heaven opened and behold a white horse; and he that sat upon him was called Faithful and True, and in righteousness he doth judge and make war./His eyes were as a flame of fire, . . ./And he was clothed with a vesture dipped in blood: and his name is called The Word of God. . ./And out of his mouth goeth a sharp sword, that with it he should smite the nations: and he shall rule them with a rod or iron: and he treadeth the winepress of the fierceness and wrath of Almighty God"; Rev. 12:5—"And she brought forth a man child, who was to rule all nations with a rod of iron: and her child was caught up unto God, and to his throne."

persons."[12] The book in question was probably the treatise on the textual corruptions of scripture (in particular 1 John 5:7, and 1 Timothy 3:16), sent by Newton to Locke, and, subsequently forwarded by Locke to Jean Le Clerc for publication.[13] There is, as far as I am aware, no record of how Locke or Lady Masham reacted to it—though the term "approbation" suggests they approved of it. Locke's part in trying to get it published implies approval. Newton's advisors at Oates do not seem to have answered his query about the "Ancient of Days" at the time of the June letter.[14]

In another letter on the same topic, probably to Locke, Newton denies the identification of "the Ancient of Days" with Christ, by claiming that the words "I am Alpha & Omega the first and last" of Rev. 1:ll are a late interpolation ("have crept erroneously into some few greek MSS.")[15] Nonetheless, he appears to have been emboldened to emphasis the role of Christ in his scheme. For among Locke's papers, there survives a draft by Newton of his apocalyptic scheme. (See illustration, p. 000).[16] In two other drafts of the scheme among the Yahuda MSS (*A Synopsis of the Synchronisms of the Apocalypse*), the variant inscriptions are inserted along the left-hand edge, in place of events in historical chronology. First:

> Christ dictates the Epistles to the Seven Churches wth reference to the following Prophecies.

In the second:

[12] Newton to Locke, 30 June, 1691, in Locke, *Correspondence*, 4:288.

[13] Dated 14 November 1690, this letter is printed in *The Correspondence of Isaac Newton*, 7 vols., ed. H. W. Turnbull, J. F. Scott, A. R. Hall, and Laura Tilling (Cambridge: Cambridge University Press, 1961), 3: 83-122. The treatise was not actually published until 1754. On the publication history, see Newton, *Correspondence*, 3: 123, note 1. This letter is not included in de Beer's edition of Locke's correspondence, presumably because the adressee is not named. The content, which links it with the published version, helps identify the recipient as Locke.

[14] Newton writes, "Concerning the Ancient of days Dan.7, there seems to be a mistake in my last letter or in your's because you wrote in your former letter that the Ancient of Days is Christ, and in my last I either did or should have asked you how you knew that." Newton to Locke, 30 June, 1691, in Locke, *Correspondence*, 4: 288.

[15] "And again in Apoc. 1. 11 the words of the son of Man, I am Alpha and Omega the first & the last, have crept erroneously into some few greek MSS out of one of wch Erasmus printed it & into the Arabick. . .God is called ye first & ye last to signify not his eternity but that it is he who sits upon the throne in the beginning and end of the prophesy: wch some not understanding have applied here to Christ to prove his Eternity." Newton, *Correspondence*, 3: 140. Turnbull heads this letter, "Newton to a Friend [?John Locke]" and suggests November 1690 as a date. Newton, *Correspondence*, 3: 140. Again, this letter is not included in the De Beer edition. But, as a follow-up to Newton's letter of 14 November 1690, it must be addressed to Locke. If these letters are to Locke, the Turnbull (et al.) edition of Newton's correspondence contains considerably more Newton-Locke correspondence than the De Beer edition of Locke's correspondence, and shows that bible interpretation was a major subject of discussion between Locke and Newton.

[16] This is preserved in the Lovelace collection of Locke MSS at the Bodleian Library, Oxford, Locke MS c.27, fol. 88r, and reproduced by kind permission of the Bodleian Library.

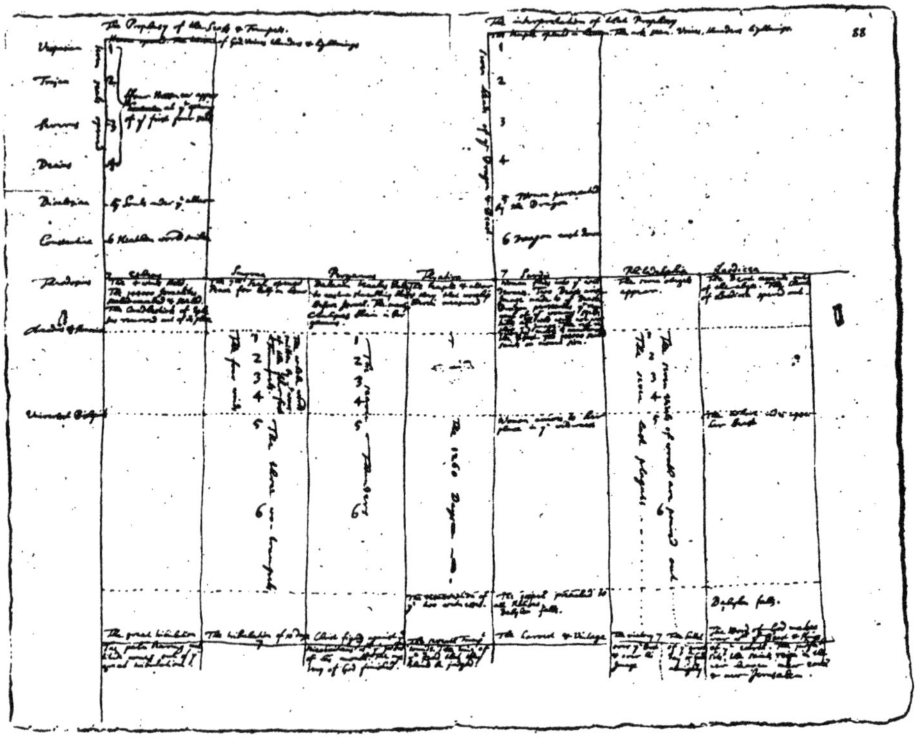

Newton's Scheme of the Prophecies of St John. MS in the possession of John Locke.
Bodleian Library MS. Locke c.27, fol.88r.
Reproduced by kind permission of the Bodleian Library.

> The Son of Man in the form of an High priest Dictates Epistles to ye Seven Churches.[17]

There are differences of detail between these two diagrammatic synopses, as well as the one sent to Locke, but the over-all scheme is the same in all three synopses, and corresponds to Newton's endless discussions of the Book of Revelation among his unpublished papers, as well as to his posthumously published *Observations upon the Prophecies of Daniel and the Apocalypse of St. John*. And, as we shall see, the prophecies have special relevance to Newton's view of Christ.

3. NEWTON'S PROPHETIC SCHEME

In his "The first book concerning the language of the Prophets,"[18] Newton notes that the narrative order of the Book of Revelation is one which doubles back on itself. Books i-vi concern the times preceding the great Apostasy. Books vii-xi describe the apostasy and its overthrow. Thereupon the whole sequence is recapitulated from the end of book xi to book xx. The last two books describe the last days after the fall of the apostasy and subsequent judgement. This scheme preserves intact the given order of the chapters of the Book of Revelation. And, in Newton's account, the second part of the Book of Revelation (from book xi) is an "Interpretation of the Prophesy" set out in the first part (that is "The Prophesy of the Seals and Trumpets.") In all, there are seven subdivisions, each corresponding (in the Bodleian MS) to one of the seven churches of Revelation 1:3 (Ephesus, Smyrna, Pergamus, Thyatira, Sardis, Philadelphia, Laodicea.)

The Bodleian version of the scheme (Locke MS c. 27) – see illustration on the opposite page – indicates the chronology of the scheme. The seven seals (Rev. Chaps. 5-7) are opened between the reigns of Vespasian and Theodosius (each vial corresponding to the reign of one emperor.) Each seal finds correspondence with the seven heads of the dragon and the beast (Rev. Chaps. 11-14.) The trumpets, thunders, 1260 days, plagues and vials of wrath (Rev. Chaps. 8-16) refer to events after the reign of Theodosius. But the only date lines which Newton gives are the reign of Arcadius and Honorius (ending the "peace for half and hour" before the sounding of the trumpets) and the "Universal Bishoprick" which co-incides with the fifth and sixth of the trumpets/thunders/ vials/plagues. The dateline of the seven

[17] Yahuda MS 8.5. One of these is reproduced in Westfall, *Never at Rest*, p. 322. The headings refer to Rev. 1:12-14—"And I turned to see the voice that spake with me ... I saw.../one like unto the Son of man, clothed with a garment down to the foot, and girt about the paps with a golden girdle/His head and his hairs were white like wool, as white as snow; and his eyes were as a flame of fire."

[18] Keynes Newton MS 5.

churches (co-inciding with Theodosius' reign) appears to be the critical turning point in the unfolding of events.

4. NEWTON AND MEDE

Newton's commentaries on the Book of Revelation are not eccentric, but fall clearly within a well-established English seventeenth-century tradition: he adopts the synchronal method for interpreting the book elaborated by the Cambridge bible scholar Joseph Mede (d. 1638) in his *Clavis apocalyptica* (1627, second edition 1632.)[19] According to Mede the visions described in the Book of Revelation should be grouped together as a series of co-terminous segments of time. For example the visions of the two witnesses and of the ten-horned beast refer to events which run concurrently with the sounding of the first six trumpets. By reading the visions as concurrent rather than as sequential, Mede was able to suggest a more coherent chronology for the widely accepted Protestant interpretation of the apocalypse as referring to post-biblical times, where anti-Christ was identified with the Church of Rome.[20] According to this, the victory over the beast denotes the final triumph of the remnant of the true faith against the papacy, that is the forces of antichrist.

Newton adopts the general principles of Mede's synchronal programme, though he differs in many particulars.[21] An example which shows this is their disagreement over the relative positioning of the visions of the seven trumpets and seven vials—and consequently over their interpretation of them. As my earlier quotation from Newton shows, Newton regarded the seven vials and seven trumpets of Revelation Chaps. 8-12 as co-inciding with one another and co-terminous one with another. This is a point where Newton can be seen to be operating a synchronal scheme similar to Mede's. At the same time, it is a point where he differs from his exegetical mentor, and says as much:

[19] On Mede, see J. van den Berg, "Continuity within a Changing Context: Henry More's Millenarianism Seen Against the Background of the Millenarian Concepts of Joseph Mede," *Pietismus und Neuzeit* 14 (1988), pp. 185-202; Katherine Firth, *The Apocalyptic Tradition in Reformation Britain, 1530-1640* (Oxford: Oxford University Press, 1979); S. Hutton, "More, Newton, and the Language of Biblical Prophecy".

[20] See B. W. Ball, *A Great Expectation: Eschatological Thought in English Protestanitsm to 1660* (Leiden: Brill, 1975); T. Bauckham, *Tudor Apocalypse: Sixteenth Century Apocalypticism, Millenarianism and the English Reformation from John Bale to John Fox* (Oxford: Oxford University Press, 1978); P. Christiansson, *Reformers and Babylon: English Apocalyptic Visions from the Reformation to the Eve of the Civil War* (Toronto: University of Toronto Press, 1978); C. Hill, *Antichrist in Seventeenth-century England* (London: Oxford University Press, 1971.)

[21] See Hutton, "More, Newton and the Language of Biblical Prophecy" for an outline of the differences between Mede and Newton.

> Mr Mede hath explained the Prophecy of the first six trumpets not much amiss: but if he had observed that the Prophecy of the pouring out the vials of wrath is synchronal to that of the sounding the trumpets, his explanation would have been yet more complete.[22]

The relationship of vials to trumpets is also one of the issues in which Newton's fellow Cantabrigian and Granthamite, Henry More, differs from Mede. In More's scheme the vials are poured out during the sounding of the seventh trumpet. Newton, like More, was working within a particular interpretative tradition, the synchronal one established by Joseph Mede, but, also like More, he differs from him in important respect—though in different ways from More. When we compare Mede, More and Newton in respect of their prophetic schemes, we can see what Newton meant when he compared the book of Revelation to the components parts of an "Engin made by an excellent Artificer," which can be fitted together in any number of ways, but only one of them will produce a clock that works.[23] Both More and Newton can be seen as calibrating Mede's original system. As I have noted elsewhere, another area where they agree to differ from Mede is over the prophetic status of the seven churches of Asia (Revelation 1:4.)

5. NEWTON'S RELIGIOUS BELIEFS

To the historically distant twentieth-century reader of seventeenth-century interpretations of the Book of Revelation, adjustments to the narrative sequencing of the visions therein contained can seem insignificant details. But in any interpretative scheme that depends on parallels for self-confirmation, re-arrangements of the order, howsoever large or small produce major alterations in the scheme of symbolic equivalence. Newton's divergence from Mede and More points to fundamental differences both of interpretation and theology.[24] Perhaps the most important areas of difference concern, on the one hand, trinitarianism, and, on the other, their respective understanding of the divine attributes. These two areas are, of course, connected, since they both concern the conception of the deity. Where Henry More asserted that the book of Revelation confirmed the doctrine of the Trinity, Newton, as is well-known, took an anti-trinitarian position. As may be expected from an anti-trinitarian, Newton's deity is single-personed.

[22] Newton, *Observations*, p. 334.

[23] Yahuda Newton MS 1, p.15

[24] I have argued that the divergent treatments of prophetic language by More and Newton were not "merely ornamental" but indicate fundamental theological differences between the two men. See Hutton, "More, Newton and the Lanuage of Biblical Prophecy."

Newton also emphasises God's will and power. God is, he writes in the *Irenicum*,

> one everliving omnipresent omnipotent omniscient invisible God the creator of heaven and earth & of all things therein. That He is what he is by the necessity of nature yt all other things are what they are by the power of his will.[25]

Newton's emphasis on the will and power of God sets him firmly within the voluntarist tradition of theology.[26] By contrast, More like Mede, emphasised the divine attributes of goodness and wisdom above divine power.[27] The most familiar face of Newton's voluntaristic God is the God of dominion described in the General Scholium to the 1713 edition of the Principia. Newton's God is the God of power, *Deus Pantocrator* whom, "we may fear and obey." God is, he writes in Yahuda MS 15:

> the *παντοκρατωρ*, Lord of all things with an irresistible & boundles power & dominion that we may not hope to escape if we rebell & set up other Gods or transgress the laws of his monarchy.[28]

As James Force justly notes, such a voluntaristic conception of the Deity has important consequences for Newton's Christology. Newton underlines the subordination of the Son to the Father: "ye son hath no power in him but w^{t} derives from ye father for he professes that of himself he can do nothing."[29] Nonetheless, Newton's denial of the divinity of Christ, does not mean that the Son of Man has an insignificant place in Newton's theology. The contrary is the case, as is apparent from the manuscript known as the *Irenicum* (Keynes MS 3).

[25] *Irenicum*, Keynes MS 3, fol. 35. Compare Yahuda MS 15.5, fol. 97v: "We must believe that he is the father Almighty, or first author of things by the almighty power of his will, that we may thank and worship him & him alone for our being and for all the blessings of this life." Quoted in Westfall, *Never at Rest*, p. 823.

[26] On voluntarism, see F. Oakley, *Omnipotence, Covenant, & Order: An excursion in the History of Ideas from Abelard to Leibniz* (Ithaca: Cornell University Press, 1986); A. Funkenstein, *Theology and the Scientific Imagination from the Middle Ages to the Seventeenth Century Theology and the Scientific Imagination from the Middle Ages to the Seventeenth Century* (Princeton: Princeton University Press, 1986). On Newton see, James E. Force, "Newton's God of Dominion: The Unity of Newton's Theological, Scientific, and Political Thought," in James E. Force and Richard H. Popkin, *Essays on the Context, Nature, and Influence of Isaac Newton's Theology* (Dordrecht: Kluwer Academic Publishers, 1990), pp. 75-90.

[27] According to More, "all the Orders of the Creation in the whole Universe" derive not from God's Will and Omnipotence, but from His "infinite Goodness, Wisdome and Power," *Henry More, Divine Dialogues* (London, 1668), vol. 2, pp. 24-5.

[28] Yahuda MS 15.3, fol. 46v. Quoted in Westfall, *Never at Rest*, p. 827.

[29] Yahuda MS 14, fol. 25, as quoted in James E. Force, "Newton's God of Dominion," in Force and Popkin, *Essays*, p. 79.

In the *Irenicum*, Newton outlines the fundamentals of religion observed since ancient times, derived through Moses from the time of Noah and preserved within the true religion, Christianity.[30] The fundamentals of this religion are the first two injunctions of the decalogue (i.e., love of God and love of ones neighbour.) From these basics are derived other fundamentals (also contained in the ten commandments.) Love of God, entails avoiding ungodly behaviour such as fornication, murder, theft and injuring others. Love of ones neighbour is an injunction to mercy and forgiveness, being merciful even to animals and bearing with those weak in faith. From the last Newton derives his latitudinarianism (such as it is):

> We are to bear with one anothers infirmities, to beware of offending our weak brethren in doing things wch in themselves are lawfull but through their weaknes give them offence & the weak must be also tender in judging their brethren in things of whose lawfulness they themselves are not yet satisfied. ffor we are the servants of God & Christ & have no authority to judge one another.[31]

The compatibility of these positions with the basics of Christianity is self-evident. But what is not immediately obvious, from this account, is how Christianity differs from other religions, that is to say, in what its uniqueness resides. According to Newton, although the coming of Christ as Messiah was predicted in ancient times, the uniqueness of Christianity was only manifested relatively late, at the time of the resurrection, which imposed a new duty to believe in Christ and to accord him honour and glory in his own right, separately from God. We worship God as our Creator, the supreme governor of the universe, to whom we owe our existence. Christ, however, is worshipped because by his act of atonement for our sins, he suffered for us and redeemed us. We have a duty to love Christ in gratitude for the atonement. And it follows that if we want our sins to be forgiven, we should forgive those who trespass against us. From this it follows that to persecute others for refusing to believe things not contained in the fundamentals of religion is to break the second and third commandment, to break the injunctions to mercy and forgiveness. So to persecute believers is to be guilty of "persecuting Christ in his mystical members," and therefore to deserve the name of an Antichristian, to be in a state of apostasy from Christ.[32] Furthermore, Christ, by virtue of his act of atonement (an act of submission to the will of God) is given power to raise all men from the dead and to judge them.

[30] See Westfall, *Never at Rest*, pp. 820ff. for an account of the *Irenicum*.

[31] *Ibid.*, fol. 33.

[32] *Ibid.*, fol.14. Cf. Yahuda Newton MS 3, fol.2 , "Introductio. Continens Apocalypseos rationem generalem": "Voluit igitur Christo, qua mystice loquitur, passiones corporis mystici per passiones membrorum exprimere, figurate accipiendo partem pro toto, ut Rhetoricis mos est."

As we have already observed, Newton himself believed that biblical prophecy was fundamental to a proper understanding of God's word. Newton's *Irenicum* confirms this. There are several points of connection between the doctrines expounded in the *Irenicum* and Newton's interpretation of the prophetic books of the bible, especially the Book of Revelation. First of all we find reiterated in the *Irenicum* Newton's view that the prophecies of the Old Testament were obscure until Christ interpreted them and that the prophecies of the second coming will be obscure until the second coming, at which time "the interpretations become the religion of Gods people till Christ shall deliver the kingdom to the father."[33] Secondly, in the *Irenicum* we are told that the Book of Revelation contains prescriptions for Christian worship, specifically the veneration of Christ separately from the God, but as honoured by God:

> For this form of worship is plainly prescribed in the Apocalyps where the scene of the visions is in the Temple & God the father is represented sitting there upon his throne that is, above the Ark between the Cherubims, and the son is represented in the form of a Lamb receiving the book of prophesy from him that sitteth upon the throne & the Presbyter & People are represented by four and twenty Elders & four Beasts worshipping in the courts of the temple.[34]

Thirdly, Newton states that the article about belief in Christ was added to the creed at the time of the resurrection, which imposes a new duty "to believe in the resurrection of Christ who will raise all men from the dead and judge them."[35] The Book of Revelation describes that coming to judgement, after the victory over the beast of apostasy.

The *Irenicum* summarises Newton's religious scheme, but the Book of Revelation is the well-spring of truth itself. Newton insisted that prophecy was a mainstay of Christianity—after all as St. John reminds us, "the testimony of Jesus is the spirit of prophecy" (Rev. 19:10)—and Newton himself calls the Book of Revelation, "the REVELATION OF JESUS CHRIST."[36] St. John's prophecies were intended as a guide to the faithful "that in the midst of Apostacies we might be able to discern the truth & be established in the faith thereof."[37] Manuel, Westfall, and others have rightly observed that Newton interprets the book of Revelation as a prophecy of the major apostasy of the Christian church: the introduction of the false doctrine of Trinitarianism. By anatomizing this apostasy, prophecy gives a graphic

[33] Keynes MS 3, fol. 14.

[34] *Ibid.*, fol. 48.

[35] *Ibid.*

[36] Newton, *Observations*, p. 308.

[37] *Ibid.*, p. 306 and Yahuda Newton MS 1, fol. 3r.

account of false faith, thereby promoting true faith illustrating its obverse. This mystery of iniquity can be read off against the record of secular history and the evidence of the transmission of the text of the Bible. But the truth contained in the Book of Revelation is not just a negative image as recorded in providential history. Rather, underlying this interpretation of history as apostasy, are fundamental beliefs about the nature of God and Christ, and Christian faith. The prophecies of the Book of Revelation are not just a forecast of future events, but designed "for effecting a recovery and re-establishment of the long-lost truth." In this, as the Apocalypse shows, Christ plays a key role. So what, on Newton's interpretation, does the Book of Revelation tell us about Christ?

Newton himself emphasises in the *Irenicum* that even though Christ is subordinate to God, he deserves worship as redeemer. His role as redeemer in the apocalyptic account, however, is embattled. Just as it is Christ who presides over the last judgement and so it is Christ who conducts the battle against apostasy. The revealed content of the Book of Revelation is the battle against the beast of apostasy, the "Babylonian Kingdom." The images of Christ in the Book of Revelation are all fearsome, whether he figures as the "Son of Man," "the Lamb," "the Word of God" or the "man child." Newton, elsewhere in his theological writings, picks up on St. John's account of Christ in a rather literal way. Christ is "The Prince of Hosts" who "at his second coming will rule the nations with a rod of iron."[38]

> We must also believe that Jesus Christ shall come to judge the quick & the dead, that is to reign over them with justice & judgement untill he shall subdue all rule & authority & power, & all enemies be put under his feet the last of w[ch] is death, & by consequence until all the dead be raised & judged. For he sits at the right hand of God not only in this world but also in that which is to come untill all enemies be put under his feet.[39]

Above all, Christ's wrath is reserved for the apostates who held the church in Babylonian captivity, corrupting the original articles of faith, and persecuting those steadfast in their original faith.

In Newton's interpretation, the "Prophesy of the Seals and Trumpets" concerns the war against and final defeat of the Babylonian kingdom. The duration of that kingdom is co-extensive with the opening of the seventh seal and with the sounding of the seven trumpets. The trumpets, as Newton insists, signify "so many Alarms to war," while "The Calamaties wch follow upon sounding ye Trumpets are all by war."[40] The weaponry of that war are

[38] *Of the Church*, 40r, as quoted by Matt Goldish, "Newton's *Of the Church*: Its Contexts and Implications," pp. 147-66, infra.

[39] Yahuda Newton MS 15.3, fol. 46. Quoted in Westfall, *Never at Rest*, pp. 823-4.

[40] Yahuda Newton MS 1.3, Prop. IV, fol. c[r], in Mamiani, p. 110.

the Vials of Wrath poured out on the Beast. These deadly missiles are equated by Newton with the seven plagues and the seven thunders. Together they constitute "ye tripple repetition of his description by the Trumpets Vials & Thunders."[41] The struggle against the Beast is an on-going process, not one final showdown. It consists in a series of offensives, each depicted as equivalent vials, plagues and thunders, not one climactic assault. The signal for delivering each set of plagues and vials (not to mention thunders, which "most probably denote the same thing with the seven Trumpets"[42]) is the sounding, in sequence, of one of the seven trumpets. The presiding genius over this mayhem is the Son of Man (alias "the Lamb," "man-child," "rider on the white horse.") Christ of the Book of Revelation is *Christus ferox*, a figure of power, not the God of love, fit son, one might say to a God conceived as *Deus Pantocrator*, the God of power whom "we may fear & obey." In literal conformity with St. John's word, Newton installs a Christ of wrath and power in place of the merciful sacrificial victim of the crucifixion. That Christ should be manifested in such fearsome guise, instead of as a figure of *misericordia*, a figure of vengeance rather than of mercy, is perhaps part of the great mystery of Revelation. This fundamentalist depiction of the Son does not by itself, explain why the seven vials have to be "synchronal" to the seven trumpets. But that synchronism certainly makes sense in terms of the battle against the beast. For this reason, though also for others, the synchronicity of vials and trumpets is not just a matter of the co-incidence of numbers. On it hangs a fundamental component of Newton's theology — his Christology. And the Christology that hangs on it is part of the great theological divide between Newton and the other Cambridge Millenarians, Henry More and Joseph Mede. It is a reminder that while Newton's Arianism lead him to deny the divinity of Christ, this did not diminish the role of Christ in the unfolding of providential history or the significance of Christ in defining the essentials of the true faith.

[41] *Ibid.*, Prop.III, fol.5, in Mamiani, p. 122

[42] *Ibid.*, fol. 4r in Mamiani, p. 120.

ESSAY 9

Interpretive Strategies in Newton's *Theologiae gentilis origines philosophiae*

KENNETH J. KNOESPEL
Georgia Institute of Technology

"What is it that Newton really thought he was doing? Is it all the usual stuff and nonsense?" The question posed by John Maynard Keynes in a letter to A.S. Yahuda in 1938 may still stand as an appropriate query regarding the puzzlement that accompanies an initial examination of Newton's *Theologiae gentilis origines philosophiae.*[1] From the earliest descriptions in David Castillejo's eccentric Cambridge dissertation, to Richard Westfall's work that presented it to a wider group of scholars; from Frank Manuel's early English transcriptions of related Yahuda manuscripts to Betty Jo Dobbs' presentation of the manuscript in her own work on alchemy, Newton scholars have received orientation and a multitude of intriguing quotations of what surely seemed to be another side of Newton.[2] Certainly anyone who has looked at the *Origines* appreciates even more the preliminary accounts

[1] Letter from J. M. Keynes to A. S. Yahuda (April 3, 1938), Box I: [Yahuda letters] in the Archive of the Hebrew National Library, Jerusalem. Quoted by permission.

[2] David Castillejo, "A Theory of Shifting Relationships in Knowledge as Seen in Medieval and Modern Times with a Reconstruction of Newton's Thought and Essays on Patronage," Ph. D. diss., 3 vols., Cambridge University, 1967. Castillejo's Cambridge dissertation contributes to his valuable but also eccentric book, *The Expanding Force in Newton's Cosmos as Shown in his Unpublished Papers* (Madrid: Ediciones de Arte Y Bibliofilia, 1981); Richard Westfall, "Isaac Newton's Theologiae Gentilis Origines Philosophicae" *The Secular Mind: Transformation of Faith in Modern Europe* (New York: Holmes & Meier, 1982), pp. 15-34; also "Newton's Theological Manuscripts," *Contemporary Newtonian Research* ed. Z. Bechler (Amsterdam: Reidel Publishing Company, 1982), pp. 129-43; and *Never at Rest: A Biography of Isaac Newton* (Cambridge: Cambridge Univ. Press, 1980); Frank Manuel, *The Religion of Isaac Newton* (Oxford: Oxford University Press, 1974); Betty Jo Dobbs, *The Janus Face of Genius* (Cambridge: Cambridge University Press, 1992); Matt Goldish, "Newton on Kabbalah" *The Books of Nature and Scriptures: Recent Essays on Natural Philsoophy, Theology, and Biblical Criticism in the Netherlands of Spinoza's Time and the British Isles of Newton's Time*, James E. Force and Richard H. Popkin, eds., (Dordrecht: Kluwer, 1994), pp. 89-103; Richard Popkin, "Newton's Biblical Theology and his Theological Physics," *Newton's Scientific and Philosophical Legacy*, ed. P. B. Scheurer and G. Debrock (Dordrecht: Kluwer Academic Publishers, 1988), pp. 81-97.

J.E. Force and R.H. Popkin (eds.), Newton and Religion, 179–202.
© 1999 *Kluwer Academic Publishers.*

which at times may strike us as probes to some other planet.[3] The manuscript tests our Latin, paleographic skills, patience, and eyesight—all our seventeenth-century interpretive strategies. Through the oral and written matter that accompanies the manuscript, one learns that it carries other stories or myths as well, stories that touch twentieth-century science and religion (Einstein and Yahuda), the early orthodoxy of the history of science (George Sarton refused to consider the manuscript as part of the history of science), and even moments in the foundation of the State of Israel. Indeed, in the account of Castellejo's unpacking and cataloging of the Yahuda manuscripts after the Six Day War, there is even a bit of Indiana Jones.[4] My objective is not to consider such twentieth-century mythologies, but rather to ask how Newton deals with the voluminous narratives he gathers in his own work. After describing the manuscript, I will turn my attention to what I have referred to as Newton's interpretive strategies. In particular, I will make several comments on Newton's synchronisms and the changing status of allegory and hermeneutics registered in the *Origines*.

1. AN OVERVIEW OF THE *ORIGINES*

Theologiae gentilis origines philosophiae is an unfinished Latin manuscript book by Isaac Newton currently located in the Department of Archives in the Jewish National and University Library in Jerusalem. As planned the book was to include at least eleven chapters devoted to showing how ancient religious practice could reveal physical truths about the universe.[5] In its current state, the manuscript exists in some five chapters including numerous fragments that probably represent research for other chapters. Entitled by Newton and identified since 1969 as Yahuda MS Var. 1, Newton MS 16, the work consists of 79 folio pages. Considering the arbitrary classification and presentation of the manuscripts in the Portsmouth Collection, it is apparent to anyone who has worked with these manuscripts

[3] I am indebted to the Edelstein Center for the History and Philosophy of Science at the Hebrew University for a 1989 fellowship that allowed me to complete a preliminary transcription of the *Origines* and other manuscripts in the Yahuda collection of the Hebrew National Library. All references to manuscripts in the Yahuda collection of Newton manuscripts are quoted by permission.

[4] The Yahuda Collection in the Hebrew National Library in Jerusalem contains a detailed collection of correspondence regarding the acquistion of the Newton manuscripts. See for example, the correspondence between A. S. Yahuda and Albert Einstein in Yahuda Var 1: Newton MS. 43. Interviews with Mordekai Nadav, former Head of the Archive, who participated in unpacking the Newton manuscripts provided a range of anecdotal information.

[5] My work on the *Origines* began in 1988 in Cambridge where I first studied the manuscript on microfilm. It continued the following year in Jerusalem where I spent some four months completing a preliminary transcription. My work has been repeatedly interrupted by my administrative responsibilities as school chair and associate dean.

that the *Origines* project extended well beyond Yahuda MS Var. 1, Newton MS 16. However, as a consequence of the chapter headings and given Humphrey Newton's transcription of some 57 folio pages, it remains appropriate to view Yahuda MS 16 as the most significant departure point for all of Newton's work on mythography.[6] Given Richard Westfall's preliminary research on the manuscript, we know that Newton was working on the project in the 1680s when he was also working on the *Principia* and that part of it still occupied him when he was working on *The Chronology of Ancient Kingdoms Amended* (1728) at the end of his life.

Since the chapter titles provided by Newton give us an idea of the book, we may look at them briefly before proceeding to a more detailed consideration of the material that we find in them (all references are to Yahuda MS. 16):

> Cap. 1: Quod Theologia Gentilis Philosophia erat, et ad scientam Astronomicam & Physicam systematis mundani apprima spectabat: quodque Dies duodecim maioram gentilis erat Planetae septem cum quatuor elementis et quintessentia terra. (incipit f. 1 r.)
> [Cap. 1: That the theology of the Gentiles was philosophical and pertains to the knowledge of astronomy and the physics of the world system; how the twelve major gentile gods are the seven planets, the four elements and the earthly quintesence.]
>
> Cap. 2: Quod Hominem post Diluvium primorum memoria in astres et elementis colebatur; quodque hominis illa aetatibus tribus primis et quarta. (incipit f. 7v.)
> [Cap. 2: That after the flood humans retained their earliest memories in the stars and elements; and how there were were first three and then four ages.]
>
> Cap. 3: Quod Noachus sit Saturnus et Janus, quodque Chamus sit Jupiter Hammon et quod Chami filii certa nepotibus sint et Dii reliqui et Gigantes qui cum Dis aetati tertia in Aegypto dimicarunt: et quomodo Chamus cum suis in Aegyptum deservit et regionen intra filias divisit. (incipit f. 53v.)
> [Cap. 3: That Noah is Saturn and Janus, and that Cham is Jupiter Hammon and that the sons of Cham are his children and how the remaining gods and giants fought in Aegypt with his family and divided the region between his daughters.]
>
> Cap. 4: Quod Chus sit Hercules, Mars, Moloc, et Belus primus (ff. 65r.-72r.)
> [Cap. 4: That Chus is Hercules, Mars, Moloc, the the first Belus]

[6] From here on references to the Newton manuscripts in the Yahuda collection will be designated simplified as Yahuda MS 16 rather than including the variorum listing. It should be emphasized, however, that the variorum listing is important considering the extensive collection of oriental manuscripts that make up the large part of the Yahuda collection. I have followed the foliation included in the manuscript.

Cap. 5: Quod Mirvaim sit Osiris et Serapis et Menatius & Dis vel Pluto quodque Mirvaimo nati sint filius Orus seu Apollo et filia Bubaste seu Diana et filius nothus Thoth seu Mercuius primus (ff. 73r.-79v.)
[Cap. 5: That Mirvaim is Osiris and Serapis and Menoetius and Dis or Pluto and that to Mirvaim were born a son Horus or Apollo and a daughter Bubaste or Diana and a son, the well-known Thoth or first Mercury.]

Cap. 5 (another version): Quod Mirvaim sit Osiris et Serapis et Menatius & Dis vel Pluto quodque Mirvaimis filius Pathros sit Orus vel Apollo et quod filiam habuit Bubastem seu Dianam et filiam nothum Thoth seu Mercurium primum (incipit f. 68v.)
[Cap. 5 (another version): That Mirvaim is Osiris and Serapis and Menoetius and Dis or Pluto and that Pathros, the son of Mirvaim was Horus or Apollo and that he had a daughter Bubaste or Diana and a daughter and a son, the well-known Thoth or first Mercurcy.]

[The following chapters headings appear on f.45r and remain undeveloped.]

Cap. 6: Quod Phut sit Typho et Antaeus et Atlas et Neptunus (f. 45r.)
[Cap. 6: That Phut was Typho and Antaeus and Atlas and Neptune]

Cap. 7: Quod Canaan sit Vulcanuus et Prometheus et quod eius exor et soror Venus fuit Astyr seu Astyrgates (f. 45r.)
[Cap. 7: That Canaan was Vulcan and Prometheus and that his wife and daughter was Venus Astyr or Astyrgates]

Cap. 8: De aliis Deorum praedictorum nominibus (f. 45r.)
[Cap. 8: Concerning the original names of other gods]

Cap. 9: Quomodo nomina praedictorum hominum sideribus imposita sunt & qualis fuit Astronomia veterum Aegyptiorum (f. 45r.)
[Cap. 9: In what ways the names of the first humans were transferred to the stars and the nature of the oldest Egyptian astronomy]

Cap. 10: Quomodo animae hominum in stellas translata sunt & stellae sic animata per Diis intelligentibus haberi ceperunt, dequod origine et progressu in idololatria et artium magicam (f. 45r.)
[Cap. 10: In what ways the souls of humans were transferred to the stars and how the stars began to be understood as containing animated spirits, and concerning the origin and progression of idolatry and the art of magic]

Cap. 11: Qualis fuit religio vera Noachidarum religio antequam per cultum falsorum deorum corrumpi coepit, et quod religio Christiana non vera fuit se exempta sunt (f. 45r.)

[Cap. 11: What was the true religion of the ancient Noachides before it began to be corrupted through the veneration of false gods, and how the Christian religion is not exempted from such corruption]

Although the book begins with a detailed account of the Egyptian synthesis of stars and elements, it proceeds in subsequent chapters to establish the manner in which human events become associated with astronomical events. According to Newton, Noah's children and grandchildren become absorbed within other mytho-histories, or pre-histories, of antiquity through a process in which they first become localized as historical figures and then memorialized as stars and planets. (In the *Chronology of Ancient Kingdoms Emended* [1728], Newton explains diffusion through trade and technology transfer.)[7] Consequently, while a proliferation of gods and goddesses appears in antiquity, the multiple pantheons are structurally related. Through his extensive historiographic research Newton factors out redundant figures until he has reduced a multitude of genealogies to rudimentary formulae. As the chapter headings show, Newton's method involves 1) formulating linkages between astronomical objects and a limited number of historical figures; and 2) establishing equivalencies between figures belonging to major mythographic pantheons. We should think of Newton as creating a translation table not only between a plethora of histories and mythographies but between eastern and estern Mediterranean mythologies. Works in Newton's library show that he was hardly alone in seeking equivalencies through a process of historical classification or taxonomy.[8]

Rather gathering up lists of gods and goddesses according to moral interpretations, Newton seeks to establish a genealogical study of multiple pantheons (Assyrian, Babylonian, Egyptian, Hebraic) in order to understand the observations they share on natural philosophy. Following seventeenth-century precedents set by Gerardus Vossius, Samuel Bochart, and John Marsham, Newton's project becomes distinctly orientalist. Inherent in such a project is an effort to situate the study of universal history away from the scholarly assumptions of the Latin west. In contrast to approaching the Old Testament through the New Testament and then viewing both through the Roman Empire, Newton's research inscribes a religious anthropology of the eastern Mediterranean. However, his objective in the *Origines*, extends well beyond the study of eastern religious practice that one would expect from

[7] See Kenneth J. Knoespel, "Newton in the School of Time: The Chronology of Ancient Kingdoms Amended and the Crisis of Seventeenth-Century Historiography," *The Eighteenth Century* 30, No. 3 (1989), pp. 19-41.

[8] Robert Morden, *Geography Rectified: or, a Description of the World* (London, 1700), [N.Q 9.29], *Introductio ad chronologiam sive ars chronologica in epitomen redacta* (Oxford, 1704) [N.Q 7.2], and Christian Helvetius, *Theatrum historicum* (Oxford, 1662) [NQ 10.57.]

Biblical archaeology. By seeking to establish an intellectual foundation for universal history that entails Biblical history, Newton not only challenges simplistic Christian typology that would authorize the New Testament but indicates that his effort to rework historical patterns is accompanied by an expectation that they may be interpreted to confirm divine revelations that not only extend beyond Biblical history but include Newton himself. From a practical vantage point, Newton's sights are set far beyond the vulgar moral readings of classical fabulae established for use in school texts or iconographic lexicons. Newton's work is distant from the scholarly traditions explored by the early iconographic work of the Courtauld and Warburg Institute. In fact, Newton's research reminds us how often twentieth-century mythographic research occurs in a Greek and Roman register. A useful snapshot of differences in the application of classical authors appears in a comparisons between Milton's rhetorical and psychological use of Ovidian poetry and the physico-historical purposes ascribed to Ovid by Newton. Where Ovid provides John Milton with psychological detail for Biblical figures, he supplies Newton with a scholarly source for universal history and alchemy. At a time that seventeenth-century classical scholarship was challenging prolix and even facile Latin of Ovid, Newton approaches Ovid far less as a Roman poet than a Latin mythographer whose verses could provide testimony to anthropological information on ancient religious practice.[9]

More than anyone else, it is Gerardus Vossius who provides Newton with a significant seventeenth-century framework for his examination of natural philosophy and its bearing on religion. The extensive dog-earing and marginal marks and notes in Newton's hand in his copy of Vossius in the Wren Library, Trinity College, Cambridge indicates its extensive use as a resource in his research.[10] More than anything else it is Vossius's extended discussion of the sun and its assimilation into religious practice that provides an orientation for Newton's research. A juxtaposition of chapters that make up Vossius' discussion entitled *Quae est de cultu corporis caelestis* [*The use of celestial bodies within religious cults*] (vol. 2: chapters 1-17) and Newton's subjects shows how Vossius marks out a structure that could facilitate Newton's research.

[9] Newton's personal copy of *Metamorphoses* ed. G. Bersmanius (London: 1655) remains in the Trinity College Library, Cambridge (NQ.9.168). Marginal notes indicate that Newton read the *Metamorphoses* with the *Fasti*. See John Harrison, *The Library of Isaac Newton* (Cambridge: Cambridge University Press, 1978), entry 1221. Ovid is also used in the *Origines* to confirm historical epochs or ages (f. 52r. and f. 53r.)

[10] My comments on Vossius are based on my examination of the copy in the Wren Library, Trinity College, Cambridge (NQ.8.46^2) as well as work with the facsimile edition published by Garland Press: Gerardus Vossius, *De Theologia Gentili*, 3 vols. (New York: Garland Press, 1976 [Amsterdam, 1641]).

Comparison of Newton and Vossius Chapter Headings

Isaac Newton, *Theologiae Gentilis Origines Philosophicae* (Yahuda MS. 16)	Gerardus Vossius, *De theologia gentili* (Amsterdam, 1641)
Newton 1: planets and religion	Vossius 1: transformation from animal gods to natural elements
Newton 2: humans into stars and elements	Vossius 2: explication of causes used by Gentiles for their transformations
Newton 3: Noah/Saturn/Janus/Chamus/Jupiter	Vossius 3: sun cults among the Hebrews
Newton 4: Chus/Hercules/Mars/Moloc/Belus	Vossius 4: sun cults among the Phoenecians
Newton 5: Mirvaim/Osiris/Serapis/Menatius	Vossius 5: sun cults among the Assyrians
Newton 6: Phut/Typho/Antaeus/Atlas/Neptune	Vossius 6: sun cults among Babylonians
Newton 7: Canaan/Vulvanuus/Prometheus/Venus	Vossius 7: sun cults in Arabia
Newton 8: Other gods	Vossius 8: Moabites
Newton 9: Astronomia verterum Aegyptiorum	Vossius 9: Persia
Newton 10: idolotry	Vossius 10: Egypt
Newton 11: Noachean religion/idolatry....	Vossius 11: Ethiopia/Ammon
	Vossius 12: Other gods
	Vossius 13: Planets translated into sun
	Vossius 14: Libero
	Vossius 15: Hercules
	Vossius 16: Janus
	Vossius 17: Celts

At the same time that Vossius provides Newton with a massive collection of detail, he supplies a taxonomic structure and multiple precedents for integrating contemporary natural philosophy into his work with ancient mythology. Beyond gathering examples regarding solar worship (2:312f), he gathers detail regarding ancient observations on the sun's characteristics from the vantage point of natural philosophy. His subsequent discussion includes reference to Copernicus (2:312), a mathematical confirmation of the sun's velocity (2:313), and to the way the sun continues to be celebrated in the Christian church (2:320f). (He shows how Jesus is associated with the sun and notices how the chancels of Christian churches face toward the east and the rising sun.) Newton's reference to Galileo in the *Origines* (which I return to below) finds multiple precedents in Vossius' discussion of seventeenth-century natural philosophy. However, even though there are similarities between the mythographic work of Vossius and Newton, the *Origines* does not imitate *De theologia gentilii* but simplifies and systematizes Vossius's rendition of mythographic material.

It is helpful to think of Newton as combining two interpretive strategies: one a euhemeristic hermeneutic and another found in physical allegory. Such strategy dismisses the layered tradition of moral interpretations linked with education and indoctrination and fixes attention on physical evidence drawn either from history or from nature. With little exaggeration, Newton's interpretive strategies may be compared to even earlier use of mythography to examine natural phenomena.[11] Even a cursory review of Newton's major sources in the *Origines* indicate that he is seeks out sources that show how serious discussion of natural phenomena could be modulated through mythological narrative.[12] For Newton, an interpretive strategy strongly linked to physical allegory is much less associated with university programs in Greek than with his own extensive alchemical research. For Newton, alchemy represents not simply a collection of arcane techniques for approaching matter but also embodies an alternative means for

[11] See Brian Stock, *Myth and Science in the Twelfth Century* (Princeton: Princeton University Press, 1972); Peter Dronke, *Fabula* (Leiden: E. J. Brill, 1974); Kenneth J. Knoespel, *Narcissus and the Invention of Personal History*, (New York: Garland, 1985); and Richard H. Popkin, "Polytheism, Deism, and Newton," in *Essays on the Context, Nature, and Influence of Isaac Newton's Theology*, ed. James E. Force and Richard H. Popkin (Dordrecht: Kluwer Academic Publishers, 1990).

[12] Newton refers to at least 50 different authors in the five chapters that make up the *Origines*. A tentative list of authors includes the following: **Chapter 1:** Clement of Alexandria, Macrobius, Plato, Cicero, Diodorus, Laertius, Porphyrius (cited from Eusebius), Eusebius, Herodotus, Lucian, Philostrius Brixiensis, Ennius, Augustine, Virgil, Plutarch, Servius, Ovid. **Chapter 2:** Galileo, Philo of Byblus, Philo of Sanchomathius (cited from Eusebius), Proclus (cited from Photius), Homer, Berosius, Manethus, Josephus, Hyginus, Hesiod. **Chapter 3:** Tibullus, Bochard, Aristotle, Plotinus, Tertullian, Strabo, Eratosthenes. **Chapter 4:** Pliny, Isagogius, Athenagoras, Apollodorus, Kircher, Pausanius, Cyril, Xenophon, Euphorus, Hesychius. **Chapter 5:** Apollonius Rhodius, Stephanus Byzantius, Apuleius. Although many of the authors are cited multiple times, I have listed them only once in this preliminary list.

understanding history.[13] Newton's alchemical work was inherently historical in the sense that he was seeking to establish what amounts to a stemma for alchemical practice. The construction of such a stemma inevitably involves an integration of textual study and physical experimentation with the expectation that their intersection will be accompanied by mutual confirmation.

Although Newton's methodology might be thought to be a blatant form of *prisca theologia* or *prisca scientia*, we must use such characterizations carefully. Newton's method is by no means identical to established Renaissance understanding of *prisca scientia*. Unlike commonly espoused hermetic teaching that situated a *prisca scientia* with Moses and even more importantly in a setting of lapsarian theology, Newton's *prisca scientia* emerges from his conviction that religious practice is related to natural phenomena and that just as natural phenomena experienced by human beings has remained relatively constant in the course of human history so has human proclivity to practice religion in certain ways. Rather than being occupied by occult practice *per se*, something characteristic of much Renaissance discourse, Newton's method works by expecting to find patterns in all natural phenomena. For Newton, meaning is not concealed in written historical texts but enacted through a correlation between the understanding of the text and the understanding of experience.

Newton's description and analysis of a sacred Egyptian procession at the beginning of the work provides a useful example for approaching Newton's strategy. Immediately at the outset, he notices that ancient philosophy was divided into practices that were either sacred or mundane. While sacred philosophy was conveyed to adherents through types and enigmas, vulgar philosophy was conveyed in an accessible, popular style: "Philosophiam antiques duplicem coluere, sacram et vulgarem, sacram Philosophis per typos et aenigmata discipulis fuit traditere: vulgarem oratores aperte et stylo populari scripserunt." [The ancients practiced philosophy in two ways, one sacred and the other vulgar; they communicated sacred philosophy to their students through types and enigmas; orators wrote vulgar philosophy in an open manner with a popular style.] (Yahuda MS. 16, f. 1r.) Newton underscores the distinction by noticing that sacred philosophy flourished in Egypt where it originated in a knowledge of the stars. ["Philosophia sacra in

[13] The work of Michael Maier and others suggests that a form of alchemical historiography becomes even more developed in the seventeenth century. Newton owns Maier and, as his notes show, used him. "In hoc enim, omnibus aliis ignorantibus exclusis concordant omnium aetatum, etiam antiquissimi, philosophi, & in praecipuis suis allegoriis & fictionibus primipoetae, omnium nationem populi differentes ingeniis, linguis, moribus, religione, legibus & aliis vitae institutis; imo omnium locorum sapientes, non solum tot regnorum europae, sed & Africae & Asiae. (f. 3r.) Michael Maier, *Symbola auriae mensae duodecim nationum* (Frankfort, 1617) I have used Newton's personal copy of the book located in the Edelstein Library in the Jewish National and University Library, Jerusalem.

Aegypto apprime florint et in scientia syderum fundata fuit." ["Sacred philosophy obviously flourished in Egypt and was founded on the science of the stars."] [Yahuda MS. 16, f. 1r.] To support his observation, he describes an ordered procession of Egyptian priests described in *The Stromata* by Clement of Alexandria.[14] Each figure in the procession carries a symbol representative of his function. The Cantor who leads the procession carries two books, one containing hymns to the gods and the other regulations for governing life. He is followed by a Seer who carries a clock in the palm of his hand which is a symbol of astrology. In front of him, he has four astrological books of Mercury. The first book contains the location of the fixed stars, the second the conjunction of the sun and the moon; the third and the fourth contain the rising and the setting of the sun and moon. Next comes the Scribe who carries a pen in his headdress, a book and a ruler in his hands. There are ten books associated with the scribe pertaining to 1) hieroglyphics, 2) cosmography, 3) geography, 4) the location of the sun and moon, 5) the location of the five planets, 6) Egyptian chorography or topography, 7) the description of the Nile, 8) the description of the equipment of the priest and its location, 9, measurement, and 10) other essentials. The Prophet follows carrying the rod of justice and a cup for drinking. He too has ten books that deal with the Egyptian religion. After all the others comes the prophet who holds a serpent in his lap.

[14] For the Egyptians pursue a philosophy of their own. This is principally shown by their sacred ceremonial. For first advances the Singer, bearing some one of the symbols of music. For they say that he must learn two of the books of Hermes, the one of which contains the hymns of the gods, the second the regulations of the king's life. And after the Singer advances the Astrologer, with a horologe in his hand, and a plam, the symbols of astrology. He must have the astrological books of Hermes, which are four in number, always in his mouth. Of these, one is about the order of the fixed stars that are visible, and another about the conjunctions and luminous appearances of the sun and moon; and the rest respecting their risings. Next in order advances the sacred Scribe, with wings on his head, and in his hand a book and rule, which were writing ink and the reed, with which they write. And he must be acquainted with what are called hieroglyphics, and know about cosmography and geography, the position of the sun and moon, and about the five planets; also the descriptions of Egypt, and the chart of the Nile; and the description of the equipment of the priests and of the places consecrated to them, and about the measures and the things in use in the sacred rites. Then the Stole-keeper follows those previously mentioned, with the cubit of justice and the cup for libations. He is acquainted with all points called Paedeutic (relating to training) and Moschophatic (sacrificial). There are also ten books which relate to the honour paid by them to their gods, and containing the Egyptians worship; as that relating to sacrifices, first-fruits, hymns, prayers, processions, festivals, and the like. And behind all walks the Prophet, with the water-vase carried openly in his arms; who is followed by those who carry the issue of loaves. He, as being the governor of the temple, learns the ten books called "Hieratic;" and they contain all about the laws, and the gods, and the whole of the training of the priests. For the Prophet is, among the Egyptians, also over the distribution of the revenues. There are then forty-two books of Hermes indispensably necessary; of which the six-and-thirty containing the whole philosophy of the Egyptians are learned by the forementioned personages; and the other six, which are medical, by the Pastophoroi (image-bearers),—treating of the structure of the body, and of diseases, and instruments, and medicines, and about the eyes, and the last about women. Such are the customs of the Egyptians, to speak briefly. Clement of Alexandria, *The Stromata, or Miscellanies* in *The Ante-Nicene Fathers*, vol. 2, ed. Alexander Roberts and James Donaldson (Buffalo: The Christian Literature Publishing Company, 1885), p. 488.

The ekphrastic description shows Newton engaged in presenting the details of an allegorical problem presented by a hieroglyphic emblem. Newton follows a widespread practice found in many ancient neo-platonic texts and which would receive even broader dissemination at the beginning of the eighteenth century through Vico's famous allegory at the beginning of *Scienza nova*. The fact that it occurs at the very beginning of his book is surely not a coincidence for it sets forth an expectation that the study undertaken in the pages that follow will engage the "types and enigmas" of sacred philosophy. In the emendations that follow Newton amplifies the interpretation already implicit in the iconography of the procession.

> In this procession the hymns of the first priest are associated with the harmony of the heavenly spheres. Next comes the astronomer with the holy books concerning the study of the stars. Next comes the sacred scribe who understands the planets, stars, and the sacred things. Finally the priest and chief appear, who know all things as a consequence of studying the sacred rites and theology, and who close the entire procession. By bringing together a knowledge of the stars and the earth with the study of that which was the most important, the Egyptians indicated that their theology concerned the study of the stars. [Indeed?], the gods of the Egyptians were stars and elements. (Yahuda MS. 16, f. 2r.)

> In hac Processione per Hymnos praefatorios a harmonia sphaerarum coelestium significatur. Deinde procedit Astronomis cum libris sacris ad scientiam Syderum spectantibus. Hunc sequitur scriba sacrorum qui descriptiones Coeli Terrae Stellarum et rerum sacrarum intelligit. Ultimo Sacerdos et Praefactus Sacrorum, ea callentiae quae ad ritus sacros et Theologiam spectant, Processionem totam claudunt. Conjungando scientiam Stellarum et Mundi cum Theologia et scientiam illam primo loco ponendo insinuabant Aegyptii Theologiam suam ad astra respicere. [Nimirum?] Dii Aegyptiorum erant astra, et elementa. (Yahuda MS. 16, f. 2r.)

Further emendations show Newton seeking sources that provide additional information about the physics inherent in ancient accounts. An addition on the reverse folio page includes consideration of the nine muses and their relation to planetary orbits as a consequence of music theory. The notes show that even the strophic and antistrophic structure of Greek meter may be related to the movement of the stars. Although I cannot expand on the relation between poetic and cosmic structure at present, it is important to recall that it is precisely such links that are explored by Richard Bentley, the eminent Greek scholar in Cambridge who was also Newton's friend, in the Boyle lectures in 1692.[15] Similar comments are found in other related manuscripts.

[15] We have formerly demonstrated that the body of a man, which consists of an incomprehensible variety of parts, all admirably fitted to their peculiar functions and the conservation of the whole, could no more

> The eastern & Egyptian nations were very much addicted to speake by figures in their language to introduce the qualities and substances of things under the character of intelligent beings or persons. So things often represented death & the grave & time & fortune & health & wealth & love & flame & the elements & planets by persons; & the Jews gave the names of evil spirits to diseases & to vices and amorous opinions & so Solomon spoke of wisdom as a person & Orpheus, Plato & Philo & some of the gnosticks gave the name λογοσ to the wisdom of God, considered as a person.... And the Idea of the Platonists, sephizoths of the Cabbalists, & Aeons of the Gnosticks are nothing else than the thoughts notions actions previous names attributes or parts of the deity turned into persons & sometimes into the souls of men." [Yahuda MS 8.1 f. 2r.]

The fact that the foregoing passage appears in Newton's manuscript commentary on the prophecies of Daniel and Revelation reminds us how the mythographic and Biblical commentaries need to be regarded as part of the same project. In a practical sense, this means that the "Rules for interpreting ye words & languages in Scripture" included in Yahuda MS 1 should be regarded as compatible with the interpretive strategies demonstrated in the *Origines*. The repeated admonishments in the Rules to approach allegory with caution and to "chose those interpretations wch are most according to ye litterall meaning of ye scriptures unless where the known circumstances of ye place plainly require an Allegory." (Yahuda MS 1 f. 12v.) actually confirms the practice of the *Origines* and reminds us that Newton practices physical interpretation or allegory in his Biblical and mythological work.

The importance which Newton attributes to scholars such as Clement of Alexandria indicates the extent to which he seeks out commentaries that aspire toward an extensive integrative structure. In scholars like Clement, Newton found extensive hermeneutic projects that repeatedly asked how one could integrate the learning of the Egyptians, Hebrews, and Greeks with the message of the New Testament. For Newton, however, the question was not accommodation between ancient wisdom and Christian revelation but the extent to which natural philosophy provided the fundamental structure for natural religion to such a degree that in the end all religious practice could be shown to be an expression of natural philosophy. Commentators like Clement, who sought to bundle massive amounts of material into patterned

---

be formed fortuitously than the Aeneis of Virgil, or any other longer poem with good sense and just measures, could be composed by the casual combinations of letters. Now, to pursue this comparison, as it is utterly impossible to be believed, that such a poem may have been eternal, transcribed from copy to copy without any first author or original; so it is equally incredible and impossible that the fabric of human bodies, which has such excellent and divine artifice and, if I may to say, such good sense and true syntax, and harmonious measures in its constitution, should be propagated and transcribed from father to son without a first parent and creator of it... Richard Bentley cited in Rudolf Pfeiffer, *History of Classical Scholarship 1300-1850* (Oxford: Oxford University Press, 1976) pp. 146-7.

structures, provide models for seventeenth-century scholars such as Gerardus Vossius.

Although the manuscript chapters include frequent euhemeristic readings, Newton's primary objective is to establish a means to interpret astronomical phenomena. At the beginning of his second chapter, Newton notices how the practice of commemorating humans by turning them into the names of stars or deities has continued into his own age: "The names of the gods are taken from human beings. Even recently we have seen that Galileo named the moons of Jupiter to honor and memorialize his Medicean benefactor; we have also seen the names of famous persons used to identify features on the moon." ["Deorum vero nomina ab hominibus desumpta sunt. Videmus enim nuper quod admodum Galilaeus sidera circum-Jovialia in honorem et memoriam benefactorum suorum Medicea nominavit, et alii nomina Clarissimorum hominum transtulerunt in Lunam."] (Yahuda MS. 16, f. 11r.) The passage indicates Newton's expectation that euhemeristic habits identified in antiquity continued to be practiced within European society. As Newton continues, it becomes apparent that euhemerism is not used simply as a strategy to demythologize traditional narratives and integrate them into historical record but also functions as a screening mechanism that permits him to discern narrative elements that pertain to physics. By surveying a multitude of religious practices, Newton shows how the worship of God in all periods has become replaced by anthropomorphic ideas of divinity. "They are accustomed to think of Osiris and Isis as the world or the elements of the earth, or the Nile river or the land of Egypt, or even the sun and the moon because the superstition of all people always seeks to exalt humans into spirits, spirits into gods, and lesser gods into greater gods." ["Osiris et Isis quandoque de Tellura et Elemento terrae quandoue de Nilo et terra Aegypti, quandoque etiam de sole et Luna exponi solent quippe ea est gentium omnium superstitio ut homines in Divos, Divos in Deos & Deos minores in Deos maximos referre et extollare semper conarantur."] (Yahuda MS 16, f. 44r.) "The name of the eternal and supreme God is Jehovah which in Greek is Iaw and in Latin is Jao-pater or Jupiter. Nevertheless, people who always have a propensity toward superstition transform humans who have died into living souls, and the souls into powerful spirits and celestial gods. From these celestial gods they always make up in time an immense and omnipotent spirit. It is a fact that each and every people cultivates that god which seems greatest of the heavens and the entire universe and calls that god Jove." ["Cum enim Jehovah quem Graci (Iaw), Latini Jao-pater seu Jupiter dixere, nomen esset Dei aeterni et supremi; gentes autem in superstitiones semper propensae, ex mortuis hominibus genios vivos, ex geniis divos potentas ac Deos coelestes, ex his Deo semper coelestes et in horum maximo animam nundi totius ac tandem Numen immensum et omnipotens gradatim configerent; factum est ut gens unaquaeque Deum illum quem maxima coluerat pro Deo coelorum ac universi totius haberet ac Jovem propterea nomnaret:" (Yahuda MS 16, f. 67r.)

The portions of the *Origines* which appear as completed drafts (Chapters 3 through 5) seek to accommodate the Noachids and other ancient pantheons. As the chapter develop they also show that Newton's method assumes the mutual interaction of natural philosophy and religious practice. The chapter titles for the portions of the books that remain incomplete suggest that Newton objective was to use his discussion of the Noachids to return to a more detailed discussion of idolatry and magic. The headnote for Chapter 10 shows that he intends to look at "how the human spirit is transferred to the stars and how, as a consequence, the stars become animated as if they have intelligence and consequently how one may approach the origine and progression of idolatry and the magical arts." (Yahuda MS 16, f. 45r?). In Chapter 11, Newton notes that he would show "[w]hat the true religion of the children of Noah was before it began to be corrupted by the worship of false Gods. And that the Christian religion was not more true and did not become less corrupted." (Yahuda MS 16, f. 45r?)

The extensive revisions of the *Origines*, as manifest in the manuscript that has come down to us, pose more than paleographic problems; they show that Newton has increasing difficulty in working out an acceptable translation table for all the pantheons. Indeed, given the material in *The Original of Religions*, another hitherto unpublished Yahuda manuscript (Yahuda MS 41), there is even reason to conjecture that Newton abandoned the *Origines* when he found how his mythographic work could be simplified through the use of the prytanean fire as an architectural synchronism. If such a conjecture is correct, the *Origines* functioned as a highly significant heuristic space for Newton's work. Rather than dismissing the manuscript because of its complexity and unfinished state, it makes more sense to see in its iterative process of inscription, emendation, and revision and exhibition of Newton's hermeneutic practice. Many scholars have commented on the disorder of the manuscript and it is undoubtedly the case that their inaccessibility has done much to keep them in the microfilm versions that we continue to use. Rather than using the state of the manuscripts as reason to discount them, it is more useful to take the writing process evidenced in them as an vehicle for further learning. Initially Newton's scholarly habits are part of seventeenth-century academic practice that involves the collection and inscription of multiple proof-texts that lead to a particular conclusion. To use vocabulary familiar to the seventeenth century, we may say that the process is syllogistic or enthememic in the way that it proceeds from thesis statements to conclusion. For example, after collecting proof texts from Cicero, Diodorus, Laertius, Porphyrius, Eusebius, Newton succinctly concludes that "These gods clearly indicate a relation between the stars and the elements and that from the Egyptians such knowledge was disseminated to all races in antiquity." The collection of copious proof-texts in Latin and Greek, often from eastern authors that our own school systems have forgotten, in a seventeenth-century secretary hand, helps explain the difficulty of the manuscript. But above all, it is Newton's compulsion for continuous revision that makes the manuscript appear inaccessible. Newton's practice of revision shows that the text was under continual surveillance or inspection to see what would emerge from its growing linkages. Combined with its interlinear and marginal commentaries, and diacritical marks of texts to be added or subtracted, the text has a fluidity or instability that Newton simultaneously seeks to control and use a vehicle for further meaning. On a fundamental level, the text shows us how Newton inscriptions are signs or premonitions of meaning that he still seeks to register. What makes Newton so important in this regard is that we can see in his writing and notes the gradual transition from one semiotic system to another. The sense that there is always something else that lies beyond any conceptual horizon, also characterizes Newton's mathematical calculations.[16] The act of writing—literally setting down of multiple versions—becomes a phenomenological exercise. The process itself becomes a kind of developing bath for it is finally not the single text which is most important but the meaning which evolves. It is a hermeneutical practice which suggests a self-conscious effort to extend beyond the hermeneutical circle. Newton is not interested in the single narrative but in the meaning that emerges from the interplay of multiple intersecting plots.

The manuscripts that surround the *Origines* provide evidence that Newton's work became increasingly focused not on the accommodation of ancient pantheons but on the analysis of an archeological structure that could provide a common but highly significant element in all the narratives Newton was working with. This structure is the prytaneum or "public building in which magistrates and committees assembled." (*Oxford Latin Dictionary*). Contrary to the impression given by Westfall, Dobbs, and others, the prytaneum is not discussed in the *Origines*.[17] In fact, one of the most significant aspects of the *Origines* may be found if we think of it as an

[16] "Now is not this very fine? Mathematicians that find out, settle & do all the business must content themselves with being nothing but dry calculators and drudges & another that does nothing but pretend & grasp at all things much carry away all the inventions as well as those that were to follow him as those that went before." Newton to Halley, 20 June 1686. *The Correspondence of Isaac Newton*, 7 vols. ed. H. W. Turnbull, J. F. Scott, A. R. Hall, and Laura Tilling (Cambridge: Cambridge University Press, 1959-77), 2:438, letter #288.

[17] Here it is important to notice that care must be taken not to confuse related projects. Westfall's 1983 article, "Isaac Newton's Theologiae Gentilis Origines Philosophicae," makes little distinction between Yahuda MS 16.1 and 16.2 and Yahuda MS 41. "Whatever the name he gave to the treatise, a description of the origins of gentile theology was not Newton's goal" (p. 24). Dobbs follows Westfall by referring to "the often reworked *Origines* manuscripts Newton carried his convictions somewhat further" (p. 151).

experimental text that permits Newton to discover the significance of the prytaneum not as a structure in itself but as an architectural synchronism. The mytho-historical project envisioned in Yahuda 16 finally comes to be controlled when it is recast, in English rather than Latin, around the Prytaneum fire.[18]

2. SYNCHRONISM

Throughout his research, Newton directs his interpretive strategy away from isolated narratives and towards clusters or bundles of narratives that suggest possible synchronisms.[19] Whether he seeks to comprehend the parables of Jesus, alchemical texts, evidence from poetry, or recorded historical events his standard practice involves an exhaustive effort to list evidence so that it may be reduced to a few fundamental principles. Often his impulse to classify leads to the formulation of tables or charts. From the vantage point of logic, Newton's approach to mythography, history, alchemy, and prophecy is similar to a mathematical operation in which complex algebraic equations are infolded into simpler symbolic notation. In each of the areas cited above it is possible to identify a particular synchronism that permits Newton to collect and then factor information. His mythographic work is structured initially by the Noachids; his mature alchemical work concentrates on the star of regulus or antimony; his work on prophecy focused on an iterative hermeneutic process that involves ever evolving temporal confirmation. A focus on the prytaneum lies at the center of his mytho-historiographic research.

The most developed discussion of the prytaneum is found in Yahuda 41 where Newton underscores its significance from the outset:

> The religion most ancient & most generally received by the nations in the first ages was that of the Prytanea or Vestal temples. This was spread over all nations before the first memory of things. Cecrops began his reign over Attica about 60 years before ye Israelites came out of Egypt & before his reign the Greeks had Prytanea in all their cities as you may understand by ye passages newly cited out of Thycydides. There was one Prytaneum in every city placed in ye principal part of ye city. And in the Prytaneum was ye Court where the Council of Senate of ye City met as the Sanedrin did in the Temple of the

[18] Although I had completed my own transcription of *The Original of Religions* (Yahuda MS 41) in Jerusalem, I am indebted to Rob Iliffe for permitting me to review his transcription which he is currently preparing for publication by Oxford University Press.

[19] See Arnaldo Momigliano, "Time in Ancient Historiography," in *Essays in Ancient and Modern Historiography* (Middletown, Conn.: Wesleyan University Press, 1977), pp. 179-204, and "The Origins of Universal History," in his *Settimo contribuo alla Storia degli studi classici e del mondo antico* (Rome: Edicione Storia E Letteratura, 1984), pp. 77-103.

> Jews. For the performance of holy rites belonged to ye chief Magistrates of ye City, the king was Pontifex maximus as appears by the instances of Melchizedech, Abraham, Ninus, the Kings of Egypt & Athens & ye Roman Emperors. (Yahuda MS 41 f. 1r.)

Newton's interest in the prytaneum permits him to think that even an English antiquity such as Stonehenge may be viewed as a prytaneum. For Newton, the universal appearance of such structures becomes an argument for what is, if not perfectly understood, a universally shared impulse toward religion. The focal point of the such religion is not the prytaneum itself but the sacral fire that it contained.

> And by this means I conceive it came to pass yt the sacred fire at ye first plantation of ye earth was to be found in every City, as an essential part of the government. For in ye first ages when ye whole world was distinguished into as many kingdoms as cities, I understand not how one & the same religion could as soon spread into them all had it not been propagated wth mankind in ye beginning. (Yahuda MS 41 f. 4r.)

After reviewing the use of sacred fire in Egyptian, Hebraic, Greek, and Roman settings, Newton considers how sacred fires were kept in ancient temples and how the temples and sacred fires together could reflect the nature of the universe. Newton's discussion in the *Original of Religions* (Yahuda MS 41) is reinforced by a similar discussion in Yahuda MS 17:3. "It becomes evident that all this pertains to significant rituals. The Prytaneum with its fire in the center was an emblem of the earth's system and was believed by everyone. Here alone they perceived that they could designate the sun and the earth by the fire (Yahuda MS 17:3 f. 10r.)." ["Nam ad spectacula tam magna habes visus est. Prytaneum cum igne in medio emblema fuisse systematis mundani concredunt omnes, et in eo solo discrepant quod aliqui Solem alii Terram per igne illum designari volunt (Yahuda MS 17:3 f. 10r.)." Newton shows that altar fires and the rituals associated with them may be interpreted as representations of cosmological systems.[20]

In its most developed form, the location of the Prytaneum fire become a manifestation of the original religion embedded both within all religions including Judaism and Christianity.

> The placing ye fire in the commong center of the Priests court & of ye outward court or court of ye people in the Tabernacle & in Solons Temple [& the framing ye Tabernacle & Temple so as to make it a symbol of the world] is part also of ye religion wch ye nations received from Noah. For they placed ye fire in ye middle of ye Prytanea. The paying of tenth to the Priests was also

[20] Gerardus Vossius provides ample material on use of sacred fire in religious ceremony and must be regarded as a source for *the Original of Religions* just as he was for the *Origines*.

> ye religion of the nations before the days of Moses. For Abraham did it to Malehizadeck & Jacob did it at Bethel & there are instances of its being done by ye Heathens to Jupiter, Apollo & Hercules. And lastly as the Tabernacle was contrived by Moses to be a symbol of ye Heavens (as St. Paul & Josephus teach), so were ye Prytanea amongst ye nations.] And as the Tabernacle was a symbol of the heavens, so were the Prytanea amongs ye nations. The whole heavens they reconed to be ye true & real temple of God & therefore that a Prytanaeum might deserve ye name of his Temple they framed it so in the fullest manner to represent the whole systeme of the heavens. A point of religion then wch nothing can be more rational. (Yahuda MS 41 f. 6r.)

Using proof texts from Macrobius, Dionysius Halicarnassus, Plutarch, Pythagoras, and Pliny, Newton elaborates his idea by suggesting that the movement of the priests symbolizes the movement of the planets. "He who worships, by turning about, becomes a symbol of ye earth. Whence the Greeks called a man microcosmus." (Yahuda MS 41 f. 7r.)

> So then was one design of ye first institution of ye true religion to propose to mankind by ye frame of ye ancient Temples, the study of the frame of the world as the true Temple of ye great God they worshipped. And thence it was yt ye Priests ancinetly were above other men well skilled in ye knowledge of ye true frame of Nature & accounted it a great part of their Theology. (Yahuda MS 41 f. 7r.)

The discussion of the cosmic significance of sacred ritual found here is related to the account of the Egyptian priests at the beginning of the *Origines*. Indeed, the iconographic opening of the *Origines* not only anticipates his explanation of the Prytanean fire but his study of Solomon's Temple. I noted above that the emblematic quality of the sacred Egyptian procession; Newton's drawing of the temple (published in his *Observations on the Book of Daniel and the Book of Revelation*) is certainly linked to the earlier procession and functions even more starkly as an enigmatic emblem that invites interpretation of different levels. While Newton gathers copious detail about many ancient structures, his research ultimately concentrates on the Temple of Solomon. Sacred ritual, liturgical practice, architectural description, iconographic detail—all contribute to the prytanean structure that most embodies the Nochaen religion. "[W]hen the priests ascended to the altar, they moved in a circle about the fire and then descended. This fire in the center of the darkness was brought out from the interior and from the center was lifted up to seven lamps (which were symbols of the planets) and it was placed in the holy of holies, in the highest heaven where God is seated among the Cherubim (Yahuda MS 17:3 f. 11r.)" ["Per conversionum illam nos hominas circa ignarum centralem in vero mundi systemate revolvi insinuabant Hierosol ymitanum ad imaginem mundi constructum esse

Josephus docet ibique sacerdotas quoties in altare ascendebant, ibant in gyrem circa ignem antequam discendebant. Ignis ille i comuni centro atiorum duorem rioris et exterioris extreabatur, et a centro illo undo ascendabatur, per lampades septem, (quae Planetarum erant symbola), in sanctum sanctorum tanquam in coelum altissimum ubi Deus inter Cherubinos sedebat." (Yahuda MS 17:3 f. 11r.)] One glimpses in these passages, Newton thinking about the ways religious ritual and religious worship enacts what is simultaneously the study and worship of the universe. Understood from such a perspective, Newton's mythographic work untangles the corrupting influence of anthropomorphism at the same time that it explores the relation between religion and physics.

It is useful to compare Newton's work on prytanea and the Temple with similar projects at the end of the seventeenth century. Olof Rudbeck's *Atlantica*, which resulted in his election to the English Royal Society, continues to be approached with a combination of embarrassment and curiosity for arguing that Sweden evolved from the ancient Atlantis and that ancient Swedish culture concealed many links to the classical world of the eastern Midterranean. Rudbeck's work, which challenged contemporary readers to view universal history from northern or Baltic vantage point rather than from a southern or Mediterrean perspective, in effect sought to turn European historiography upside down. Especially in his efforts to focus attention on the relation between written records and physical remains, Rudbeck's work invites comparison to Newton's mythohistorical project.[21] (There is an autograph letter by Newton, in Carolina Redivia, the University of Uppsala library, in which he requests a copy of Rudbeck's research.)[22] But while they also share an interest in measurement and technology transfer, they resonate above all in the way each provides a context for their other work. Just as the *Origines* provides a historico-religious grounding for Newton's *Principia* (and for his alchemical experimentation), *Atlantica*

[21] My comments on Olof Rudbeck are part of a book I am writing on Baltic relations during the seventeenth and early eighteenth centuries. See Kenneth J. Knoespel, "Reshaping the Earth: Olof Rudbeck and the Transformation of Sweden," *Cultural Exchange between European Nations during the Renaissance* ed. Gunnar Sorelius and Michael Srigley (Uppsala: Almqvist & Wiksell, 1994), pp. 193-212. See Gunnar Eriksson, "Historia och naturalhistoria hos Olof Rudbeck" in *I Idèhistoriens virvlar: Festskrift till Rolf Lindborg* (Stockholm: Brutus Östlings Bokförlag, 1991), pp. 29-60; also Gunnar Eriksson, "Olof Rudbeck d. ä" in *Lychnos* (Lärdomshistoriska Samfundets Årsbok 1984 (Stockholm: Almqvist & Wiksell, 1984), 77-119; also "Rudbeck och Linné: Enmansinstitutioner" in *Kunskapens trädgårdar: om institutioner och institutionaliseringar i vetenskapen och livet* (Uddevalla: Atlantis, 1988), pp. 17-39; See also Gunnar Eriksson's *The Atlantic Vision: Olaus Rudbeck and Baroque Science* (Canton, Massachusetts: Science History Publications, 1994). See also Matts Malm, *Minervas äpple* (Stockholm: Brutus Östlings Bokförlag Symposion, 1996). Research on Rudbeck and seventeenth-century Sweden continues to be indebted to the seminal work of Sten Lindroth, *Svensk lärdomshistoria: stormaktstiden* (Stockholm: Norstedts, 1975).

[22] See Marco Beretta, *A History of Non-Printed Science: A Select Catalogue of the Waller Collection* (Uppsala: Acta Universitatis Upsaliensis, 1993), pp. 138-9.

supplies an historical explanation for the industry not only of Rudbeck himself but also for his Swedish contemporaries. Each author establishes historical-scientific parameters for ongoing physical experimentation. Even more importantly, it is possible to see how each author also locates their research in a diligent study of practical arts. There is no better sign of this than the sites that are central for their research. Both Newton and Rudbeck sought to establish foundational narratives based on studies that directed their attention to artifacts and architecture. For Rudbeck, the mounds of Old Uppsala and the physical remains of Viking history provided a means to reorient the study of written record. For Newton, the study of mythology lead to the conviction that the Temple in Jerusalem held within its record of ritual and within its architecture, information about the structure of the universe.

Although the *Origines* is not an alchemical text, it also provides a setting for understanding Newton's alchemical study. That there is a constancy of matter and that it can be confirmed through natural and human history constitutes a fundamental thesis for all of Newton's research. Such a thesis, however, is by no means self-evident but requires Newton to interpret continually the relationships between multiple representations of matter that occur in history and the universal principles used to explain them. The interpretation of the representation of matter is absolutely central to Newton's alchemcial work and reminds us that the significance Newton gives to the prytaneum fire should not be limited to its symbolic function as a sacral represetation of the universe. Inherent in his research is the liklihood that the sacred fire is also connected to the transforming fire of the alchemist. Newton's extensive work with alchemcial texts cannot be imagined to involve the study of matter alone but needs to be viewed as a study of the narrative forms that enable him to approach matter. For Newton the study of these narrative forms encompasses an historical project that surveys centuries of chemical research. The *Origines* challenges us to see that alchemical texts are also a manifestation of sacred ritual and liturgical practice. His efforts to understand the "types and aenigmas" of the Egyptian procession are linked closely to his efforts to interpret and practice alchemical texts. Similarly, his efforts to justify and understand prophetic texts resonate with his work to comprehend the systematic nature of matter through interpretive intervention and manipulation. We must expect to find Newton working simultaneously on multiple levels. The configuration of fabulae form an equation which supports or guides the application which involves repeated experiments. The experiments in turn, are controlled on another level by the patterns which emerge and which in turn are explained not by the fabulae but by the relationship which they are seen to emplot.

3. Reading and Writing Under Surveillance

Allegory's importance as an interpretive strategy lies in its capacity to permit multiple levels of understanding to exist simultaneously. Since the time of the earliest Homeric commentary, it has also been recognized as a practice that is most effective in systems that are closed and in which meaning can be shared by a community.[23] Newton approaches allegory with caution both because it could be hermeneutically confining and because it reinforced the communal structure of meaning. It is even more to the point to say that Newton found allegory an instrument that could be used to mislead.

> But the world loves to be deceived, they will not understand, they never consider equally, but are wholly led by prejudice, interest, the praise of men, & authority of the church they live in: as is plain because all parties keep close to the religion they have been brought up in, & yet in all parties there are wise & learned as well as fools & ignorant. There are but few that seek to understand the religion they profess, & those that study for understanding there in, do it rather for worldy ends, or that they may defend it, than to examine whither it be true with a resolution to chose & profess that religion which in their judgment appears the truest. (*A Treatise on the Apocalypse* [Yahuda MS 1 f. 5r.])

In contrast to the deception, self-interest, and authority that reinforces the interpretive community described here, Newton's strategy seeks an interpretive mode that would allow his contemporaries "to understand the religion they profess." As demonstrated through the research inherent in the *Origines* and its surrounding manuscripts, Newton practices strategies associated with euhemerisim and physical allegory to reconfigure the way that he approaches history. There is more to Newton's practice, however, than the stellification of heroes and the formulation of a set of correspondences between words and physical phenomena.

One consequence of spending a long time working with Newton's manuscripts is a growing awareness that he watches his own writing as a natural phenomenon that can be surveyed for evidence. As we follow his literal inscription of multiple versions of a paragraph in the *Origines*, it is finally not the single text alone which carries the most importance but the expectancy that meaning will continue to evolve among the emended sentences, crossed-out lines, and marginal additions. Of course, such hermeneutical surveillance may be compared to the process invoked by Newton's alchemical experimentation. The continually rereading or recalculation comprises an interpretive mode that is itself experimental.

[23] See Jean Pépin, *Mythe et Allégorie: Les origines grecques et les contestations judéo-chrétinnes* (Paris: Études Augustiniennes, 1976).

From such a vantage point, there is an affinity between the experimental mode that accompanies Newton's natural philosophy and the deductive space of synchronisms. With Newton the study of myth becomes another way to understand the study of science. In his book *The Kabbalah in a New Perspective*, Moshe Idel refers to the combinatory hermeneutics of the cabbala.[24] Although it would·be misleading to seek a detailed understanding of the Jewish cabbala in Newton's work, I find Idel's term useful because it helps define Newton's strategy as a means to excite or provoke meanings that extend beyond closed interpretive systems. In effect the power of magic comes not simply from the manipulation of natural phenomena but in challenging or opening fixed orthodox meanings. For the seventeenth century, where reading was conducted in a field of socially accepted meaning, the exfoliation and infoliation of multiple mythographic pantheons offered a license to broader speculation. Newton seeks to negotiate meaning by engendering disruptions that are hermeneutically controlled through a process of reading writing that are continually under surveillance.

The question of interpretive strategies and the coherence of interpretation is fundamental to Newton's undertaking. For him the Bible is affirmed not through external commentary but by its own internal cogency. Newton found demonstration of such cogency not in the church's interpretation of the New Testament but in the way that Jesus came to understand his own relation to the Old Testament. "[T]he prophecies of the old Testament remained in obscurity till Christ's final coming & then were *interpreted* by Christ & the *interpretations* became the religion of the Christians: so the Prophesies of both Testaments relating to Christ's second coming may remain in obscurity till that coming & then be *interpreted* by divine authority & the *interpretations* become the religion of God's people for the future (my emphasis)." (*Drafts on Daniel and the Apocalypse* [Yahuda MS 7.2 f. 1r.]) As the passage indicates, for Newton interpretation is the very ground of religion. His position inevitably raises questions about his own hermeneutical activity for just as the interpretive work of Jesus confirms the meaning of the Old Testament, so it ultimately falls to interpreters beyond the New Testament to confirm its prophecies. Is it possible that Newton found in his own work the creation of a new interpretive instauration that would lay a foundation for a reformed religion integrating the moral teachings of Jesus with a knowledge of the coherence of creation? Such questions surely occur to Newton:

> And if natural philosophy in all its parts, by pursuing this method shall at length be perfected, the bounds of moral philosophy will be also enlarged so far as we know by Natural Philosophy what power the first cause has over us

[24] Moshe Idel, *Kabbalah: New Perspectives* (New Haven: Yale University Press, 1988); see also Harold Bloom, *Kabbalah and Criticism* (New York: Seaburgy Press, 1975).

> & what benefits we receive or may expect from him; so far our duty toward him as well as that towards one another will appear to us by the light of nature. And no doubt, if the worship of fals Gods had not blinded the Hethens, their Moral Philosophy would have gone further than to the four Cardinal Vertues, & indeed instead of teaching us to worship the sun & moon & dead men's souls, they would have taught us to worship ye true Benefactor. (ULC Add MS. 3970 f. 244v. Quoted by permission)

The *Origines* establishes experimental grounds for understanding how religious practice may reveal physical truths about the universe. As we have seen, the project involves the collection of a vast amount of historical and mythographic information with the expectation that they will reveal patterns that may be discerned through his own interpretive intervention. In the work surrounding the *Origines*, it is ultimately the prytanean fire that provides the common denominator for his mythographic research by confirming the continuity of religious practice across millennia. For Newton such constancy becomes a sign that natural philosophy lies at the very basis of religious revelation. Even more, the prytanean fire provides a basis for Newton's belief that an original religion was grounded on a physical understanding of the universe that had become corrupted through the vulgar practice of all religions. Viewed with the *Principia*, the *Origines* the provides historical justification for Newton's own momentous discoveries. From such a vantage point, Newton's discoveries in the *Principia* may even be regarded as rediscoveries of ancient revelations. The linkage between ancient religion and his own discoveries has considerable bearing on his status as an interpreter as well. While it is all too common to separate Newton's interpretive work as a natural philosopher from his interpretive work in history and religion, we gain much drawing them together. Finally, such integration affirms Newton's status as an interpreter/prophet. The *Principia* is not only a book about natural philosophy but, seen together with the *Origines*, is work of prophecy because of the ways it confirms God's patterned presence in history. It is entirely appropriate to compare Newton's role at such an interpretive juncture to Jesus for just as the cogency of Jesus' interpretation established a new religion, so the cogency of Newton's work could ground religion on renewed physical truths.

Although Newton never answers questions about the personal status of his own interpretive revelations, there is ample reason to see in the instauration of a Newtonian science a practice thoroughly antithetical to his own envisioned union of science and religion. Considering Newton's own abiding interest in detecting the human proclivity to deify its own accomplishments in a false religion, it is possible to view the subsequent glorification of Newton as a manifestation of the very human inclination he sought to criticize. If Newton ever truly wondered whether his work in natural philosophy could be seen as a divine relation, the subsequent reception of his work would surely make him conclude that his was a truly failed messianic science. At the beginning of my comments I noticed how the *Origines* continues to touch foundational narratives of the twentieth century. In conclusion, I would like to notice how the *Origines* touches our own work. As we integrate Newton's ignored manuscripts with the heroic legends so established by his published works, we participate in a process of demythologizing that restores his vision of the fusion of science and religion. In the vocabulary of the philosophy of science, manuscripts such as the *Origines* challenge the positivistic presentation of Newton and reveal him, as a demythologized figure, far closer to our own late twentieth-century discussions of realism and constructionism.

ESSAY 10

Newton's Apocalypse

MICHAEL MURRIN
University of Chicago

Isaac Newton was not an original interpreter of the Apocalypse. He himself stated that he followed Joseph Mede, and he imitates Mede in his lexicon of symbols with due attention to dream lore like the *Oneirocriticon* of Ahmad ibn Sirin, in his synchronisms, and in his attention to Hebrew ritual.[1] Seen in the long history of Apocalypse interpretation, however, even Mede might not be original. Few in fact are those who have changed the whole discourse on this book. Mede, while he may have had a new method, did not alter the standard Protestant interpretation of Revelation. Commentators and critics who might meet such tough criteria would be Dionysius of Alexandria, Tyconius, Joachim of Fiore, Nicholas of Lyra, and perhaps Ribera and Alcazar, whom Mede read.

Bishop Dionysius provides an example. In his book *On Promises* he answered the charges of Bishop Nepos of Arsinoë, a literalist millenarian, whose *Refutation of the Allegorists* had caused much local dissension.[2] For

[1] Newton, *Trattato sull' apocalisse*, ed. Maurizio Mamiani, 1994 (Turin: Bollati Boringhieri, 1995), 1, 8r, 14-16. In citations from this text I refer first to the manuscript number (Newton MS 1), then to the folio number (8r), and finally to the page numbers in Mamiani's edition (14-16). Frank Manuel notes Newton's debt to Mede in *The Religion of Isaac Newton* (Oxford: Clarendon Press, 1974), p. 90; and Sarah Hutton argues that Newton is closer to Mede in his interpretive practice than More, who was Mede's own student. See "More, Newton, and the Language of Biblical Prophecy," in *The Books of Nature and Scripture: Recent Essays on Natural Philosophy, Theology, and Biblical Criticism in the Netherlands of Spinoza's Time and the British Isles of Newton's Time*, ed. James E. Force and Richard H. Popkin (Dordrecht: Kluwer Academic Publishers, 1994), p. 49. She also argues that Ahmad ibn Sirin gave Mede his basic idea of the analogy between the natural and the political world, p. 43. Newton developed his interest in Hebrew ritual in the late 1670s. See Richard Westfall, *Never at Rest: A Biography of Isaac Newton* (Cambridge: Cambridge University Press, 1982), pp.344, 346.

[2] For the basic account see Eusebius, *Ecclesiastical History*.7.24-25. The early witnesses to the Apocalypse were millenarians: Papias, Justin, and the heretic Cerinthus. See also Wilhelm Bousset's remarks in the Introduction to his edition, *Die Offenbarung Johannis*, (Göttingen: Vandenhoeck and Ruprecht, 1906), pp. 19-20.

J.E. Force and R.H. Popkin (eds.), Newton and Religion, 203–220.
© 1999 *Kluwer Academic Publishers.*

Dionysius every verse of the Apocalypse conceals something and is more wonderful than each interpretation, a statement Jerome would repeat and which would justify the Western tradition of allegorical readings, developed soon afterwards by Victorinus and Tyconius.[3] Dionysius' second set of arguments, however, had very different results, ones the Bishop did not intend.

Dionysius argued that John the Presbyter not John the Evangelist composed the Apocalypse, and he proved his point with careful philology. The author of the Gospel and the First Epistle attributed to John the Evangelist never names himself, while the John of the Apocalypse frequently does so. The Gospel and the First Epistle have the same kind of metaphysical beginning and share common terms and concepts not found in the Apocalypse.[4] The Gospel and the First Epistle also differ radically from the Apocalypse stylistically. The author of the first two texts wrote faultless and eloquent Greek, while the John of the Apocalypse fell into grammatical errors.[5] Many modern scholars agree with Dionysius and with his attribution of the book to John the Presbyter.

Dionysius' arguments killed the Apocalypse in the Levant, though it survived as a canonical text in Egypt. The Syrian churches never had it in their canons, and now the Greek-speaking Christians of Palestine and Antioch dropped it from theirs. Antioch had always favored a literal and historical approach to Scripture, and its clerics would presumably not have been impressed by Bishop Dionysius' allegorical defense but would have reacted to the question of authorship. The Apocalypse still had influential defenders in Cyprus and Anatolia, but not in Greece, and the book never became part of the Greek canon. In fact, the book survived only in areas like Egypt and the Latin West and among theologians like Athanasius who especially aroused Newton's suspicion.[6]

[3] Alexandria, of course, was a center for allegorical interpretation. Dionysius was also answering those who complained that the book was not an unveiling (*apocalypsis*) but veiled its content with a heavy, thick curtain of unintelligibility. For Jerome see his letter to Paulinus in the *Opera*, ed. Dominicus Vallarsius (Venice, 1766), 1.1.270-81.

[4] Dionysius has a long list: life, light, turning from the dark, truth, grace, joy, flesh, the blood of the Lord, judgment, the forgiveness of sins, love by God of us and the command to love one another, condemnation of the world, of the Devil, of Antichrist, the proclamation of the Holy Spirit, and the adopted sonship to God (Eusebius, *Ecclesiastical History*, 7.25.21).

[5] Newton was still trying to refute this argument in his *Observations upon the Prophecies of Daniel and the Apocalypse of St. John* (London, 1733), pp. 243-7. For a discussion of Newton's argument, see Richard H. Popkin, "Newton as a Bible Scholar," in *Essays on the Context, Nature, and Influence of Isaac Newton's Theology*, ed. James E. Force and Richard H. Popkin (Dordrecht: Kluwer Academic Publishers, 1990), pp. 107-8.

[6] For a discussion of the Apocalypse and the canon see Bousset, Introduction, pp. 19-31; Bruce M. Metzger, *The Canon of the New Testament: Its Origin, Development, and Significance* (Oxford: Clarendon Press, 1987), pp. 113-14, 148, 150, 155, 203-06, 209-20, 224-5.

In his *Annotations* Erasmus revived Dionysius' arguments, which in turn influenced the early Reformers. In fact, for a while it seemed likely that Apocalypse would not make the Protestant canon. Among the Lutherans, Carlstadt put Revelation in the bottom class of texts, along with Second and Third John, Second Peter, Jude, James, and Hebrews, all texts the authenticity of which had been called into question previously. Luther himself argued in the following year that the book was neither apostolic nor prophetic, and Lutherans refused to have it numbered among the canonical books, something which marked Lutheran Bibles throughout the sixteenth and into the seventeenth century.[7] Among the evangelicals Zwingli and Oecolampadius rejected it, and Calvin ignored it.[8] In the seventeenth century doubts about the textual accuracy of biblical books would lead Newton to devote a whole chapter of his *Observations* to the issue, and in that chapter he would attempt a point by point refutation of Dionysius.[9]

Yet even then a crucial hesitation was evident. In Luther's New Testament Revelation was the *only* book initially to have woodcuts. Cranach adapted Dürer's famous series to polemic, putting the papal tiara on the whore who rides the beast in Chapter Seventeen. And it was polemic which soon gave the book such prominence in Protestant circles.

Luther himself shows the change. He wrote a second preface for the book in 1530, providing an interpretation which set Protestant standards for the

[7] Luther made his arguments in the Preface to Revelation which he composed for his translation of the New Testament. See his "Vorhede auf die Offinbarung Sanct Johannis," in *Luther's "September Bible" in Facsimile* (Ann Arbor: Ann Arbor Publishers, 1972).

[8] For a general discussion of the issue, see Bousset's Introduction, pp. 31-3, and Metzger, *The Canon of the New Testament*, pp. 241-3.

[9] For Newton the issue was crucial. Already in the 1670s he believed the essence of the Bible was prophecy and had reduced that essentially to Daniel and the Apocalypse. See Westfall, *Never at Rest*, pp. 319, 826-7, and Popkin, "Newton as a Bible Scholar," p. 106. Popkin cites an unpublished note of Newton's, written in the early eighteenth century, to the effect that the historical parts of Scripture were written with the ordinary assistance of the Spirit, but the prophetic sections had special impulses of the spirit (p. 109 & n. 66). In the *Observations*, Newton argues that John the Evangelist composed the Apocalypse at the time of the Christian dispersal, caused by the Jewish Revolt, and this fact explains his faulty Greek. Later, after he had lived long in Asia Minor, he could write the excellent Greek which marks the Fourth Gospel. See especially *Observations*, pp. 235-47. Father Richard Simon had recently made the text of the Bible a major issue for English Protestants, and Newton knew his work. He owned the English translation of his *Histoire critique du Vieux Testament*, which came out in 1682. See Manuel, *The Religion of Isaac Newton*, 84, and Popkin, "Newton as a Bible Scholar," pp. 103-4. Already, however, in the early 1670s, Newton was collating textual variants of the Apocalypse. They survive in Yahuda MS 4 and were complete by 1694. Newton looked at twenty-two versions and arranged the variants verse by verse. See Manuel, *The Religion of Isaac Newton*, pp. 93-4, and Westfall, *Never at Rest*, pp. 327-8. For a careful and detailed discussion of Simon and the English response see Justin A. I. Champion, "'Acceptable to inquisitive men': Some Simonian Contexts for Newton's Biblical Criticism, 1680-1692," infra.

next 150 years and more.[10] He borrowed his methods from Nicholas of Lyra, whose *Postille* (1329) had changed previous approaches to Apocalypse and had wide circulation in the later Middle Ages. As Denis the Carthusian complained, Lyra imposed a rigorous chronological order on the book and made each symbol refer to a specific historical event.[11] Lyra paralleled the text with ecclesiastical history, arguing for example that the woman clothed with the sun in Chapter Twelve referred to the feast of Candlemas or the Purification of Mary, instituted after a plague at Constantinople, but more often he stressed the persecutions and heresies which the church suffered and the war and politics which affected it. The effect of his interpretation, however, was to remove prophecy from the Apocalypse, since he traced a continuous history at least as far as Chapter Twenty, in which he found the reign of Pope Innocent III.[12]

Luther took over Lyra's methods, substituting a Protestant for a medieval view of church history. At the same time he allowed for contemporary events in the second half of the Apocalypse, and this part would become the locus for a lively Protestant discussion of the present and near future of the church. Newton would still be using this approach when he claimed that prophecies are histories of things to come.[13]

Despite its importance for the social and political history of the North, Protestant commentary on Revelation lacked originality and never really came out of Lyra's shadow. Furthermore it suffered from two major weaknesses. The first is the history itself. Catholics had argued that the Reformers left no room for a church in the last thousand years, and Protestant historians answered with the story of the hidden church. The Lutherans started and soon had *Carion's Chronicle* (1531), which relied heavily on the prophets,[14] and then the *Magdeburg Centuries* (1559-74), while Foxe provided the same service for the English. In the interpretations of the Apocalypse, however, this new history appeared unfortunately as confused and uncritical. Commentators lumped together all who opposed or criticized the Papacy in the Middle Ages, even writers like Dante and Petrarch. Moreover the Waldensians and Albigensians formed a pair, beginning the first reformation, despite the fact that the Albigensians were

[10] For a translation of this preface in its latest form, that of 1546, see the text in *Word and Sacrament*, vol. 1, ed. E. Theodore Bachmann, who also revised the translation of Charles M. Jacobs, vol. 35 in *Luther's Works*, ed. Helmut T. Lehmann, 1960 (Philadelphia: Fortress Press, 1987).

[11] Denis wrote his commentary shortly before his death in 1471. For a discussion see Henri de Lubac, *Exégèse médiévale: Les quatre sens de l'écriture*, Part II, vol. 2 (Paris: F. Aubier, 1964), pp. 363-7.

[12] De Lubac, *Exégèse médiévale*, II.2.346-7.

[13] Newton said this more than once. See his remarks under Rules 13 & 14 at *Trattato*, 1, 16r, 32.

[14] See Robin Bruce Barnes, *Prophecy and Gnosis: Apocalypticism in the Wake of the Lutheran Reformation* (Stanford: Stanford University Press, 1988), pp. 51, 106-7.

really Manichaeans not Christians. The two regularly appear together in Apocalypse commentary. More has them, so does Newton.[15]

Such a method also reversed normal Protestant modes of interpretation. The commentator usually explicated the literal sense of the text, but, as Bishop Dionysius had argued, the Apocalypse has no plain sense, veiling its discourses instead. Luther had to allegorize it, so did Mede and his followers. When Mede claimed he gave only the literal sense of the text, he meant that each symbol had if possible only one meaning, one application to history. Newton said the same.[16] This is still allegory, though in the more restricted form introduced by Nicholas of Lyra. The language and terminology might suggest Protestant norms, but the practical criticism of this book was still medieval.

It was rather a Jesuit who made the first original commentary in the period. Francisco de Ribera is known more often as the confessor and biographer of Teresa of Avila, but around 1578 he gave lectures in Salamanca which eventually became *In sacram beati Ioannis apostoli, et evangelistae Apocalypsin commentarii*, which was published, however, only after his death in 1591. Wilhelm Bousset, the great editor of the Apocalypse in the last century, argued rightly that Ribera broke completely with Lyra and even with the older spiritualized tradition which came from Tyconius. Relying on patristic sources, Ribera tried to grasp the thought world implied by Revelation.[17] As a result he assumed most of the visions still referred to the future,[18] and read the book literally, as science fiction. The trumpets and vials provide examples. In the second trumpet something like a burning volcano drops into the sea. Ribera assumes that some day a great fiery globe will fall into the sea.[19] He likewise reads the star Wormwood which hits the rivers in the third trumpet and the first six vials literally.[20] Now literalism is a possible response to descriptions of the beginning and the end of the world, since in such extraordinary situations, which must remain outside the experience of the commentator, it is difficult to determine what would be literal, what allegorical. By his approach Ribera exposed the fact that Protestants, whatever the terminology they might use, did not read the Apocalypse plainly and literally but allegorically.

[15] Henry More, *Apocalypsis Apocalypseos; or The Revelation of St. John the Divine Unveiled* (London, 1680), pp. 16-17, 128; Newton, *Trattato*, 1, 3v, 10.

[16] Under Rule 2. *Trattato*, 1, 12r, 20.

[17] Bousset, Introduction, pp. 91-2.

[18] See, for example, his remarks at *In sacram beati Ioannis apostoli, et evangelistae Apocalypsin commentarii* (Lyon: Iuntae, 1593), pp. 2, 9.

[19] Cited via David Pareus, *A Commentary upon the Divine Revelation of the Apostle and Evangelist John*, tr. Elias Arnold (Amsterdam, 1644), p. 159.

[20] *Ibid.*, p.162, 375, 377.

Ludovicus ab Alcazar was even more disturbing. For his *Vestigatio arcani sensus in Apocalypsi* (1614) he used methods normally associated with the "High Criticism" of the nineteenth century.[21] He applied the first half of Apocalypse to the Jewish Revolt and the second half to early Roman persecutions of Christians.[22] While he owed his chronological order to Lyra, he dropped ecclesiastical history and eschatology. The whole book concerned events long ago and no longer served a prophetic function. He supported his argument with the first complete survey of Apocalypse criticism, from antiquity to the present, and Bousset still used Alcazar to date many medieval works.[23] In the North his followers were Grotius, Hammond, and Bossuet. Newton cites approvingly his reading of the wilderness where the woman hides,[24] but attacks Alcazar's approach, saying that those who apply Apocalypse to the Apostolic Age must explain why their interpretations were not expressed then.[25]

Most learned or academic English readings in the seventeenth century responded to Ribera and Alcazar. There was a flurry of new methods and approaches but always the same interpretation. Commentators wished to defend the continuing prophetic function of Apocalypse against Alcazar, but they also wanted to support their allegorical and polemical reading against both Spaniards: the attack on the Papacy or, at mid-century, by Protestants on each other. Mede devised the most distinguished of these approaches and initiated what came to be a Cambridge tradition through his pupil Henry More and then Isaac Newton. Yet Mede showed that, of the two parallel lines, the textual order and the new Protestant history, the second took priority. The whore and the beast had to symbolize papal not pagan Rome, this despite Newton's protest that the history should rather be fitted to the Apocalypse.[26]

Mede nevertheless can still interest modern readers, who can learn much from him in their own studies of Apocalypse, mostly because he became so excited over his method that he put all his emphasis there, particularly in his study of prophetic language and Hebrew ritual. Both are striking, but the

[21] See Richard Bauckham, *Tudor Apocalypse* (Appleford, Abingdon: Sutton Courtenay Press, 1978), pp. 236-7.

[22] For a discussion of Alcazar and his antecedents see Bousset, Introduction, pp.93-4.

[23] Bousset also implies that the erudition in Pareus' Variorum commentary comes from Alcazar, Introduction, pp. 96.

[24] *Trattato*, 1.1, 11r, 70. Alcazar reads it as paganism.

[25] *Ibid.*, 1, 9r, 16. Because Alcazar, like Ribera, does use patristic sources, Newton's criticism loses much of its force.

[26] Under Rule 12. *Trattato* 1, 15r-16r, 30. For a discussion of Mede see Michael Murrin, "Revelation and Two Seventeenth Century Commentators," in *The Apocalypse in English Renaissance Thought and Literature*, ed. C. A. Patrides and Joseph Wittreich (Manchester: Manchester University Press, 1984), pp. 126-9, 132-3, 139-40.

jury must still be out on the question of their originality. Ribera had already noted the similarity in the phrases, metaphors, and tropes between Revelation and the Hebrew prophets, and in fact he had been studying the twelve prophets when he began to lecture on Apocalypse at Salamanca.[27] Published with his commentary was also his treatise on Hebrew ritual, festivals, and the Temple. Mede may have had a major source for his erudition, and a final judgment on his originality must wait on an adequate study of the Jesuit scholar.

Despite his new methods Mede showed the usual strains of Protestant polemic. He falls into internal contradiction, once he reaches the monster stories, mixing up the dragon and the beast in his attempt to fit the text to the preestablished historical pattern.[28] We shall see that Newton experienced the same difficulty.

Mede succeeded well enough, however, to have a wide, deep, and lasting influence, not just at Cambridge, but among the millenarians of the mid-century and among Germans. He still had followers in the eighteenth century.[29] Yet he could not stop the spread of Spanish methods in the North, particularly the historical-critical approach of Alcazar, which Grotius and others adopted.[30]

Newton then belongs to a conservative movement to maintain the old Protestant readings of Revelation. We know he owned the third edition of Mede's *Works* (1672),[31]and Mede may have given the impetus to the studies which Newton had finished by 1675, now in the Yahuda collection and recently edited by Maurizio Mamiani.[32] That influence continued throughout Newton's life and still marks his *Observations upon the Prophecies of*

[27] Ribera, *In...apocalypsin commentarii*, Proem, 3.

[28] Mede had used the angel's gloss on the scarlet Beast of Chapter 17 for the red dragon of Chap. 12. He then said the Dragon, not the sea Beast, had a wounded head, although Rev.13:12 syntactically refers to the Beast, and John never calls the Dragon *to thērion*. See his *Commentarius* in *The Works* (London, 1677), pp. 500-1. Mede's correspondents all corrected this confusion: Lawenus (Mede, Appendix to the *Clavis* in *The Works*, pp. 541-2) and De Dieu (p. 566). Mede tried to answer these objections but in the course of his argument allowed his interpretation to determine his presentation of the literal sense, a tendency he criticized in others. See the Appendix in *The Works*, pp. 555, 570 and *The Remains*, also in *The Works*, p. 596.

[29] See Murrin, "Revelation," pp. 125, 138-9.

[30] Bousset, Introduction, pp. 93-4.

[31] Mamiani, Introduction, p. xiv & n. 32.

[32] This is implied in Mamiani's chronology (Introduction, p. xiii). Richard Westfall pointed out that Yahuda MS 1.1 had to be complete by 1675. See *Never at Rest*, pp. 319-20 & n. 114. Richard Popkin dates it as early as 1671. See his "Newton as a Bible Scholar," p. 103. Alan Shapiro, however, notes that Yahuda 1.1 by its watermarks is dated to the mid-1670s. See Rob Iliffe, "'Making a Shew': Apocalyptic Hermeneutics and the Sociology of Christian Idolatry in the Work of Isaac Newton and Henry More," in *The Books of Nature and Scripture*, p. 63 & n. 29. What Mamiani now lists as Newton 1 contains rules for interpretation with a brief lexicon of symbols. Newton 1.1 has a detailed lexicon with proofs, and 1.3 has his synchronisms or the application of the symbols to the Apocalypse.

Daniel and the Apocalypse of St. John, published posthumously in 1733.[33] And it is as a student of Mede that Newton should be analyzed.

Like Mede Newton thinks that the prophets shared a common symbolic discourse, resembling a code which could be broken or a forgotten language which could be recovered.[34] The system in both authors turns on a simple identification, that between the cosmos and politics. Mede and Newton claimed that the prophets normally talked of a political entity as if it were a world with its sky, stars, and land symbolizing parts of that entity.[35]

Newton also accepted Mede's synchronisms, a logical corollary of the method. If a symbol normally means the same thing, then its occurrence in different places signals a return to the same topic or event. Apocalypse, however, repeats many symbols, so it invites a synchronic analysis of its famous sequences: the seven letters, seals, trumpets, and vials.[36] Newton, therefore, could synchronize the seven trumpets and seven thunders, which reply to the cry of the storm angel (Rev. 10:3).[37] Mede had claimed that such synchronisms, like the symbol system he had rediscovered, were not the caprice of an interpreter but *objective*.[38]

A further development of the same method led both authors to a ritual reading of Apocalypse. Newton succinctly expresses the logic: "Temples were anciently contrived to represent the frame of the Universe as the true Temple of the great God." (*Theological Manuscripts*, 125) Since prophetic language expressed political events through cosmic imagery, it could also evoke Temple ritual, which referred to the same things. Mede, for example, read the vials as part of a dedication ceremony, alluding to that of

[33] Westfall says that the *Observations* combines two different manuscripts. The base text was an unfinished manuscript from Newton's old age, supplemented by an old, completed manuscript (*Never at Rest*, pp. 321, 818-9). Iliffe says that Keynes MS 5 has this same base text. See "'Making a Shew'," p. 63 & n. 28. Also relevant is the tract on the Last Judgment which Manuel transcribed as Appendix B in his *Religion of Isaac Newton*, pp. 126-36. Manuel dates a parallel version to the 1680s (pp. 99-100). Finally there is the material on prophetic language in Keynes MS 5, which H. McLachlan transcribed in *Theological Manuscripts* (Liverpool: University Press, 1950), pp. 119-26.

[34] For examples in Mede see Murrin, "Revelation," p. 136. Newton expresses his version of this view in McLachlan, *Theological Manuscripts*, pp. 119-20.

[35] Mede, *Commentarius*, in *The Works*, 448-9; Newton, *Observations*, p. 16; *Theological Manuscripts*, pp. 120-1.

[36] Murrin, "Revelation," pp. 131, 137.

[37] *Observations*, p. 312. He added the seven vials and classed them all as temporal divisions of the seventh seal.

[38] Mede makes this claim on the title page of the *Clavis*. See also the Appendix to the *Clavis*, in *The Works*, pp. 550, 558, and *The Remains*, also in *The Works*, p. 581. Earlier commentators had called such repetitions recapitulation, and Victorinus and Tyconius had already assumed recapitulation in the fourth century. The approach still has its modern proponents. See the discussion in André Feuillet, *L'apocalypse: État de la question* (Paris & Bruges: Desclée de Brouwer, 1962), pp. 10-11.

Solomon.[39] This approach to Apocalypse continues to thrive in the twentieth century. André Feuillet provides many examples and himself supports it.[40]

Newton gives his own slant to the ritual approach. Mede had stressed the Tent in the wilderness encampment,[41] but Newton prefers the Jerusalem Temple.[42] The difference here is slight, since biblical authors assumed virtually the same ritual for both,[43] and Mede and Newton draw when necessary on both Tent and Temple.[44] Newton makes his own contribution to the tradition when he gives the ritual reading of the sequences a specific focus, that of the Feast of Tents or Tabernacles. Newton assumes that Israelite festivals were prophetic and that this feast in particular pointed to the Second Coming (*Observations*, p. 255.) He especially uses the feast to explain the seals sequence.[45] He says that the opening of the seals imitated the seven days of the seventh month, during which the High Priest read the Law. John has the sealed book kept on the right side of the throne or Ark, and Newton claims that the Israelites put the Torah in the right side of the Ark.[46] On the last day, that of the Atonement, the High Priest offered incense, carrying it in a gold censer into the Holy of Holies. While he burned it there, he sprinkled the mercy seat seven times with the blood of a bullock and then with that of a goat. Meanwhile the people outside waited silently. This is the half-hour silence when the seventh seal is broken (Rev. 8:1). Having made these sin offerings, the High Priest sent the goat Azazel into the wilderness, dressed himself in new clothes, and had the two sin offerings burnt outside the camp with altar fire carried in the censer. In Apocalypse the angel who offers the incense takes the censer and fills it with fire from the altar, which he tosses on the earth (Rev. 8:5). Connected to this ceremony is the sealing of the servants of God (Rev. 7:2-17). The Israelites had opened three books at the beginning of the month, one for the

[39] Mede, *Remains*, pp. 590-1. See also Murrin, "Revelation," p. 137.

[40] See *L'apocalypse*, pp. 61-2, 71-3. Most modern scholars, however, stress correspondences to early Christian liturgical practice rather than to the Jerusalem Temple.

[41] Mede, *Commentarius* in *The Works*, pp. 437-8; Appendix in *The Works*, pp. 566-7; *Fragmenta*, in *The Works*, pp. 916-7; Murrin, "Revelation," p. 137.

[42] See the way Newton reads the vision which begins the seals sequence (*Observations*, pp. 257-61).

[43] There were, however, differences. There was probably, for example, no incense altar in the desert encampment. See Roland de Vaux, *Religious Institutions*, vol. 2 of *Ancient Israel*, trans. into English (New York & Toronto: McGraw Hill Book Company, 1965), pp. 409-10.

[44] Mede, for example, talks of the Temple courts at *Commentarius*, pp. 438-9, and Newton on the four animals uses Mede's explanation of the tribal banners in the desert encampment (*Observations*, pp. 258-60).

[45] For Newton's use of this feast to explicate the letters to the Asian churches see *Observations*, pp. 255-7; for the trumpets, pp. 265, 268-9.

[46] *Observations*, pp. 261-4, 309-10. De Vaux, citing Deuteronomy, observes, however, that the High Priest read the Law at the feast only once every seven years. See *Ancient Israel*, 2.498.

saved, one for the damned, and one for those not yet judged. On this day they sealed them up (*Observations*, p. 265-7.)

Like other followers of Mede, Newton had to struggle with contradictions at the heart of the synchronic method. Also like Mede he failed, but his efforts did eventually produce his one original specific reading, that of the woman in the wilderness.

As we have seen, Mede had to misread John's stories of women and monsters which he presents in the second half of his prophecy. Mede mixed up the dragon and the beast, and such a confusion was critical for Protestant polemic, since commentators normally assumed that the beast and the whore of Chapter Seventeen together signified papal Rome but had other readings of the Dragon. Contradiction here would undermine an identification which had been basic to Protestant thinking since Martin Luther's time. More struggled with this confusion, so did Newton, but he eventually looked more to the women in the stories than to the monsters.

John has two or three women take part in these sequences. He first presents the woman with child who flees to the wilderness in Chapter Twelve. Some chapters later he sees in vision the whore who rides a scarlet beast and is ultimately destroyed by its horns (Rev. 17-18). Finally, he compares the New Jerusalem, which descends from heaven after this beast has been destroyed, to a bride (Rev. 21: 9-11). None of these figures overlap.

Protestant commentators had a standard reading of the first two women. They opposed them as the true church or faithful remnant, chased into the wilderness during the Middle Ages, and her enemy, the false church or Rome. They, therefore, read the two women as contemporary with each other, though they are not presented together in the Apocalypse. Spenser made this interpretation familiar to English readers when he contrasted Una in the wild, traveling about with a lion and trying to teach aniconic religion to satyrs and fauns, and the whore Duessa, who rides the seven-headed monster in Orgoglio's castle.[47] The young Newton subscribed to this opposition. In his symbol system and in his specific readings he equated the woman with the true church and the whore with the adulterate church, fallen into idolatry.[48]

In the Apocalypse a monster links the two women. In Chapter Thirteen the dragon, frustrated in its pursuit of the woman, who took refuge in the wilderness, calls a composite beast out of the sea. It has the body of a panther, the feet of a bear, a lion's mouth, seven heads with names of blasphemy, and ten horns (Rev.12:17-13:2). This beast wars with the followers of the exiled woman. Later John sees the second woman or whore

[47] For Una, see *The Faerie Queene* 1.3 & 6; for Duessa *TFQ* 1.7.16-18; 8.6.12-15, 25.

[48] *Trattato*, 1, 27r, 42; 1.1, 20r-22r, 86-88; 1.1, 23v, 92; 1.3, 8r, 128; 1.3, 40r, 208.

ride a scarlet beast, which likewise has seven heads, ten horns, and names of blasphemy (17:3). By the method Mede and Newton used to decode images, the two beasts must have the *same* signification. Now John identifies the scarlet beast with Rome, therefore the sea beast must have the same meaning.[49]

This Protestant reading had ancient roots. Already at the beginning of the fourth century, the martyr bishop Victorinus had identified the woman in exile with the faithful remnant.[50] Later in the same century Tyconius made an analogy to Exodus.[51] The woman who flees from the dragon to the desert resembles the Israelites who left Egypt for the Sinai.[52] Many Protestant commentators, including Mede, accepted this old analogy,[53] and even their enemy, the Jesuit Ribera, allowed it, since he stressed patristic understandings of the Apocalypse.[54]

This Protestant application of these symbols soon became fixed. Commentators took the earlier part of the woman's story, her birth pangs, the menace of the dragon in heaven, and equated them with events which immediately preceded the Middle Ages. Mede, for example, connects the birth of the child to Constantine, who stopped the persecution of Christians, and sees the war in heaven as the battle between paganism and Christianity in the Roman Empire.[55]

Newton accepted the essentials of this reading. The heavenly war represents persecution, and Michael's defeat of Satan the Christian victory over paganism and its idol worship.[56] In the *Observations* he made his agreement with Mede exact. The woman in heaven is the true church, and the birth of the child signifies Constantine's victory (*Observations*, pp. 279, 316).

[49] Richard Bauckham notes that these two chapters, Rev. 13 and17, became especially popular by the middle of the sixteenth century and would often get monographs by themselves. See *Tudor Apocalypse*, pp. 102-3.

[50] Pierre Prigent, *Apocalypse 12: Histoire de l'exégèse* (Tübingen: J. C. B. Mohr {Paul Siebeck}, 1959), p. 8.

[51] *Ibid.*, p. 16. Ambrosius Autpertus carried this analogy into the medieval period. He composed his commentary between 758 and 767 (pp. 24-5).

[52] Ambrosius Autpertus pointed out that the dragon was a type of Pharaoh (Prigent, *Apocalypse 12*, p. 25). Isaiah had compared the dragon with multiple heads to Egypt at 27:1 as did Ezekiel 29:2-4.

[53] In his *Eicasmi* (1596), Foxe used the Exodus analogy for the faithful remnant in the Middle Ages, as did David Pareus in 1618. See Prigent, *Apocalypse 12*, pp. 12, 59, 62. For Mede see his *Commentarius* in *The Works*, pp. 496-7.

[54] Ribera, *In...Apocalypsin commentarii*, pp. 244-5.

[55] Mede, *Commentarius* in *The Works*, pp. 493-6. Foxe and Brightman read the heavenly war in the same way, and Pierre Prigent cites many Continental parallels: Artopeus (1549), Osiander (1580), and Kromayer (1662). See *Apocalypse 12*, pp. 58-9, 64.

[56] *Trattato*, 1.3, 10ar, 132; 10r-11r, 136; 11ar, 138; *Theological Manuscripts*, p. 125; *Observations*, p. 23, 279-80.

With the beast, which Newton identifies in conventional fashion with Rome,[57] he nevertheless falls into the old confusion between the dragon and the beast which had affected Mede's reading. He begins even more mixed up than Mede. Since in his reading the beast is a kingdom (the Western Roman Empire), derived from the dragon's kingdom (initially the whole Roman Empire), the beast becomes one of the heads of the dragon. More exactly it is part of the seventh head, which the dragon and the beast share together.[58] Meanwhile the dragon, once it is limited to the Eastern Roman Empire, becomes one of the beast's horns![59] In the *Observations* Newton dropped these claims, though he kept the basic analysis which required them, but he fell into another confusion when he claimed that the dragon, trying to drown the woman, must be the Western not the Eastern Empire.[60]

His reading of the whore on the other hand remained stable, as one would expect in Protestant polemic. The angel identifies the beast she rides with Rome (Rev. 17:9), and both Newton and Protestant commentators generally applied the figure to papal rather than to pagan Rome (*Observations*, pp. 282-3, 323). In his image series Newton had said riding a beast signified rule and that the purple and scarlet clothes which the whore wears (Rev. 17:4) meant royalty (*Observations*, p. 22; *Theological Manuscripts*, p. 124.)

Newton changed his reading of the woman in the wilderness, however, fundamentally between his tracts composed in the 1670s and the posthumous *Observations*. He had begun, as we have seen, making the standard identification of the woman with the true church, but already polemic and his system of synchronisms had made problems. He originally equated the woman clothed with the sun, who had the moon at her feet, not only with glory, the meaning required by his lexicon of symbols,[61] but also with righteousness. He has the moon stand for this because it shines with borrowed light. Such a reading fits Protestant theology *exactly* but has no precedent in his discussion of symbols, where he has the moon symbolize the person next in dignity to the king or sun.[62] He later opted for Mede's

[57] William Fulke had identified the Babylon of Rev. 14:8 and 17 with Rome in his Hampton Court sermon of 1570, which went through many editions (1570, 1571, 1572, 1574, 1579, 1580) and solidified in England an already common identification. See Bauckham, *Tudor Apocalypse*, pp. 103, 106, 146.

[58] *Trattato* 1.3, 9r-v, 128-30; 1.3, 24r & 22v,174-76.

[59] *Ibid..*, 1.3, 31r-33r, 188-90. Since the horns of the beast are kingdoms ultimately derived from the dragon, and the dragon and the beast coexist at Rev. 16:13, then the dragon must become one of the beast's horns.

[60] *Observations*, pp. 277, 281, 315-7.

[61] *Trattato*, 1.1, 23v, 92.

[62] *Ibid.*, 1, 24r, 38. He got this interpretation from Ahmad ibn Sirin. See 1.1, 2r, 52-54. In McLachlan, *Theological Manuscripts*, p. 123, Newton has the moon symbolize the common people of a state.

reading of the moon as Israelite ceremonies,[63] but this too violated his lexicon of symbols.

More troubles came with his attempt to fit the woman's exile into his synchronisms. She flees from the dragon into the wilderness, but the monster vomits a flood, vainly trying to destroy her. It then calls up from the sea a beast, which wars with the seed of the woman and which the whore eventually rides.[64] John gives temporal measurements for the operation of these characters. The woman remains in the wilderness for a time, and times, and half a time (Rev. 12:14), and the beast, cured of its head wound, has power for forty-two months (Rev. 13:5). In Newton's system both measurements refer to the same period. Textual order, however, undercuts this standard equation. The woman remains in the wilderness only during the beast's rule,[65] but, since John puts her there before the dragon calls up the beast, Newton must insist that the woman was *still going* into the wilderness, when the dragon sent the flood, and that afterwards she went still further in. Only then can the time count begin, once the beast has come and survived a mortal wound. Newton in effect rewrites the text to fit his system.

The older Newton tried to resolve this dilemma but still could not avoid adding to the text. He begins by identifying the woman of Chapter Twelve with the whore of Chapter Seventeen, an equation which would have shocked English readers. Newton argues that, when she flees, she leaves behind the faithful remnant (*Observations*, p. 279), but in the standard reading she *is* that remnant. Then Newton substitutes for the Exodus analogy a journey across the desert to *Babylon*, the name of the whore (Rev. 17:5). So the woman becomes the whore. Newton here reads his interpretation into the text, since John has the woman remain in the desert, not cross it to Babylon on the Euphrates.

To make this reading Newton drew on the most influential of the late medieval commentators, Nicholas of Lyra, who equated the desert with Greece during the wars between Heraclius and Chosroes. The church fled west because the Persians devastated Syria and destroyed Christian churches.[66] Newton adapts this equation to Protestant tradition, substituting Babylon for Greece. Since Babylon on the literal level signifies Rome on the

[63] Newton, *Observations*, p. 279. For Mede see his *Commentarius* in *The Works*, p. 492. The equation of the moon with Israelite cult is very old. Oecumenius had already made this identification in the sixth century. See Prigent, *Apocalypse 12*, pp. 12, 29.

[64] *Trattato* 1.1, 19r, 82; *Theological Manuscripts*, p. 124; *Observations*, p. 22.

[65] *Trattato* 1.3, 40r, 208.

[66] Nicholas in turn simply adapted the reading of Alexander of Bremen, who applied the story to the same war but narrowed the symbolism of the desert to Constantinople. See Prigent, *Apocalypse 12*, pp. 45, 47.

allegorical level, he argues that the wilderness symbolizes the Western Empire, which is spiritually barren (*Observations*, pp. 273, 280.)

Newton thus makes out of old ingredients a completely new and strange reading. Yet it grows out of his method. He identifies all three women with the church, but they represent different stages in its history. The woman in heaven is the true, primitive church. In the desert she becomes the corrupt Roman church of the medieval period. She will end presumably as the bride or be replaced by her, coming down as the New Jerusalem from heaven.

This reading indicates that despite his early protestations Newton let the external history determine his reading of the text. The early Newton had warned the commentator not to force the text to fit the history but to choose things from history suitable to the text.[67] Here, however, Newton does just the opposite. His new reading shows admirably how the early uncorrupted church became the adulterate medieval church, a transformation aptly symbolized by the woman becoming the whore. At the same time it violates the text more clearly than did his earlier reading. John never made Babylon the goal of the fleeing woman.

Newton's most important contribution to apocalyptic commentary, however, did not depend so much upon an original interpretation as on the way he used Mede's synchronisms. That in turn had its basis probably in the problems millenarians faced after the Restoration.

Millenarianism had already been a main cause for the disrepute Apocalypse had received in late antiquity. Dionysius of Alexandria made his influential analysis of the book as part of an argument with a millenarian, a fact Newton recognized. In the *Observations* he connected the bad reputation of Apocalypse in later antiquity to early millenarians like Papias and Justin Martyr (*Observations*, pp. 247-49.)[68] Even a recent literalist like Franciscus Ribera backed away from a millenarian reading, contradicting his own method.[69] The matter was especially important, since Mede and his followers were millen-arians.[70] After 1660 Mede's millenarianism connected him to revolution.[71] Millenarianism had already moved to the social fringes

[67] Rule 12 in the *Trattato* 1, 16r, 30.

[68] Mede had based his argument for a millennial reading on Justn Martyr. See his *Commentarius* in *The Works*, pp. 533-4.

[69] He instead followed Augustine and argued that the millennium was the whole time until Antichrist comes. David Pareus points out the confusion, showing that Ribera's interpretation of the first resurrection, which begins the millennium, as the immortality of souls departs from the literal (*Commentary ... upon Revelation*, pp. 507, 520).

[70] Sarah Hutton has said that Mede could be called the founder of a Cambridge school of millenarianism. See "More, Newton, and the Language of Biblical Prophecy," p. 39.

[71] Murrin, "Revelation," p. 125. Katharine Firth remarks that his editor and biographer, John Worthington, tried in vain "to disentangle the thought of Joseph Mede from his reputation as a prophet of the Revolution." See her *Apocalyptic Tradition in Reformation Britain 1530-1645* (Oxford: Oxford University Press, 1979), p. 245. I do not enter here into the refinements of millenarian theory. For those

after 1653[72] and was connected with the Fifth Monarchists, who advocated violent social change.[73] Newton, accordingly muted his own millenarianism in the tracts he wrote on Apocalypse in the 1670s.[74] He has the millennium, only he never calls it that. He argues that the palm-bearing multitude plays its role in history from the resurrection of the witnesses to the utter ruin of the beast.[75] They make up, therefore, part of the millennial kingdom, but a casual reader might miss the fact. Newton's strategy here is rhetorical, since he could be quite explicit about his millenarianism, when he wished. The discussion of the New Jerusalem, which he composed in the following decade, a version of which Frank Manuel has edited, provides a good example.[76]

The old Newton, however, while he still believed in the millennium, no longer expected it to come soon. He put it off to the twenty-first century[77] and made no mention of it in the *Observations*. Glancing at other millenarians, he remarks: "The folly of Interpreters has been, to foretel times and things by this Prophecy, as if God designed to make them Prophets." (*Observations*, p. 251.) Newton thinks that such rashness has brought the Apocalypse into contempt. God wished the prophecies to serve a different purpose: "He gave this and the Prophecies of the Old Testament, not to gratify men's curiosities by enabling them to foreknow things, but that after they were fulfilled they might be interpreted by the event, and his own Providence, not the Interpreters, be then manifested thereby to the world." (*Observations*, p. 251.) So commentators must content themselves "with interpreting what hath been already fulfilled." (*Observations*, p. 253.) In effect, as Westfall observes, the apocalyptic had vanished from the Apocalypse.[78] Despite his millenarianism Newton sounds more like the orthodox Lutheran reaction against millenarianism which developed

---

interested, William Lamont succinctly explains the differences between pre- and post-millenial expectation. See *Godly Rule: Politics and Religion, 1603-60* (New York: Saint Martin's Press, 1969), p. 7.

72 Paul Christianson, *Reformers and Babylon: English Apocalyptic Visions from the Reformation to the Eve of the Civil War* (Toronto: University of Toronto Press, 1978), p. 244.

73 B. S. Capp, *The Fifth Monarchy Men: A Study in Seventeenth-century English Millenarianism* (Totowa, N. J.: Rowman and Littlefield, 1972), pp. 14, 56, 131, 231. He points out, however, that they only engineered two minor risings (p. 132). Their violence was rhetorical (p. 134). Lamont thinks that the Fifth Monarchy men and the ranters were symptoms of the decline of millenarianism (*Godly Rule*, pp. 106-7).

74 Manuel indicates that Newton shared with Henry More and John Spencer at Cambridge a revulsion for the fanatical enthusiasts of the 1640s and 50s. See *The Religion of Isaac Newton*, p. 87.

75 *Trattato* 1.3, 56r-57r, 236-38.

76 See Appendix B in Manuel, *The Religion of Isaac Newton.*

77 To the year 2060 or later. See Westfall, *Never at Rest*, pp. 816-7.

78 Westfall, however, is talking of Newton's early works. See *Never at Rest*, pp. 329-30.

especially during the Thirty Years' War, [79] though in English circles this viewpoint was commonly attributed to Calvin.[80] Newton's decision had consequences.

Misgivings about millenarian readings may account for Newton's treatment of the vials series. Since Luther the vials had been the focal point for Protestant speculation about the present and future.[81] Interpreting the vials was linked to expectations of the Final Judgment, soon to come. Commentators differed among themselves under which vial they were living, though the fourth was one of the most popular choices.[82] Brightman expected it to come soon,[83] and Mede hoped they were living under it.[84]

Newton in contrast connected them to past history. Already in the 1670s he synchronized the vials and the trumpets, showing by a careful set of parallels how the one series duplicated the other.[85] This was new, as Robert Iliffe has pointed out,[86] and, since the trumpets referred to heathen wars against the Christian Roman Empire, the vials too no longer concerned the present and future. In the *Observations* they almost disappear, like the millennium, but what Newton gives reaffirms his old reading.[87] He reads the third vial, which turns rivers and fountains to blood, for example, as Honorius' attack on heretics in the early fifth century (*Observations*, pp. 298-300.)

[79] Robin Bruce Barnes traces the Lutheran attack on millenarian expectations back to Alexandrus Utzinger (1589) and argues that orthodox Lutherans in effect destroyed apocalyptic speculation during the Thirty Years' War. See *Prophecy and Gnosis*, pp. 232, 241-7, 254-60.

[80] Calvin stressed the hiddenness of the Last Day, and the Second Helvetic Confession (1566) expressly rejected millenarianism. See Barnes, *Prophecy and Gnosis*, pp. 32-3; Lamont, *Godly Rule*, pp. 22-3. Bodin attributed to Calvin a statement which he may or may not have made but which circulated widely: "I thoroughly approve the reply of Calvin, not less polished than sagacious, when he was asked his opinion about the book of the Apocalypse. He candidly answered that he was totally at a loss regarding the meaning of this obscure writer, whose identity was not yet agreed upon among the erudite. Similarly, I do not see how we are to relate the wild beasts and the image discussed by Daniel to these empires which flourish everywhere now-a-days and have flourished for so many centuries." (Firth, *Apocalyptic Tradition*, p. 197). Richard Bauckham considers Calvin's remark probably apocryphal (*Tudor Apocalypse*, p. 40, n. 9).

[81] Luther, for example, compared the frogs during the sixth vial to some of his attackers (Second Preface, pp. 407-8). Luther's reading was decisive for the tradition. See Barnes, *Prophecy and Gnosis*, pp. 41-4.

[82] Bale, however, classed the English Reformation and the dissolution of the monasteries under the sixth vial. See Christianson, *Reformers and Babylon*, pp. 17-8. Writers on the Protestant left similarly liked to move the time schedule ahead, stressing the fifth and sixth vials (pp. 64-5, 93, 140-1, 231, 238). Tillinghast, though he thought the angels had reached only the third vial by 1654, nevertheless expected the millennium in 1656! See Capp, *Fifth Monarchy Men*, pp. 105, 192-3.

[83] Firth, *Apocalyptic Tradition*, p. 170.

[84] *Commentarius* in *The Works*, p. 483; Murrin, "Revelation," p. 128.

[85] *Trattato* 1.3, 1r-4r, 2v & 3v, 112-20.

[86] Iliffe, "'Making a Shew'," p. 63.

[87] *Observations*, pp. 275, 295.

Henry More sensed what was at stake and after some conversations with Newton attacked the synchronization of the vials and the trumpets.[88] The implications of this synchronization were indeed far-reaching. If the vials concern events which occurred more than a thousand years previously, then almost the whole book no longer involves the present, and what applies to the future Newton has already stated cannot be interpreted. Newton here replicates, though with a different method and set of interpretations, the anti-prophetic readings started by the Jesuit Alcazar at the beginning of the seventeenth century and reworked for the Protestant north by Grotius and his disciple Hammond.[89] Now Mede had specifically devised his interpretation to counter such readings and keep alive the prophetic status of Apocalypse. The student took the master's method, applied it to the same text, but realized finally that the method itself required him to kill the prophecy.

With the disappearance of prophecy polemics also vanish. While Newton assumes the old exterior historical schema, devised by Protestants to attack first the Pope and then each other, and while he expresses violent anti-Catholic sentiments elsewhere, in his commentaries on Revelation Newton refrains from name-calling.[90] David Pareus, the influential Calvinist commentator on Revelation in the early seventeenth century, provides a contrast. Pareus considered himself a moderate and yet insisted that the Pope and his Cardinals would be tossed into hell fire *alive* after Armageddon, though he allowed their followers simply to die.[91] Newton is content to leave such applications to inference.

This reticence again relates to his method. The equation of the cosmos with the state which underlies Newton's approach to prophetic symbols and which he develops at great length and in detail[92] and which informs his interpretation of Apocalypse leads him despite his love of prophecy and his fervent Christian beliefs to a basically political and secular reading of the

[88] In his *Exposition of Daniel* (1681). See Iliffe, "'Making a Shew,'" pp. 69-71.

[89] Like modern academic critics Alcazar read Apocalypse in terms of events which occurred in the first century of our era. Grotius and Hammond, who adopted Alcazar's approach, in effect destroyed the Protestant way of reading the book. They limited Revelation, except for prophecies of the millennium, to the times before Constantine. For them the vials signify the victory of Christianity over pagan Rome. See Firth, *Apocalyptic Tradition*, pp. 246-7. Mede's pupil, Henry More, attacked Grotius constantly and bitterly. As he complained, in Grotius' approach prophecy disappears, since the events signified all happened long ago. See, for example, *An Explanation of the Grand Mystery of Godliness; or, A True and Faithful Representation of the Everlasting Gospel of our Lord and Saviour Jesus Christ, the Onely Begotten Son of God and Sovereign over Men and Angels* (London, 1660), pp. 182-4.

[90] For the schema see Murrin, "Revelation," pp. 132-3. Bale may have been the first to use it in England. See Bauckham, *Tudor Apocalypse*, pp. 69-70, 95; Christianson, *Reformers and Babylon*, pp. 15-6. Christianson stresses the polemical function of this historical pattern, p. 5.

[91] Pareus, *Commentary upon Revelation*, pp. 494-5.

[92] *Observations*, pp. 17-23; *Theological Manuscripts*, pp. 120-6.

book.[93] Such an exegesis, of course, would fit the emerging Enlightenment culture of the eighteenth century, but some Protestants might have said that Newton paid too high a price for it. So many in the previous two centuries had read Apocalypse with fascination and hope, wrote about it, and argued heatedly over its interpretation. Newton's Apocalypse, secularized and cut off from the present, would arouse no such excitement. What Alexander Pope said of the scholastics could as well apply to Revelation and its interpreters:

> Once *School-Divines* this zealous Isle o'erspread;
> Who knew most *Sentences* was *deepest read*;
> Faith, Gospel, All, seem'd made to be *disputed*,
> And none had *Sense enough to be Confuted*.
> *Scotists* and *Thomists*, now, in Peace remain,
> Amidst their *kindred Cobwebs* in *Duck-Lane*.
> (*An Essay on Criticism*, 440-45)

Newton unwittingly sent Apocalypse off to join the scholastics, collecting dust in used book stores.[94]

93 There are exceptions. Newton argues that the seven letters relate to church conditions in the period covered by the fifth and sixth seals (*Observations*, pp. 256, 286). The fifth seal concerns the Great Persecution under Diocletian and Maximian (pp. 256-7, 278) and the apostasy begins under the seventh seal (pp. 256-7).

94 In his note, Pope indicates that Duck-Lane was a place near Smithfield, where old and used books were sold. I cite from *The Poems of Alexander Pope*, ed. John Butt, (New Haven: Yale University Press, 1974).

ESSAY 11

Newton and the Guaranteeing God

G. A. J. ROGERS
University of Keele

INTRODUCTION

It used to be thought, and often still is thought, that the Problem of Induction was discovered by Hume and nobody since has solved it. The problem, as characterised by one philosopher, is that of the "validity of judgements about future or unknown cases."[1] Famously it and similar problems reduced Hume himself to intellectual despair. Bertrand Russell in one of his short stories even preserved a special place in hell for all those philosophers who have claimed to solve the problem. In this inductivist's hell all predictions as to how things would behave made on inductive grounds are always falsified. Until, that is, the person in desperation, attempts to infer counter-inductively, at which point objects behave according to inductive principles, and so on.[2] Since induction lies at the heart of the scientific enterprise, and since Hume has called it into question, the argument goes, then science itself verges on the edge of incoherence or, perhaps more correctly, crashes ineluctably into a chasm of irrationality.

Hume also called into question with similarly devastating results, or so it has often been alleged, all causal inferences. To infer from constant conjunctions to necessary connections is always flawed. But causal connections, if they exist at all, are necessary. So our belief that A is the cause of B can never be known. Hume is not now credited with having originated such insights into causal inference. Popkin and others have pointed out that Glanvill some fifty years earlier was aware of some of the

[1] Nelson Goodman, *Fact, Fiction and Forecast* (Indianapolis: Bobbs Merrill, 2nd edition, 1965) p. 59.

[2] *Nightmares of Eminent Persons, and Other Stories* (London: Allen and Unwin, 1954).

J.E. Force and R.H. Popkin (eds.), Newton and Religion, 221–235.
© 1999 *Kluwer Academic Publishers.*

problems.[3] But it is Hume who is rightly credited with having seen further and deeper into the issues than any of his predecessors. So, without taking anything away from Hume's achievement, we may wonder why the problems that Hume finds, or seems to find, in induction and causation only came to the attention of philosophers in the early eighteenth century. For, somewhat in line with Locke's comment about Aristotle and the syllogism that "God has not been so sparing to Men to make them barely two-legged Creatures, and left it to *Aristotle* to make them Rational,"[4] so we did not have to await Hume's arrival to infer inductively. But apparently we did have to await Hume to make us realize that there might be serious problems with such inferences. In this paper I wish to explore the question why it was only with Hume that this issue arose so effectively. I shall do so by focusing in particular on the place of Newton in this story, because, for reasons which will soon emerge, I believe he plays a special role in it, a role closely connected with his theological beliefs. Let us, therefore, begin with a preliminary look at the place of induction in Newton's thinking. We can do so with a further consideration also in mind. It is that the place of induction in Newton's philosophy of science is but part of his wider response to sceptical worries about the possibility (or perhaps even the desirability) of knowledge. It therefore fits with the wider response to sceptical issues that is an important feature of late seventeenth century empirical science. I shall want to say something about those wider issues as we proceed.

There were other important contributory reasons why the problem of induction did not feature in the pre-Humean world which I shall not discuss but which I shall just mention, at least partly to indicate that the complete story is much more complicated than I shall consider here. These other factors include the two following.

One important change that occurred in the world of learning in the late sixteenth and early seventeenth century was the gradual abandonment of the idea that academic institutions existed only to preserve and consolidate a tradition of learning already in existence. In looking to new ways in which to expand human knowledge new questions about methodology and their epistemic implications were soon generated.

Another important factor that interacted with the search for new methods was the impact of the new or newly recovered sceptical arguments against claims to knowledge and the varying attempts made to defeat them or set them aside.

[3] Cf. Richard H. Popkin, "Joseph Glanvill: A Precursor of David Hume," *Journal of the History of Ideas* 14 (April, 1953), pp. 292-303, reprinted in Popkin's collection *The High Road to Pyrrhonism*, (San Diego: Austin Hill Press, 1980, reprinted Indianapolis: Hackett Publishing Co., 1993), pp. 181-95. Popkin considers earlier writers who have noticed similarities between Glanvill and Hume.

[4] *An Essay Concerning Human Understanding*, ed. P. H. Nidditch (Oxford: Clarendon Press, 1975), Book IV, Ch.XVII, §4, p. 671.

A third important consideration is linked to the recovery of ancient atomism and the deep epistemic problems which appeared to flow from them. One central one in perception was the acceptance of the atomist's distinction between the primary and secondary qualities of objects and the supposed lack of resemblance between the experiences or ideas of the observer of that object and the causes of those ideas in the object itself, the colourless particles of classical atomism. Although the argument of this paper is often closely related to all three of these further considerations, I shall not, for reasons of time, attempt to elaborate on them further.

NEWTON AND INDUCTION

Newton's philosophy of science is embodied in the four *Regulae Philosophandi* at the beginning of Book III of the third edition of the *Principia*. It is now well known that it took some time for these rules to take their final form and it would be inappropriate to rehearse that story now.[5] I do, however, wish to stress that for all the variation in their draft formulations, throughout their history there is clear commitment to certain methodological considerations. Those considerations are based upon a certain, clear view about the nature of the world which I will shortly characterise.

In a draft version of Rule IV we have the following sentence:

> ...if arguments from hypotheses are admitted against inductions, then the arguments of inductions on which all experimental philosophy is founded could always be overthrown by contrary hypotheses. If a certain proposition drawn by induction is not yet sufficiently precise, it must be corrected not by hypotheses but by the phenomena of nature more fully and more accurately observed.[6]

Without entering further than we need into the complex waters of Newton's understanding of "hypothesis," we can surely draw from these few words the following implications:

1. Experimental philosophy (i.e. natural philosophy based upon observation) only advances through induction.

[5] Some of the now classic sources here are: Alexandre Koyré, *Newtonian Studies* (London: Chapman and Hall, 1965), Chapter VI, "Newton's Regulae Philosophandi"; J. E. McGuire, "Atoms and the Analogy of Nature: Newton's Third Rule of Philosophizing," *Studies in the History and Philosophy of Science*, I, 1970, pp. 3-58, reprinted in J. E. McGuire, *Tradition and Innovation. Newton's Metaphysics of Nature* (Dordrecht: Kluwer Academic Publishers, 1995, pp. 52-102; and M. A. Finocchiaro: "Newton's Third Rule of Philosophizing: A Role for Logic in Historiography," *Isis* 65, March, 1974, pp 66-73.

[6] I quote from the texts given by Koyré in "Newton's Regulae Philosophandi."

2. The phenomena of nature are in some sense already fixed. What they are is independent of our minds.

3. Our knowledge of them may be improved by our observational techniques.

4. We should not make use of hypotheses in determining the phenomena.

When we combine these commitments with the draft and never published Rule V we see that for Newton the implication is that there can be no legitimate kind of natural philosophy other than the experimental one. In Rule V he writes:

> Whatever is not derived from things themselves, whether by the external senses or by the sensation of internal thoughts, is to be taken for a hypothesis.... And those things which neither can be demonstrated from the phenomenon nor follow from it by the argument of induction, I hold as hypotheses.[7]

And in the "General Scholium" we have the definitive judgement on hypotheses in natural philosophy:

> ... whatever is not deduced from the phenomena is to be called an hypothesis, and hypotheses, whether metaphysical or physical, whether of occult qualities or mechanical, have no place in experimental philosophy. [8]

In the course of advising Cotes on the wording of the "General Scholium" added to the second edition Newton drafted a letter to him which was never sent. Instead he wrote a shorter and much less revealing one. The longer, unsent, letter, whilst mostly reenforcing familiar methodological points, also offers insight on some issues that rarely feature in Newton's public pronouncements on method. In particular it makes a point of contrasting the "manner of Philosophizing made use of in the Principia.... from the method of others."[9] The method Newton uses, he says, is one that proceeds by deducing things mathematically from principles derived from Phænomena by Induction. He goes on to illustrate the rejected, hypothetical, method:

> One may suppose that bodies may by an unknown power be perpetually accelerated & so reject the first law of motion [the principle of inertia]. One may suppose that God can create a penetrable body & so reject the

[7] Koyré, "Newton's Regulae Philosophandi," p. 272.

[8] *Mathematical Principles of Natural Philosophy*, Andrew Mott's translation revised by Florian Cajori (Berkeley and Los Angeles: University of California Press, 1962), p. 547. The "General Scholium" was added to the second edition of 1713.

[9] *The Correspondence of Isaac Newton*, 7 vols., ed. H. W. Turnbull, et al., (Cambridge: Published for the Royal Society at the University Press, 1959-77), 5:398. The draft letter is undated but the letter which was to replace it is dated 28 March 1713.

> impenetrability of matter. But to admitt of such Hypotheses in opposition to rational Propositions founded upon Phænomena by Induction is to destroy all arguments taken from Phænomena by Induction and all Principles founded upon such arguments. ... Experimental Philosophy reduces Phænomena to general Rules & looks upon the Rules to be general when they hold generally in Phænomena.[10]

Newton's method contrasts with hypothetical philosophy which "consists in imaginary explications of things & imaginary arguments for or against such explications." This latter, he says is followed "too much by Cartes, Leibnitz & some others."[11]

Newton, then, leaves us in no doubt about his commitment to induction as a method. But he also readily admits it is a method that does not generate certainty. As he explains it, induction moves from the particular to the general and looks upon the conclusion so drawn as a general truth even though the generality cannot be demonstrated a priori.[12] It is, therefore, as he put it in the last of the Queries to the *Opticks*, only probably true and "no Demonstration of general Conclusions."[13] So we see that Newton was prepared to settle for mere probabilities in much of natural philosophy. But he does so without any sense of the despair at the human condition that was later to characterise the views of Hume. And my question is why was this so? The answer, in short if vague terms, is because of Newton's understanding of the way things are with the world and with human beings. It was a way which did not leave space for the kind of sceptical worries that characterise much of Hume's thought. Precisely how Newton saw those is intimately connected with his religious beliefs, as we shall see. Before we turn to those, however, we need to explore the background to inductive and a priori inference as it was understood by earlier seventeenth-century figures of whom the most important are Bacon and Descartes.

BACONIAN INDUCTION

Bacon's vision for the advancement of learning was fired by the thought that *discovery*, not justification, was the objective of human enquiry into the natural world. It was new knowledge that he sought and it was a method to generate new knowledge that he expounded. That method was a new form of induction "which shall analyse experience and take it to pieces, and by a due

[10] *Ibid.*

[11] *The Correspondence*, 5: 399.

[12] *Ibid.*

[13] *Opticks*, Dover Edition, based on fourth edition, 1730, (New York: Dover Publications, 1952), p. 404.

process of exclusion and rejection lead to an inevitable conclusion."[14] The objective is not "argument," but "arts," and the method of induction "is that form of demonstration which upholds the sense and closes with nature." It begins from the particular and only gradually moves toward the general. It promises certainty instead of probability but only through a progressive process of investigation. Bacon recommended steering between overconfidence and a paralysing scepticism. "Those who have taken upon them to lay down the law of nature as a thing already searched out and understood ... have therein done philosophy and the sciences great injury," he wrote.[15] Equally the sceptics have taken things too far. There was a happy medium between the two extremes of "pronouncing on everything and the despair of comprehending anything." The empiricist base for this programme is combined with a confidence in its results not always echoed in the empiricist tradition which arose at least in part from Bacon's lead.

How, then, does Bacon justify his appeal to induction? Does he, indeed, offer any justification at all? In the *New Organum* his response is given in largely secular terms. Bacon offers no appeal to theological premises to justify his confidence. Rather, it at first seems to be based on the assumption that once the causes of error had been identified they can be overcome. Error was generated by, for example, the failure of the senses to give information (e.g. if the objects investigated are too small or too remote to see) or if the senses mislead us (e.g. illusion). But it is possible to overcome such difficulties by devising observational aids (e.g. telescopes) or experiments which force nature to reveal her secrets. Error, in other words is eliminable, not inherent in the human condition. This lack of overt theological connection in Bacon's account is something for which he is justifiably noted. He sounds distinctly modern.

Further reading, however, might give us pause from turning Bacon into a figure of the eighteenth century. To appreciate this we need only turn to *The Advancement of Learning*. We might begin by recalling its full title: *The Proficience and Advancement of Learning, Divine and Human*. It is soon apparent that in that work Bacon offers a far from secular justification for our confidence not only in the desirability of his programme but also of his method.

Bacon begins his justification of the objectives and method which he will propose with a rejection of a theological argument against the pursuit of knowledge based on Biblical texts. They are all ones that suggest that learning is corrupting or dangerous drawn mainly from Solomon or St. Paul.

[14] *The Great Instauration*, "The Plan of the Work," in *The Philosophical Works of Francis Bacon*, reprinted from the texts and Translations with the Notes and Prefaces of Ellis and Spedding, edited by John M. Robinson (London: Routledge, 1905), p. 249.

[15] Author's preface to *The New Organon*, Robinson ed., p. 256.

An example is the often cited phrase "knowledge puffeth up" from Corinthians I. Bacon responds with counter quotations. Solomon, he points out, also says that "The spirit of man is as the lamp of God, wherewith he searcheth the inwardness of all secrets."[16] The Bible is not advocating a rejection of the search for knowledge, Bacon argues. Rather it is warning us to contain that search within particular limits. These, he says, are three: we must not forget our mortality; we should use our knowledge for "repose and contentment and not distaste or repining"; and, thirdly, and perhaps most significantly, "that we do not presume by the contemplation of nature to attain to the mysteries of God."[17]

Bacon offers a variety of considerations drawn from theology and the history of the church to argue the desirability of searching for knowledge. They are summarised by him in the familiar form of the two books of knowledge which are he says given authority in the New Testament: "For our Saviour saith, *You err, not knowing the Scriptures nor the power of God*; laying before us two books or volumes to study, if we will be secured from error; first the Scriptures, revealing the will of God, and then the creatures expressing his power; whereof the latter is the key unto the former."[18] It is the "omnipotency of God, which is chiefly signed and engraven upon his works."

There is no reason to doubt Bacon's sincerity in thus characterising the theological justification for the activities of the natural philosopher. Granted the datum of Christianity as revealed through scripture then the activity of the natural philosopher was at least in part justified by its insight into God's power and thus reinforced the Christian understanding of the creation. Bacon made considerable use of the story of creation and Adam's first activities in support of his case. It was "not the natural knowledge of creatures but the moral knowledge of good and evil" that caused the fall.[19] Adam had indeed been set in the garden to delight in "exercise and experiment." Eden was the Cavendish Laboratory writ large!

Granted this theological framework for the activity of the natural philosopher it is scarcely surprising that the method is itself regarded as unproblematic. Why else has God created us with eyes and ears if not to contemplate his creation and to draw from that contemplation some understanding of the general truths which govern its behaviour?

There is a sense in which Bacon did secularise knowledge. He encouraged us to believe that knowledge of nature could not in the particular be inferred from knowledge of God. But he did not doubt that the activity of

[16] Robinson, p. 44.

[17] Robinson, p. 45.

[18] Robinson, p. 64.

[19] Robinson, p. 61.

searching for knowledge was sanctioned by the nature of God's relationship to his creation, and particularly in his relationship to human beings. It was more or less our duty to seek God's wisdom in his works. But nothing can be a duty that is impossible for us to do. If the study of God's works can reveal his wisdom then knowledge of his works must be possible. There is no theological space on which the sceptic may gain a foothold.

DESCARTES, GOD, AND NATURAL PHILOSOPHY

Descartes's invocation of the Deity to defeat the sceptic about the reliability of our senses is too well known to require much comment. It is, however, worth underlining that Descartes is committed to the position that we often do make false judgements about the physical world. For example when we are young or unreflective we assume that material objects have just the properties which they appear to have, such as colour, taste, and smell. Such beliefs are the prejudices of childhood, he tells us, but they can be avoided if we apply the criteria of clear and distinct ideas in the prescribed way.[20]

Like Bacon, Descartes is keen to stress that the fact that error is possible and indeed often occurs is not itself a reason to doubt that truth may be known as long as we recognise our limitations as finite beings. Descartes shares with Bacon, too, a clear desire not to be seen as claiming knowledge about the Deity and His intentions which exceed our capacities. It is in natural philosophy that both Bacon and Descartes see the main possibility of knowledge, and not nearly so much in natural theology.

If Descartes's God is the guarantor *par excellence* in the history of philosophy, we must remember also that according to Descartes's account error is still a very common, perhaps normal, feature of ordinary perception. Truth is something that has to be worked for and worked for hard. The important point is that it is obtainable because of the nature of God's relationship to his creation. The key to obtaining it is the application of the right method. Although Descartes differed from Bacon in what that method was, that we had access to such a method was central to the thought of both.

That there was such a common theme between the thought of Bacon and Descartes allowed Joseph Glanvill to classify them together as "Free Philosophers" who sought "Truth in the Great Book of Nature" but did so:

> with wariness and circumspection without too much forwardness in establishing Maxims, and positive Doctrines: To propose their Opinions as *Hypotheses* that *may probably* be the true accounts without peremptorily affirming that they are. This, among others, has been the way of those Great

[20] The most important texts here are probably those in the later sections of Part I and the early sections of Part II of the *Principles of Philosophy*.

> Men, the Lord Bacon and Des Cartes; and is now the method of the Royal Society of London.[21]

Whether every other Englishman would have recognised Descartes in that characterisation may be doubtful. Newton for one, as we saw earlier, seems not to have shared that vision of at least some of Descartes's work. But there can be little doubt that both Bacon and Descartes shared a common objective and thus a commitment to the possibility of achieving knowledge of nature which linked in the ways we have seen with their understanding of God's relationship with his creation.

GLANVILL AND RATIONAL BELIEF

Joseph Glanvill was one of those who found Descartes's theistic appeal unconvincing (Gassendi was another) even though he took over Descartes's terminology to set out the epistemic issues. Turning from scepticism to certainty—"a subject of both difficulty and importance"—Glanvill distinguishes between "indubitable" and "infallible certainty."

The first includes intuitive certainties such as "I think," "I exist." It also includes various ideas that I have, for example those of God, heaven, earth, etc., the traditional laws of thought, and so on. We assent to these as soon as we understand them.

Glanvill also includes within this list the Cartesian criterion of knowledge, viz, "That what our understandings declare of things *clearly* and *distinctly perceiv d* by us, is truly so, and agreeing with the realities of things themselves" (p. 48). But he offers no proof of this: "This is a Principle that we *believe* firmly but *cannot prove*, for all proof and reasoning supposeth it." Descartes in attempting a justification is "out in his method; when from the *Idea's* he endeavours to prove that *God is*, and from *his* Existence that *our Faculties are true*... So that, that great Man seems to argue in a Circle" (p. 48). So Glanvill finds it no problem at all to allow that our proofs come to a halt at this point. It is, of course, not that Glanvill doubts that God does indeed guarantee our intuitions here. But he also thought that we could not establish that by appeal to a demonstrative proof of his existence.

Many claims made on the basis of experience are also included in this first kind of indubitable certainties. Amongst these are the existence of matter and motion in a great variety of modifications. We learn of these and their like by sense, he tells us, and he continues in words which may remind us of David Hume many years later: "... and we nothing doubt here;

[21] *Essays on Several Important Subjects in Philosophy and Religion* (London, 1676), Essay II, "Of Scepticism and Certainty," p. 44.

although the Theory of Speculative considerations of those matters be full of difficulty, and seeming contradiction" (p. 48). In other words although we cannot find a theoretical justification for our sense-based beliefs about particular objects around us we nevertheless cannot doubt their existence even though we have no proof that things are as they appear: "... this fullness of assent is all the certainty we have, or can pretend to; for after all, tis possible our Senses may be so contrived, that things may not appear to us as they are: But we fear not this, and the bare possibility doth not move us" (p. 49). In other words the speculative hypothesis that things may be otherwise fails to disturb our belief that they are not. *Hypotheses non fingo!*

Glanvill then tells us about his second kind of certainty—"Certainty Infallible." To have this we have to be assured that it is impossible things should be otherwise. But this state of affairs is impossible for human beings to attain: "our Faculties may be so contrived as always to deceive us in the things we judg most certain and assured." And here Glanvill plays the religious card in a way quite different from Descartes. This latter form of assurance, he tells us, "... perhaps is proper only to Him, who made all things what they are; and discerns their true natures by an infallible and most perfect knowledge" (p. 50). In other words, only God can have such infallible knowledge. But this, he says, should not worry us. The bare possibility of being mistaken moves us not, for nobody can really doubt of the first principles that he has illustrated. Glanvill never actually justifies this by an appeal to the Creator but the implication is absolutely clear and in a later essay he claims that the being of a God is proved by reason, and we know what is true because of principles of reason that God has implanted. This might tempt us to accuse Glanvill of also being committed to the same circular argument for which he had dismissed Descartes.[22] For if we ask Glanvill why we may trust our reason it looks as though his answer is ultimately a kind of Platonic one. And indeed, for all his links with scepticism this is what he gives us. For although we must be careful not to fall into error, which it is very easy to do, we can avoid this if we make use of "Those principles of Truth which are written upon our Souls."[23] We are reminded that Glanvill was an admirer of Henry More. And the author of those truths is of course the deity: the light of reason he tells us which has a great congruity with "those Laws, that God hath placed in our Souls ."[24] Of course this last stage in the argument looks circular. But it is important to notice that the structure of Glanvill's argument is different from that of Descartes's. For, given Descartes's demand for the highest standards of

[22] Cf. *Essays on Several Important Subjects*, Essay 5, "The Agreement of Reason and Religion," especially pp. 6-7.

[23] "Agreement of Reason," p. 17.

[24] "Agreement of Reason," p. 6.

proof before he will take one step into the external world, the proof of God's existence must achieve that standard. Glanvill, on the other hand, only requires that there is evidence that there are physical objects, or whatever, around us, not infallible proof. Whilst the standard is less demanding the chances of success are that much the higher and, since Glanvill does not need to settle for absolute certainty in his conclusions, that they do not meet those absolute standards is of no consequence.

So the claim is that Glanvill's thought has the following construction: the senses provide us with empirical evidence about how things behave. It is perfectly possible that they will not continue to behave in the future as they have in the past. But the bare possibility is not to be taken seriously unless there is any evidence—inductive evidence—to suggest that things are going to change. Although nothing is infallibly certain for human beings we have no reason to suppose that objects will behave differently in the future from the way they have behaved in the past. We can, for good measure, add the fact that both reason and empirical evidence provide overwhelming argument for the existence of a Deity. Such a Deity would not deliberately set out to deceive us.

BACK TO NEWTON

That natural religion took enormous support from the developments in seventeenth century science is a commonplace, and we are not surprised that Newton saw his system of the world as powerful argument for Providence. As he wrote to Richard Bentley, busy preparing the first set of Boyle lectures in 1692, "When I wrote my treatise about our Systeme I had an eye upon such Principles as might work with considering men for the beliefe of a Deity & nothing can rejoyce me more than to find it usefull for that purpose."[25]

Newton, like many of his English contemporaries, constantly draws religious conclusions from what he takes to be the facts of nature, whether in the form of the Argument from Design as exhibited in the workings of the eye, or argument that the infinite nature of God may be drawn from the infinity of space. Nor is it just the existence of the deity that is implied by these facts of nature. For from them we may also infer our moral duty. As he put it at the end of the Queries in the later editions of the *Opticks*, the true method in natural philosophy, the method of "arguing from Experiments and Observations by Induction," can lead to moral insight: "And if natural Philosophy in all its Parts, by pursuing this Method, shall at length be

[25] *The Correspondence*, 3:233.

perfected, the Bounds of Moral Philosophy will also be enlarged." And he goes on to explain how this will be:

For so far as we can know by natural Philosophy what is the first Cause, what Power he has over us, and what Benefits we receive from him, so far our Duty towards him, as well as that towards one another, will appear to us by the Light of Nature.[26]

It is clear from this that Newton sees the moral truths as supervenient on the facts of the creation and therefore capable of being discovered by empirical investigation. Once again, the suggestion is that natural philosophy provides a universal key to knowledge and presumably shares with that knowledge the kind of fallibility that Newton was prepared to concede that, in theory at least, the method of natural philosophy contained: it was, he allowed, no demonstration of its conclusions.

As we noted earlier, Newton's most famous remarks on method are the *regulae philosophandi*. Although these have generally been treated as remarks about investigating the natural world, we would do well not to assume that they have no message for theology. Seventeenth-century classifications of knowledge (Locke s, for example) often include in the "Knowledge of Things " both earthly and heavenly objects.[27]

Rule I reads: "We are to admit no more causes of natural things than such as are both true and sufficient to explain their appearances."

And in explication: "To this purpose the philosophers say that Nature does nothing in vain, and more is in vain when less will serve; For Nature is pleased with simplicity, and affects not the pomp of superfluous causes."[28]

There is much that has been written about this Rule, and no doubt much more that could be said. But I wish to concentrate primarily on Newton's appeal to Nature's simplicity.

We can begin first by noting that sometimes in the seventeenth and eighteenth centuries *Nature* is used as a synonym for God, or, if not as a synonym, then as something like the female counterpart of God. Locke, for example uses it in this way in the *Second Treatise of Government*, Sections 26 and 27. So that we can be sure that when Newton says that "Nature is pleased with simplicity", he believes that God is also.

It is not at all surprising that the Lincolnshire Puritan, which Newton undoubtedly was in many important respects, should admire simplicity. Indeed, it is perhaps not too far fetched to hear in that denunciation of "the pomp of superfluous causes" the full Calvinist hostility to the Church of Rome. But whether or not such explicit theological connections can be made

[26] *Opticks*, p. 405.

[27] See, for example, the last chapter of Locke's *Essay Concerning Human Understanding*, "Of the Division of the Sciences."

[28] *Mathematical Principles of Natural Philosophy*, p. 398.

with the Rule, there can be no doubting the importance of simplicity as a methodological principle in Newton's intellectual landscape and of its theological connections.

Evidence for this is to be found in Newton's theological manuscripts. In one set of these unpublished papers, he sets out rules for theology comparable to those of the *regulae philosophandi*. They are, however, much more extensive. Under the heading "Rules for interpreting the words and language in Scripture" he gives in considerable detail fifteen principles of guidance. One of these is on the place of simplicity. The relevant passage, somewhat shortened, is the injunction that we should set out:

> To choose those constructions which without straining reduce things to the greatest simplicity. ....Truth is ever to be found in simplicity, and not in the multiplicity and confusion of things. As the world, which to the naked eye exhibits the greatest variety of objects, appears very simple in its internall constitution when surveyed by a philosophic understanding, and so much the simpler by how much the better it is understood, so it is in these visions. It is the perfection of God's works that they are all done with the greatest simplicity. He is the God of order and not of confusion....[29]

Simplicity, then, is not for Newton some secular principle plucked from nowhere as a defence against the extravagances of alternative scientific hypotheses. It rather leads us directly back to Newton's understanding of God. It is indeed a bulwark against methodological scepticism, as Newton makes clear in later passages. Just as we are entitled to generalize by induction in the natural sciences, even though "the arguing from Experiments and Observations by Induction be no Demonstration of General Conclusions",[30] so we must follow a similar methodology in theology. And we must, because if we do not, conclusions reached by inductive generalization are liable to be "evaded by hypotheses".[31] The methodology which is appealed to here has its counterpart, even its origin, in the correct method for investigating the natural world, i.e. the natural sciences. Thus Newton argues that those who regard his construction of the Apocalypse as uncertain "on pretence that it may be possible to find out other ways" then "he is not to be regarded unless he shall show wherein what I have done may be mended".[32] It would be, in other words, to allow the inductively supported conclusion to be evaded by hypotheses. Newton's own interpretation is compared with a recognition that the parts of an "Engin made by an excellent Artificer" are correctly assembled "when he sees them

[29] The text is published in Frank E. Manuel, *The Religion of Isaac Newton* (Oxford: Oxford University Press, 1974), Appendix A. The passage quoted is on page 120.

[30] *Opticks*, p. 404.

[31] *Mathematical Principles of Natural Philosophy*, p. 400.

[32] Manuel, *The Religion of Isaac Newton*, Appendix A, p. 121

joyn truly with one another notwithstanding that they may be strained into another posture." Or, in another analogy, we take it that we have the sense of a person's words when we understand them according to the rules of grammar, even though they could in theory be given "some other harsher construction".[33]

Clearly the methodology of theology and natural philosophy meet here in one. But there is more to Newton's position than this. For its strength, if we may call it that, is that in it the theological premise of his thought provides him with a defence against the inductive sceptic—against Hume if you like. For the appeal to simplicity as a rational ground for the acceptance of an explanation about the physical world is given a theological sanction. Nature and God are inherently on the side of simplicity. To object to inductive inference is to deny the place of the deity in the scheme of things. It is interesting to remember that Descartes, in justifying the laws of nature, the principle of straight-line inertia, had appealed to exactly the same thing, the immutability and simplicity of God and his creation.[34]

Newton, then, was not at all dismayed by the fact that his method of natural philosophy did not lead to the discovery of any necessary truths of nature. Indeed, it was rather the reverse, for Newton's voluntarist theology allowed no such easy route to the identification of God's will.[35] But there was nothing seriously problematic about the laws of nature he had identified for all that, because the truths in question were the product of a God whose essence we could and did, at least in part, know, e.g., that God was omniscient and omnipotent. And our knowledge of his simplicity, and the simplicity of his hand-maiden, Nature, itself provided justification for inductive reasoning. So, on this reading of things, with Newton the method of science is itself justified through our theological understanding of the nature of God.

CONCLUSIONS

So we now have at least part of the answer to the question from which we began: why did the problem of induction not seem like a problem prior to

[33] *Ibid.*

[34] Cf. *Principles of Philosophy*, Part II, Principles 37-9.

[35] For something of the relevance of the voluntarist theology to natural science, see: G. A. J. Rogers, "Locke, Law and the Laws of Nature," in *John Locke*, ed. Reinhardt Brandt (Berlin and New York: de Gruter, 1981) and the papers cited there in relation to voluntarism from Ockham to the seventeenth century; James E. Force, "Hume and the Relation of Science to Religion Among Certain Members of the Royal Society," *Journal of the History of Ideas* 45, No. 4 (Oct.-Dec., 1984), pp. 517-36; and Margaret J. Osler, *Divine Will and the Mechanical Philosophy* (Cambridge: Cambridge University Press, 1995), passim.

Hume? Given the theological framework which the empiricist natural philosophers, of whom Newton is the paradigm, conducted their enquiries, it is because there was and could be no such insoluble problem. The story does not quite end there, as we all know. For lurking in the Newtonian picture of the world were several other primed bombs which required only the touch of Hume's torch to ignite them. One was Newton's assumed views on causation and another was his account of matter. These are bombs as yet not wholly defused, but they are another story.

ESSAY 12

Newton, the "Ancients," and the "Moderns"

JAMES E. FORCE
University of Kentucky

> But you, Daniel, shut up the words, and seal the book, until the time of the end. Many shall run to and for and knowledge shall increase.
>
> Daniel 12:4.[1]

1. THE "TWO NEWTONS" OF HISTORIOGRAPHY AND THE BATTLE OF THE BOOKS

In the judgment of many 18th- and 19th-century historians, Newton's life and work epitomize modern scientific progress. In this century, despite enormous contributions to our understanding of Newton's extensive knowledge of, and interest in, the ancient world, for many interpreters and laymen alike, Newton remains either the inaugural progenitor of "an unassailable primacy in mathematics, enlightened rationalism, and experimentation" or the "dramatic climax" of the modern scientific enterprise.[2]

Whether we rank Newton as the "First Mover" of the Scientific Revolution or as the teleological climax of the scientific revolution—the primary interpretations, as Professor Dobbs reminds us, of most historians today—we mean it as a compliment to Newton.[3] Allan Bloom unhesitatingly

[1] For Newton's view of the centrality of this text, see his *Observations upon the Prophecies of Daniel, and the Apocalypse of St. John.* In Two Parts (London, 1733), Part II, pp. 249-51.

[2] James E. Force, "Biblical Interpretation, Newton, and English Deism," in *Scepticism and Irreligion in the Seventeenth and Eighteenth Centuries*, ed. Arjo van der Jagt (Leiden: E. J. Brill, 1993).

[3] B. J. T. Dobbs, "Newton as Final Cause and First Mover," *Isis* 85, No. 4 (Dec., 1994), pp. 635-6. This was Professor Dobbs' "Distinguished Lecture" which she delivered to the Annual Meeting of the History of Science Society on Nov. 13, 1993, in Santa Fe, New Mexico. For a survey of this tradition, see (in addition to Dobbs) Richard Olson, *Science Deified & Science Defied: The Historical Significance of Science in Western Culture*, 2 vols., (Berkeley and Los Angeles: University of California Press, 1982), 2:Chap. 3.

J.E. Force and R.H. Popkin (eds.), Newton and Religion, 237–257.
© 1999 *Kluwer Academic Publishers.*

enlists Newton in his personal cult of the "greats" of our modern civilization while simultaneously explaining his idiosyncratic, classicist's view of the differences between ancients and moderns. In Bloom's view, essentially modern scientists such as Newton and Einstein share a rational, scientific, essentially modern "common universe of discourse." Newton is one of the intellectually respectable elite because of his "modern" scientific rationality. Color the right stuff of Newton's steely-eyed, essentially modern, scientific rationalism the blue of the wild blue yonder or the jet black of deep space. One refracted image of Newton as seen through the prism of anachronism, then, is Newton the heroic modern scientist who, with other scientific hall-of-famers (Darwin, Einstein) "charted the course" of modern thought.[4] Interpreting Newton as either the First or the Final cause of the world of modern science extends a tradition begun in the late 17th-century during the Battle of the Books when the "scientific" Newton is first chosen as the paradigm of the sort of progress achieved by the Moderns within the context of Royal Society apologetics.

A long line of defenders of the new experimental science, authoritatively established by Royal Patent in the Royal Society in 1662, actively proclaim both the usefulness of the new science for establishing religion and, most importantly, the absolute superiority of the scientific knowledge of the Moderns.[5] In response to "scientific" propagandizing by writers such as

[4] Allan Bloom, *The Closing of the American Mind* (New York: Simon and Schuster, 1987) p. 264, writes that: "The moderns looked to and disagreed with the Greek philosophers and their heirs, the Roman philosophers. But they shared the view that philosophy, and with it what we call science, came to be in Greece and had never, so far as is known, come to be elsewhere. Philosophy is the rational account of the whole, or of nature. Nature is a notion that itself is of Greek origin and requisite to science. The principle of contradiction guided the discourse of all, and the moderns presented reasoned arguments against those of their predecessors with whom they disagreed. The moderns simply took over a large part of ancient astronomy and mathematics. And they, above all, agreed that the philosophic life is the highest life. Their quarrel is not like the differences between Moses and Socrates, or Jesus and Lucretius, where there is no common universe of discourse but more like the differences between Newton and Einstein. It is a struggle for the possession of rationalism by the rationalists." Consider the sort of modern Valhalla into which modern scientific rationalists seek to enshrine Newton. In the Introduction by Zev Bechler to a collection of essays entitled *Contemporary Newtonian Research* (Dordrecht: D. Reidel, 1982), p. 2, Bechler writes that "This belief in the overall rationality of the scientist is the unbreakable tie that unifies historians of scientific ideas into one big loving family in which disputes can't really be fundamental. Here everyone works for the common good and deviations are negligible, and the common good is an exhibition of true rationality silently throbbing wherever science exists."

[5] From the date of its second charter on April 22, 1663, one of the announced goals of the Royal Society of London for Improving Natural Knowledge is to illustrate the providential glory of God manifested in the works of His Creation. The Charter states the Society's purposes: "studia ad rerum naturalium artiumque utilium scientias experimentorum fide ulterius promovendas in Dei Creatoris gloriam & generis humani commodum applicanda sunt." Thomas Birch, *The History of the Royal Society*, 4 vols. (London, 1756-57), 1:221-30. The first Royal Charter was granted in July, 1662, when the title of "Royal Society" was conferred. This official sanctioning was preceded by a year and a half of Wednesday afternoon meetings at Gresham College, London, attended by twelve people including John Wilkins, who was appointed Chairman, and Robert Boyle.

Thomas Burnet and Bernard Fontenelle and in accordance with his own jaded cosmopolitanism and sceptical literary tastes, in 1690, in his *Essay upon Ancient and Modern Learning*, Sir William Temple asserts the superiority of the Ancients in every field of inquiry while consigning "all science, both Ancient and modern, to the dustbin."[6] William Wotton returns fire on behalf of the Moderns in his *Reflections upon Ancient and Modern Learning* of 1694, a work in which he draws heavily upon the early defenders of the Royal Society in advocating the superiority of the Moderns in their knowledge of natural philosophy. With these two works, the long-simmering Battle of the Books in England is heatedly rejoined. Wotton eventually enlists Isaac Newton specifically in the cause of the Moderns in his *A Defense of the Reflections upon Ancient and Modern Learning* of 1705, when he singles out Newton's recently published *Opticks* as the latest example of the progress made by modern scientists.[7]

Back on the continent, the work of making Newton the quintessential scientist of the Modern camp continues when Voltaire dates the beginning of the modern age from the publication of Newton's *Principia*. In his notebook, Voltaire writes: Before Kepler, all men were blind. Kepler had one eye, Newton had two.[8] David Hume, in his "Of the Rise and Progress of the Arts and Sciences," an essay first published in 1742, writes:

> What checked the progress of the CARTESIAN philosophy, to which the FRENCH nation shewed such a strong propensity towards the end of the last century, but the opposition made to it by the other nations of Europe, who soon discovered the weak sides of that philosophy? The severest scrutiny, which NEWTON's theory has undergone, proceeded not from his own countrymen, but from foreigners; and if it can overcome the obstacles, which

[6] Joseph M. Levine, *The Battle of the Books. History and Literature in the Augustan Age* (Ithaca: Cornell University Press, 1991), p. 17.

[7] Wotton's citation of Newton's *Opticks* as an example of the clear superiority of Modern science is pointed out by Joseph M. Levine, *The Battle of the Books.*, p. 352

[8] Cited in Piyo Rattansi, "Newton and the Wisdom of the Ancients," in *Let Newton Be! A New Perspective on his Life and Works* ed. John Fauvel, Raymond Flood, Michael Shortland, and Robin Wilson (Oxford: Oxford University Press, 1988), p. 185. As a young man, Voltaire unequivocally sides with the Moderns and emphasizes the experiments of the French Academy of Science to test Newton's hypothesis that the earth is an oblate spheroid. In his 1738 poem lauding the progress made by the scientists of the Academy of Science, Voltaire upbraids the Greeks for their limited scientific achievements and concludes, "Your century is vanquished by ours." [Voltaire, "A Messieurs de l'Academie des Sciences."] By 1765, in his dialogue entitled "Ancients and Moderns," Voltaire had come to a more moderate position, similar to that of Fontenelle, in which he still held that the Moderns were superior in natural science while admitting that the Ancients were the masters of rhetoric. See the article by A. Owen Aldridge in *Dictionary of the History of Ideas. Studies of Selected Pivotal Ideas*, 5 vols. (New York: Charles Scribner's Sons, 1968), I:86, s.v. "Ancients and Moderns."

it meets with at present in all parts of Europe, it will probably go down triumphant to the latest posterity.[9]

Unlike this vision of an essentially modern (because "rational") progenitor of modern physics who was the first to reveal "nature and nature's laws" which had long lain hidden in the dark "night" of human ignorance, there is a "second Newton" known to some historians but often ignored or tailored to fit Whiggish conceptions. This Newton, whose personal theology is distinctly heterodox, is revealed in his vast unpublished store of manuscripts. Newton, the heterodox theologian, considers the doctrine of the Trinity to be profoundly anti-Christian and believes ardently in the literal fulfillment of the apocalyptic scenario in the book of Revelation. This Newton studies the works of Josephus, Philo, Maimonides, and the Talmudic scholars—occasionally in his own direct confrontation with these texts, but often through the lens of English scholars such as John Selden, John Lightfoot, John Spencer, Sir John Marsham, Joseph Mede, and Edward Pococke or the work of continental scholars such as Samuel Bochart, Johann Buxtorf, Denys Petau, Richard Simon, and Gerard and Isaac Vossius.[10]

Newton's interest in ancient religious and scientific thought is as wide-ranging as the topics analyzed in the vast compendia of the Christian Hebraists.[11] It includes a fascination with ancient chronology, ancient

[9] David Hume, "Of the Rise and Progress of the Arts and Sciences," in *The Philosophical Works*, ed. Thomas Hill Green and Thomas Hodge Grose, 4 vols. (London, 1882), 3:183. In *The History of England*, first published in 1756, Hume repeats his opinion about Newton's unique genius:

In Newton this island may boast of having produced the greatest and rarest genius that ever arose for the ornament and instruction of the species. Cautious in admitting no principles but such as were founded on experiment; but resolute to adopt every such principle, however new or unusual: From modesty, ignorant of his superiority above the rest of mankind; and thence, less careful to accommodate his reasonings to common apprehensions: More anxious to merit than to acquire fame: He was, from these causes, long unknown to the world; but his reputation at last broke out with a lustre, which scarcely any writer, during his own life-time, had ever before attained. While Newton seemed to draw off the veil from some of the mysteries of nature, he shewed at the same time the imperfections of the mechanical philosophy; and thereby restored her ultimate secrets to that obscurity in which they ever did and ever will remain. He died in 1727, aged 85. [David Hume, *The History of England. From the Invasion of Julius Caesar to the Revolution in 1688*, 8 vols. (London, 1782), 8:332-4.]

[10] Richard S. Westfall, *Never at Rest. A Biography of Isaac Newton* (Cambridge: Cambridge University Press, 1980), p. 346.

[11] One of Newton's most important theological manuscripts is his *Theologiae Gentilis Origines Philosophicae* which dates probably from the period between 1683 and 1689. This work exists, in its most mature form in Yahuda MS 16.2 (at the Jewish National and University Library, Jerusalem), written largely in the hand of Newton's amanuensis, Humphrey Newton, as well as in two earlier drafts, Yahuda MSS 16.1 and 17. Yahuda MS 13.3 consists of reading notes related to the complete manuscript. Similarly the manuscript in possession of the American Philosophical Society consists of notes on the religion of the ancient Egyptians and so is related to the *Theologiae Gentilis Origines Philosophicae*. In Yahuda MS 16.2, Newton describes how he has examined the compendia of such Christian Hebraists as Samuel Bochart, Sir John Marsham, and Gerard Vossius in order to learn from them their opinions about the relationship between the religion of Moses and that of the Egyptians. He then goes on to describe his

geography, ancient religious ritual, and the relationship between Old Testament prophetic writings, such as the book of Daniel, and New Testament prophetic writings, such as the Apocalypse of St. John. From his earliest exercise sheet of the mid- to late-1660s in which he translated Hebrew words into Latin, through his lifelong preoccupation with the language of the Hebrew prophets and the properly literal meaning of Old Testament prophecies, prefigurements, and rituals to his discussion over Christmas, 1725, with William Stukeley regarding the exact shape of the Jewish temple, Newton ransacks the original sources and commentators on ancient religion and religious practices.[12]

The most common approach up—until thirty years ago—in dealing with Newton's historical/religious interests has been to ignore those parts of Newton's intellectual makeup which do not fit the pattern of Newton as the first and/or moving cause of modern science. While most modern-day Newton scholars are well aware of Newton's emphatic commitment to a Lord God of supreme dominion and of his intense interest in the knowledge of the Ancients, scientific and religious, they generally argue that, while these facts are important to Newton's Puritan psychology, they imply no necessary internal connection in Newton's thought between Newton's private God and Newton's quintessentially Modern science. Thus, for Frank E. Manuel and Richard S. Westfall, who were the first to begin to illuminate the details of Newton's profoundly held and complicated religiosity, the *Principia*, remains the *Principia* and stands—for Newton—logically apart from Newton's conception of the Lord God of Dominion discovered through his historical research into the religious wisdom of the ancients.[13]

---

own views about that relationship. Westfall has given a full account of Newton's views in Yahuda MS. 16.2 in "Isaac Newton's *Theologiae Gentilis Origines Philosophicae*," in W. Warren Wagar, ed., *The Secular Mind: Essays Presented to Franklin L. Baumer* (New York: Holmes & Meier Publishing, Inc., 1982), pp. 15-34. For a different reading of this manuscript, see James E. Force, "Biblical Interpretation, Newton, and English Deism," in *Scepticism for and against Religion in the Seventeenth and Eighteenth Centuries*, ed. Arjo van der Jagt (Leiden: Brill, 1993), pp. 282-305.

12 On Newton's Hebrew exercise sheet, see Derek Gjertsen, *The Newton Handbook* (London, 1986), p. 208. On the complex question of Newton's "literalism," see James E. Force, "Samuel Clarke's Four Categories of Deism," in *Scepticism in the History of Philosophy*, ed. Richard H. Popkin (Dordrecht: Kluwer, 1996); "Newton, the Lord God of Israel, and Knowledge of Nature," in *Jewish Christians and Christian Jews from the Reformation through the Enlightenment*, ed. Gordon Weiner (Dordrecht: Kluwer, 1994), pp. 131-58; "Science, Deism, and William Whiston's 'Third Way'," in *Ideas and Production. A Journal in the History of Ideas, Issue Seven—History of Science* (Cambridge: Cambridgeshire College of Arts and Technology, 1987), pp. 18-33.

13 Frank E. Manuel, *The Changing of the Gods* (Hanover: Univ. Press of New England, 1983), pp. 15 and 23. See, too, Manuel, *The Religion of Isaac Newton*, (Oxford: Oxford University Press, 1974) passim, and p. 49; and Richard S. Westfall, "Newton's Theological Manuscripts," in *Contemporary Newtonian Research*, ed. Zev Bechler (Dordrecht: D. Reidel, 1982), pp. 139-40. Westfall's work has been particularly valuable both in its gargantuan scope and in its absolutely meticulous precision. Westfall's glorious biography of Newton, *Never at Rest: A Biography of Isaac Newton* (Cambridge: Cambridge University Press, 1980) establishes a new paradigm of excellence and comprehensiveness in

Tracing the influence of Newton's knowledge of ancient historical textual sources both upon his religious views and upon his view of scientific "progress" are my main goals in this paper. Simultaneously, I hope to knit together the "two Newtons" of historiography into one historical person whose apparently diverse historical and scientific works are unified by Newton's persistent thematic focus upon the power of the Lord God as that power is revealed not only in scripture but also as it is manifested in the book of nature and even in the "bare stone choirs" of long-ruined "Druid Temples" such as Stonehenge.

2. The Battle of the Books and Newton's Vision of the New Jerusalem and Scientific Progress: Synthesizing the "Two Newtons" of Historiography

A. Newton and the Battle of the Books

Where does Newton see himself in the Battle of the Books, one of the defining controversies of his remarkable era? A suggestive but ambiguous answer comes from his remark to Hooke regarding Hooke's claim to priority in the theory of color announced by Newton in his paper before the Royal Society in 1672. Newton writes in an extremely famous passage that "If I have seen further it is by standing on ye sholders of Giants."

Newton's use of this metaphor is undoubtedly suggestive of a sympathy toward the Ancients while simultaneously claiming that he has, in fact, seen "further." Disputants on both sides of the Ancient-Modern debate, even before Sir William Temple's work, had made use of the metaphor of a giant

---

the study of Newton. Westfall was the first scholar who knew intimately and in detail all the many aspects of Newton's thought. But, in considering the relationship between Newton's theological interests and Newton's science, for example, Westfall holds to the importance of studying each strand of the seemingly disparate strands of Newton's thought separately. It is highly significant when a scholar of Westfall's attainments and stature writes: "Having studied the entire corpus of his theological papers, I remain unconvinced that it is valid to speak of a theological influence on Newton's science. I say specifically 'theological influences,' not 'religious influence.' The second can, I believe, be readily shown and is generally admitted. A theological influence, by which I mean the influence of Newton's central Arian position and his allied view of the prophecies, is another matter. As I indicated earlier, perhaps we can find the source of the God of the General Scholium in his Arianism. It is not clear which came first, however, his view of God or his Arianism; and even if we grant the influence we remain still on a plane of high generality from which it is difficult if not impossible to demonstrate an influence on some concrete element of his science." Westfall concludes that he "is inclined to examine the relation [between Newton's "theology" and his science] from the other side. At the end of the 17th century, theology was the study with a firmly established, long dominant role of [sic] European civilization, a role then beginning to be challenged by the early success of modern science. It appears to me that we are more likely to find the flow of influence moving from science, the rising enterprise, toward theology, the old and (as we know from hindsight) fading one." Westfall, "Newton's Theological Manuscripts," in Bechler, *Contemporary Newtonian Research*, pp. 139-40.

with a dwarf sitting astride its shoulders. This metaphor goes back at least to John of Salisbury's *Metalogicon* and was quickly adapted by such Renaissance writers as Juan Luis Vives and Ben Jonson. Proponents of the "Ancients" emphasize how big the giants of antiquity were in comparison to the puny dwarfs of the modern world and so utilize the image as an illustration of the superiority of the knowledge of the Ancients. Proponents of the Moderns emphasize, in their replies, how much further modern dwarves can see from their positions astride the Ancient giants and, thus, utilize the image as an illustration of the superiority of modern knowledge.[14] But when one reads the entire passage to get a sense of the context of Newton's remark, it seems clear that by "Giants," Newton here refers to Descartes and Hooke, i.e., to the immediately preceding generation of natural philosophers and not to the Ancients of the classical world.[15] As suggestive as Newton's use of this metaphor might be, it is too ambiguous to claim, solely on its basis, that Newton sides with the proponents of the Ancients.

Joseph Levine has discovered a suggestive hint about how one astute reader located Newton in the controversy between the ancients and the moderns. When the self-exiled high-flying Tory Bishop, Francis Atterbury, read Fontenelle and Newton in France, he recorded with some surprise his discovery that Newton thought that the Ancients were, "men of great genius and superior minds who had carried their discoveries (particularly in Astronomy and other parts of Mathematicks) much farther than now appears from what remains of their writings."[16]

But the most telling and abundant evidence for Newton's view about the progressive march of knowledge and his profound sympathy with the Ancients lies in his many unpublished manuscripts. In Newton's voluminous manuscripts, it becomes quite clear that Newton has fixed his intellectual gaze upon the future in a way not calculated to class him as a typical progress theorist. Clearly, he understands that the Moderns in general, and he in particular, have advanced the modern state of knowledge. But the

[14] "Ancients and Moderns," in *Dictionary of the History of Ideas*, p. 77.

[15] "What Des-Cartes did was a good step. You [Hooke] have added much several ways, & especially in taking ye colours of thin plates into philosophical consideration. If I have seen further it is by standing on ye sholders of Giants." Newton to Hooke, 5 Feb. 1676, in *The Correspondence of Isaac Newton*, 1:146. Cf. Westfall, *Never at Rest*, p. 274, and Frank E. Manuel, *A Portrait of Isaac Newton* (Cambridge: Harvard University Press, 1968), p. 145. Manuel has disputed the degree of irony and the psychological tone of this passage. Westfall rightly, in my opinion, rejects Manuel's suggestion that the non-explicit, veiled allusion to a dwarf is a reference to Hooke's posture. What is certain is that Newton is making use of a metaphor long in currency in the controversy between the Ancients and Moderns. Manuel, with his usual perspicacity, has directly questioned whether Newton's statement "can be taken as proof of his generous appreciation of the work of his predecessors." (Manuel, *Portrait*, p. 145).

[16] Atterbury to Thiriot, n.d., in Francis Atterbury, *Epistolary Correspondence*, 5 vols. (London, 1783-90), 1:182. Cited in Levine, *The Battle of the Books*, p. 40n.

conclusion which Newton achieves in his analysis of modern scientific progress is not the conclusion of modern-day Whig historiographers. Furthermore, part of his theory of progress is a self-conscious revival, through painstaking historical research in the ancient writings, of what was known to be true in ancient times. The only Moderns who count among the learned for Newton are those few—he calls them a "remnant"—who are capable of understanding this ancient wisdom as it has been transmitted to the modern world (in whatever form.) At issue for Newton is both the content of the ancient wisdom as well as its means of preservation and transmission through the ages.

B. Prophetic Language and Newton's Rules of Interpretation

In his *Rules for Methodizing the Apocalypse*, Newton describes Rule 9 for interpreting millennial and apocalyptic language in prophetic history. The exegete must:

> ... choose those constructions which, without straining, reduce things to the greatest simplicity.... Truth is ever to be found in simplicity, and not in the multiplicity and confusion of things. As the world, which to the naked eye exhibits the greatest variety of objects, appears very simple in its internal constitution when surveyed by a philosophic understanding, and so much the simpler, the better it is understood, so it is in these visions. It is the perfection of all God's works that they are done with the greatest simplicity.

The next sentence in Rule 9 is the most crucial point of connection between the rational methodology of Newton's empirical method—one aspect of which is to reveal in the design of nature a creator "well-skilled in geometry"—and the realm of fulfilled prophecies recorded in ancient histories, profane as well as sacred. Newton's God is supremely:

> the God of order and not confusion. And therefore as they that would understand the frame of the world must endeavor to reduce their knowledge to all possible simplicity, so must it be in seeking to understand these [prophetic] visions.[17]

[17] Newton, Yahuda MS 1.1, reprinted in Frank E. Manuel, *The Religion of Isaac Newton*, p. 120. Newton's Rule 9 in his *Rules for Methodizing Scripture* clearly echoes the famous "principle of simplicity" with regard to physical phenomena stated in Rule 1 of the second edition of the *Principia*. Cf. Richard Olson, *Science Deified & Science Defied. The Historical Significance of Science in Western Culture*, 2 vols. (Berkeley and Los Angeles: University of California Press, 1982-90), 2:125, where Olson argues that Newton's utilization in this passage of the "principle of simplicity" shows the extent to which he "depended on 'metaphysical' principles of the kind Newton found so repugnant when they were expressed by Leibniz." For a slightly different interpretation of the meaning of the "prinicple of simplicity," especially as expressed in Rule 1 of the "Rules of Reasoning" in the second edition of the Principia, see James E. Force, "Newton's God of Dominion: The Unity of Newton's Theological, Scientific, and Political Thought," in James E. Force and Richard H. Popkin, *Essays on the Context,*

Just as God has left the impression of his attributes of generally provident power and omnipotence on physical creation in the "book" of nature so, too, he has revealed abundant evidence of his specially provident intercession in his created natural order in the ancient histories of the Bible and the ancient secular historians.[18]

Newton seeks especially to understand the as yet unfulfilled scriptural prophecies regarding the Day of Judgement and the nature of the prophesied millennial kingdom to come. For Newton, such interpretation is an act of pious scholarship in which he traces God's manifest and continuous involvement in his creation from its ancient beginnings to its prophesied future end.[19] Newton is a premillennialist who believes that the second

---

Nature, and Influence of Isaac Newton's Theology (Dordrecht: Kluwer Academic Publishers, 1990), pp. 83-90.

[18] William Whiston sums up the Newtonian view of God's providence, so perfectly illustrated in the above text, when he describes the task of the Newtonian biblical scholar: "Since it has now pleased God, as we have seen, to discover many noble and important truths to us, by the Light of Nature, and the System of the World; as also, he has long discovered many noble and important Truths by Revelation, in the Sacred Truths; It cannot be now improper, to compare these two Divine Volumes, as I may well call them, together; in such Cases, I mean of Revelation, as related to the Natural World, and wherein we may be assisted the better to judge by the knowledge of the System of the Universe about us. For if those things contained in Scriptures be true, and really deriv'd from the Author of Nature, we shall find them, in proper Cases, confirm'd by the System of the World and the Frame of Nature will in some Degree, bear Witness to the Revelation." William Whiston, *Astronomical Principles of Religion, Natural and Reveal'd* (London, 1717), p. 133. Newton (and such Newtonians as Clarke and Whiston) believe in "miracles," a term they use in two senses. First, God's continuous act of preserving the natural world is itself an instance of special providence. Second, when pressed by Leibniz about their reduction of God to the status of an inferior clock repairman or when considering scriptural reports of phenomena for which there is no apparent (or conceivable) mechanical explanation, they accept, however reluctantly, standard miraculous accounts and view them as proof of God's directly interactive special providence. Finally, God's dominion is not limited to that of a purely creative deity active once long ago.

[19] Again, Newton explicitly derives the doctrines which constitute his idiosyncratic Christianity from his own literal reading of scripture. Newton writes that: "We are commanded by the Apostle (1 Tim 1.13) to hold fast the form of sound words. Contending for a language which was not handed down from the Prophets and Apostles is a breach of the command and they that break it are also guilty of the disturbances and schisms occasioned thereby. It is not enough to say that an article of faith may be deduced from scripture. It must be exprest in the very form of sound words in which it was delivered by the Apostles. Otherwise there can be no lasting fixity nor peace of the Church catholick. For men are apt to vary, dispute, and run into partings about deductions. All the old Heresies lay in deductions; the true faith was in the text." Newton, Yahuda MS 15.1, fol. 11r. (Cited in Manuel, *The Religion of Isaac Newton*, pp. 54-5). Newton echoes, in his literalism, the literalism of John Milton who also adheres to the word of Holy Scripture as his criterion for what is true and false in matters of doctrine. Milton's literalism is expressed in many places, for example: "For my own part, I adhere to the Holy Scripture alone— I follow no other heresy or sect. I had not even read any of the works of heretics, so called, when the mistakes of those who are reckoned for orthodox, and their incautious handling of Scripture first taught me to agree with their opponents whenever those opponents agreed with Scripture." Milton, *The Works*, ed. F. A. Patterson, 18 vols. (New York: Columbia Univ. Press, 1931-8), 14:15. Cited in Norman T. Burns, *Christian Mortalism from Tyndale to Milton* (Cambridge: Harvard Univ. Press, 1972), p. 165. Newton's literalism, itself an echo of Puritanism, is echoed by his disciple William Whiston. Whiston argues in his separately paginated "Introductory Discourse" to his *New Theory of the Earth* (London, 1696) that: "...'tis evident that the Holy Books ought not to be tormented and eluded, as to their obvious

coming of Christ will precede his thousand-year reign on earth.[20] In Newton's view, following the ruin of the wicked nations, following the "end of weeping and of all troubles," and following the "return of the Jews," then Christ will return for his millennial reign. And, then, Newton writes that:

> In the Apocalyps where tis said that they bring the glory and honour of the nations into the new Jerusalem those nations are certainly mortals, for they are the nations whom the Dragon deceived no more till the thousand years expired and who being at the end of those years again deceived by him did compass the beloved city and were devoured by fire from the throne, that is by war. Thus is there an end of those rebellious nations but not of the beloved city.... The state of this new Jerusalem you may see further described in Isa. 60 namely how it is a city of mortals assembled from captivity and rules over the nations and *continues for ever* and how (as in the Apocalyps) the Gentiles come to her light and the Kings to the brightness of her rising and her gates are open continually that...they may bring unto her the riches of the Gentiles and *the Sun is no more her light by day nor the moon, but the Lord her everlasting light.*[21]

During the initial millennium of Christ's reign in the New Jerusalem (which will, after all, continue "*for ever*"), Newton literally believes that, in addition to the ordinary mortals who compose it, there will be interspersed "children of the Resurrection." This kingdom of "ordinary mortals" and individually resurrected saints is the literal fulfillment of the Lord God's covenant with Abraham "when he promised that his seed should inherit the land of Canaan *for ever*, and on this...covenant was founded the Jewish religion as on that is founded the Christian."[22]

For Newton, then, the millennial New Jerusalem is the inheritance of the mortals then alive and individually resurrected saints specifically chosen by God to help Christ rule during the millennium. During Christ's initial thousand years of dominion over this earth and the New Jerusalem, the earth will be at peace in a manner unknown since the primeval paradise of Eden. Newton goes on to describe the interaction of the immortal "children of the

---

sense, on every occasion, under pretence that some particular Texts are to be construed another way. That SACRED RULE ought for ever RELIGIOUSLY to be observed, That we never forsake the plain, obvious, easie and natural sense, unless where the nature of the thing it self, parallel places, or evident reason, afford a solid and succinct ground for so doing." Whiston, "Introductory Discourse," p. 26.

[20] See James E. Force, s.v. "Millennialism," in *The Blackwell Companion to the Enlightenment*, ed. John Yolton et al. (Oxford: Basil Blackwell, 1991).

[21] Newton, *The Synchronisms of the Three parts of the Prophetick Interpretation*, "Of the [world to come,] Day of Judgment and World to come," published as Appendix B in Manuel, *The Religion of Isaac Newton*, pp. 126-8. I have emphasized the phrases "continues for ever" and "the Sun is no more her light by day nor the moon, but the Lord is her everlasting light" because they illustrate how this earth, once it becomes a wandering comet, will still be illuminated as prophesied.

[22] *Ibid.* Emphasis added.

resurrection" with the mortals who share their abode during the millennium in most striking terms:

> ...we are not to conceive that Christ and the Children of the resurrection shall reign over [mortals] the nations after the manner of mortal Kings or convers with mortals as mortals do with one another; but rather as Christ after his resurrection continued for some time on earth invisible to mortals unless [when] upon certain occasions when he thought fit to appear to [mortals] his disciples; so it is to be conceived that at his second coming he and the children of the resurrection shall reign invisibly unless they shall think fit upon any extraordinary occasions to appear. And as Christ after some stay in or neare the regions of this earth ascended into heaven so after the resurrection of the dead it may be in their power [also] to leave this earth at pleasure and accompany him into any part of the heavens, that no region in the whole Univers may want its inhabitants.[23]

At the end of this thousand year reign of Christ and his specially chosen, individually resurrected assistants, the Children of the Resurrection, Satan will rise up from his pit and make apocalyptic war. Satan's final defeat will be followed by the general resurrection of martyrs and saints not resurrected for the millennium, i.e., those who "sleep in the dust,"[24] and the day of judgement. Newton writes:

> For when the Martyrs and Prophets live again they may reign here with Christ a thousand years till all the nations Gog and Magog be subdued and the dominion of the new Jerusalem be established and death be vanquished by raising the rest of the dead (those who do not live again untill [the end of] the thousand years be finished,)....[25]

[23] *Ibid.*, pp. 135-6.

[24] Newton is a psychopannychist who does not believe that the soul maintains a continuously conscious existence after the death of the body. He quotes this famous "soul-sleeping" text from Daniel 12.2 and most of the other texts which literally support such an interpretation (e.g., "The dead know not any thing .....There is no work nor knowledge nor wisdom in the grave. Eccles.9.5,10") in the Clark Library Manuscript "Paradoxical Questions concerning ye morals & actions of Athanasius & his followers" in the Question entitled "Quest. Whether Athanasius did not set on foot the invocation of saints." See James E. Force, "The God of Abraham and Isaac (Newton)," in *The Books of Nature and Scripture: Recent Essays on Natural Philosophy, Theology, and Biblical Criticism in the Netherlands of Spinoza's Time and the British Isles of Newton's Time*, ed. James E. Force and Richard H. Popkin (Dordrecht: Martinus Nijhoff, 1994), pp. 179-200.

[25] *Ibid.*, p. 136.

3. Newton, the Wisdom of the Ancients, and Stonehenge

A. Antiquarianism and the Proponents of the "Ancients"

One of the most suggestive, if partial, analyses of Newton's interest in the wisdom of the ancients has been that of McGuire and Rattansi who have utilized the "classical scholia" of the Principia to suggest that Newton's work in ancient history is closely allied with other aspects of his thought.[26] A promising new line of inquiry may now further cement Newton's view of the nature of scientific progress with the acrimonious outburst of the debate between the Ancients and Moderns precipitated by the exchange between Temple and Wotton in the 1690's. Not coincidentally, I hope to show that Newton's interest in the Ancients extends far beyond the bounds of standard textual interpretation of the extant writings of the ancients. Of course, Newton is keen to show how the ancient writers anticipate his own scientific discoveries and how these writings thus validate his scientific views as well as his theories regarding the development of idolatry within the true religion of the Lord God as revealed to the Ancients and made manifestly accessible to some few Moderns, the "wise" remnant mentioned in prophecy. Nevertheless, Newton's interest in the Ancients extends far beyond the writings of the Ancients and into the field work of the Antiquarians of his

[26] While many writers have discussed Newton's interest in the writings from antiquity, they have most often done so from the perspective of what I have called the "disconnectedness" thesis in which, acknowledging the serious import of Newton's inquiry into the physical and religious knowledge of the ancients, they have maintained that interest was fundamentally disconnected from his Modern scientific inquiries. A handful of interpreters such as Piyo Rattansi and B. J. T. Dobbs have insisted that we take Newton's relentless pursuit of the prisca sapientia of the ancients seriously. In her book, *The Janus Faces of Genius: The Role of Alchemy in Newton's Thought* (Cambridge: Cambridge University Press, 1991), p. 7, B. J. T. Dobbs rightly insists that: "With the post-Newtonian diminution of interest in divinity and heightened interest in nature for its own sake, scholars have too often read the Newtonian method narrowly, selecting from the breadth of his studies only mathematics, experiment, observation, and reason as the essential components of his scientific method. For a science of nature, a balanced use of those approaches to knowledge suffices, or so it has come to seem since Newton's death, and one result of the restricted interests of modernity has been to look askance at Newton's biblical, chronological, and alchemical studies: to consider his pursuit of the prisca sapientia as irrelevant. None of those was irrelevant to Newton, for his goal was considerably more ambitious than a knowledge of nature. His goal was Truth, and for that he utilized every possible resource." J. E. McGuire and P. M. Rattansi first drew our attention to the "classical" scholia once considered by Newton as a significant accompaniment to his Principia in what must be now itself be considered a classic article on the topic of Newton's ransacking of antiquity for those ancients who had anticipated his own scientific discoveries. [Cf. "Newton and the 'Pipes of Pan'," *Notes and Records of the Royal Society* 21 (1966), pp. 108-43.] More recently, Rattansi, in an excellent article entitled "Newton and the Wisdom of the Ancients," (in *Let Newton Be!*, p. 191, has asked the penetrating question: "How seriously should we take these attempts by Newton to find anticipations of the *Principia* among the ancients? It may at first appear like one of the aberrations of genius on which it would be kinder to draw a veil. In doing so, however, we may be depriving ourselves of a key that may help to unlock the enigma of Newton's total intellectual endeavour." See, too, James E. Force, "Newton's God of Dominion," pp. 75-102.

day. The great ruins of antiquity are as much a text to Newton as any piece of writing. His intense interest in the exact size and shape of the Jewish temple has long been known. Less well known is the fact that Newton is equally interested in Stonehenge and speculates in a distinctive fashion about that great monument's origin and what its ancient origin tells the modern world about the wisdom of the Ancients.

It is Francis Bacon who first carefully maintains a strict division between "Antiquarianism," or the mere study of physical artifacts, and "Perfect History" in the *Novum Organum*.[27] In the seventeenth century, however, a significant and extremely dramatic shift occurs in the intellectual status of, for example, coins or burial urns, the vital historical artifacts of Antiquity and the objects of focus for the Antiquarians. In England, in the seventeenth century, curiosities dug out of the ground become evidence useful in reconstructing, first, the Roman past and, later, the Celtic past, a consequence of the fact that the early history of England is simply a blank without such artifacts as a guide. Antiquarian scholars such as William Camden, Ole Worm, John Twyne, John Stow, Richard Verstegan, Jan Goropius Becanus, William Dugdale, and, finally, William Stukeley, manifest this new regard for the physical remnants of antiquity as important evidence for reconstructing the past. Such writers are simultaneously interested in "the early geological history of the earth, the settlement of the New World, the chronology of Egyptian and Chinese history, the collection of fossils, [and] the early history of Christianity."[28]

In addition, these writers forge a close alliance between the artifacts they excavate from the fields and the writings of the ancients. Roman coins, inscriptions, burials, metalled Roman roads, and buried Roman walls become the criterion for judging the claims found in Roman literary texts about life in Roman Britain. Interestingly, this entire tradition of scholarship has suffered a parallel fate to that of Newton in the critical hands of Whig interpreters. Too often, interpreters have simply been determined to examine the writings of these Antiquarians with an eye toward what is "Modern" about them. So intent are these Whiggish modern interpreters to uncover anticipations of Modern archaeology that they simply ignore what the Antiquarians thought they were doing.[29]

[27] J. G. A. Pocock, *The Ancient Consititution and the Feudal Law: A Study of English Historical Thought in the Seventeenth Century* (Cambridge: Cambridge University Press, 1957), pp. 6-7. See, too, William B. Ashworth, Jr., "Natural History and the Emblematic World View," in *Reappraisals of the Scientific Revolution*, ed. David C. Lindberg and Robert S. Westman (Cambridge: Cambridge University Press, 1990), pp. 320-5.

[28] Ashworth, "Natural History and the Emblematic World View," p. 324.

[29] In a recent BBC documentary devoted to Stukeley, "Romance in the Stones: William Stukeley's Curious Journeys," the narrator (and script writer) David Drew, interviews modern archaeologists about how much of Stukeley's field work remains adequate today. Julian Richards of the Trust for Wessex Archaeology states that Stukeley "obviously hadn't got access to all the techniques that we have today for

B. Antiquarianism in the Service of Whig Historiographers

It is not accidental that Newton's first biographer, William Stukeley, is one of the foremost Antiquarians of Newton's day. William Stukeley, a student of Newton's at Cambridge, knows exactly what he is doing in his excavations at Avebury and Stonehenge.[30] He also knows what he is NOT doing. Stukeley is not, except incidentally, attempting to invent modern archaeology. Stukeley is looking for evidence— literary or archaeological—which connects the Britons of Celtic times with the people and chronology of the Bible. He writes:

> Can we then think that this great being, the author of all, has been less careful of the moral than of the material world? has he less regard to the far most worthy part of it, us, the thinking substances, that like him (tho' in a very inferior way) can pervade the boundless space, and traverse all his works, in whom he has sown sparks of an immortal fire, deriv'd more immediately from his own fountain? therefore I wonder at the incogitancy of our modern free-thinkers, people of learning, that will not see the beauty of that profound scheme of religion, which was carry'd on from the beginning of the world (I mean from the fall of man) to its completion and perfection in the christian revelation.[31]

---

dating and scientific analysis but as an observer of broad landscapes and as an interpreter even given the problems that he had with his interpretation, I think he's of enormous value to us in the twentieth century." Drew also interviews Stukeley's biographer Professor Stuart Piggot who contributes to the theme of Stukeley as a Modern who was different in kind from those antiquaries who pursued historical inquiry solely through literary texts and whose interest in the Druids served as the foundation of a patriotic and nationalistic view of ancient British achievements. The best modern interpretation of Stukeley's work as an "archaeologist" is found in Peter J. Ucko, Michael Hunter, Alan J. Clark, and Andrew David *Avebury Reconsidered: From the 1660s to the 1990s* (London: Unwin Hyman, 1991).

[30] In 1752, Stukeley published his *Memoirs of Sir Isaac Newton's Life*.

[31] William Stukeley, *Paleographia Sacra: or, Discourses on Monuments of Antiquity that Relate to Sacred History. Number I. A comment on an Ode of Horace, shewing the Bacchus of the Heathen to be the Jehovah of the Jews* (London, 1736), Dedication. The rest of this passage of Dedication to Sir Richard Ellys, Bart., reads: "To open our eyes a little in this age, wherein we seem to be willing to overlook it, was the purport of this small publication. We might well think this the Augustan age reviv'd; among other reasons, because our poet Horace is in every hand. He is thought unpolite and out of fashion that has not a taste for his beauties. I have endeavour'd too to throw in my mite that way, tho' somewhat out of the vulgar road; and to put a celebrated piece of his work into a dress, which probably he would not have mislik'd, had he liv'd a little longer, and had those happy opportunities which we have long enjoy'd: to which his excellent good sense and discernment in human and divine things would have secur'd his ready approbation. However, what I have endeavour'd at, will shew, I hope, that the heathen were not left out of the general view of providence in that great scheme, which was more intimately deposited among the Jews. They had such notices of the matters belonging to it, either originally imparted to them by God Almighty, or at least borrow'd from the greater splendor of those people in very early days, as enabled them upon its full manifestation to see, this was what was ultimately meant in all their perplex'd theology. And we find it so by experience, from the heathen world so soon and universally embracing the christian faith, notwithstanding persecutions and difficulties of old habit and custom, of long deprav'd passions and appetites. This, the comment I have made upon the Ode, which is but a very little of what might be said in this view, will render sufficiently clear."

Stukeley's fascinated focus upon, first, the stone circle at Avebury and, later, upon Stonehenge arises because of the perfect suitability of these bare stone circles, in his eyes, to combat Modern "freethinkers" who refuse to see the support which these monuments provide to revealed Christian scripture. As Stukeley explains, in his *Stonehenge, a Temple restored to the British Druids* (1740), that his purpose is to utilize his archaeological researches in an apologetic polemic aimed at the deists:

> God, which is no where upon earthe done, in my judgement, better than in the Church of England. My intent is (besides preserving the memory of these extraordinary monuments, so much to the honour of our country, now in great danger of ruin) to promote, as much as I am able, the knowledge and practice of ancient and true Religion; to revive in the minds of the learned the spirit of Christianity...to warm our hearts into that true sense of Religion, which keeps the medium between ignorant superstition and learned free-thinking, between enthusiasm and the rational worship of God, which is no where upon earthe done, in my judgement, better than in the Church of *England*.[32]

Stukeley's modern-day biographer, Stuart Piggot, absolutely cannot resist the temptation to link what he regards as Stukeley's unscientific use of "objective" archaeology for religious polemicizing with what Piggot regards as the "dramatic" change in Newton's mental life following a "mental

[32] Stukeley, *Stonehenge: A Temple Restor'd to the British Druids* (London, 1740), Preface. In this Preface, Stukeley carefully explains that this volume devoted to Stonehenge was only the seventh and last of a projected seven-volume work the main theme of which was to be "Patriarchal Christianity, or a Chronological History of the Origin and Progress of True Religion, and of Idolatry." Stukeley's Modern biographer, Stuart Piggot, cites this text and laments that "We have travelled a long way from the accurate scholarship of the Restoration historians, in the last decade of which tradition Stukeley began his work, and are already in the age in which Bolingbroke could express his 'thorough contempt for the whole business' of the great historical achievements of earlier scholars, and not far from that of Gibbon...." Finally, Piggot concludes his animadversion on Stukeley's stated intentions by comparing the two published volumes on Stonehenge and Avebury: "In earlier pages we have examined the *real* [emphasis added] archaeological contribution contained in the eccentric pages of *Stonehenge* and *Avebury*, where it is entangled with the most curious mixture of Biblical and classical allusions and flights of the wildest etymological fancies. The Stonehenge book is the less diffuse, and comes nearer to an objective report on field-work and excavation, but in the three years which elapsed between its appearance and that of Avebury, Stukeley's mind seems to have become perceptibly more obsessed with theory, and in the second book the *sound field-work only peeps out here and there among the nonsense* [emphasis added.] When 'PHUT son of CHAM' begins to dominate the scene (' personn of eminence, tho' not taken notice of so much as he deserves'), Druids, Phoenicians, Chaldeans, and the Tyrian Hercules are all confusedly worshipping in a Dracontium in imminent expectation of the Messiah." Piggot notes that the manuscript notes on the other projected volumes (1. A Mosaic Chronology; 2. Melchisedec, or a delineation of the first and patriarchal religion; 3. Of the mysteries of the ancients, and of idolatry; 4. A dissertation on writing and the origin of the alphabet; and, 5. The patriarchal history of Abraham and the Phoenician colonization of Britain) "was written and remains in deservedly obscure manuscript form." Finally, Piggot concludes his animadversion on Stukeley's stated reasons for pursuing archaeology ("It is time to stop," he writes after three pages) and gets back to chronicling the events of Stukeley's life in a suitably objective, modern manner. See Stuart Piggot, *William Stukeley. An Eighteenth-Century Antiquary* (Oxford: Clarendon Press, 1950), pp. 127-30.

disturbance" in 1693. After Newton's "breakdown," according to Piggot (who follows L. T. More's Whiggish biography on this point), Newton turns away from serious science and toward religion. Piggot joins the ranks of those who claim that Newton is the First and/or Final cause of modern science and laments the effects of Newton's "breakdown" while drawing a parallel with Stukeley's intellectual career. According to Piggot, Stukeley starts out as a good field observer but, sadly, winds up taking holy orders and perversely utilizes archaeological evidence in the service of religious apologetics. Piggot has no sympathy for such romancing of stones. Piggot draws this Whiggish parallel between the intellectual careers of Newton and Stukeley starkly:

> In both men there had been an early phase of scientific and objective work followed by a more or less sudden change to one of religious enthusiasm in which the disciplines and results of the earlier work were utilized in vindication of Protestant Christianity, however Latitudinarian a form at least it may have taken in Stukeley's mind. And the connexion can be brought even closer, since the lives of the two men overlapped, so that the antiquary in youth had known the ageing mathematician. In 1720, when Newton was seventy-six and Stukeley thirty-three, they had indeed discussed Solomon's Temple together—'He says it was older than any other great temple', wrote Stukeley, describing the conversation, and since Newton and Stukeley both shared a delight in draughtsmanship, it is not surprising that both made imaginary drawings of this notable building in antiquity.[33]

[33] Piggot, *William Stukeley*, p. 131. In the conclusion to the passage cited in the text, Piggot really takes Whiggish flight when he writes that: "Stukeley had therefore been interested in Biblical antiquities as early as 1720, and Newton's conversations may well have had their part in influencing his ideas a few years later, while the Mosaic Chronology Stukeley compiled in 1731 may betray an even closer connection. He had know Newton, and had collected details of his life with assiduity for Conduitt, Newton's nephew by marriage, and so could not escape a knowledge of the crisis in his hero's life and the manner in which he had turned his mathematical genius to religious ends. It is hardly likely that any conscious comparison can have existed in Stukeley's mind after Newton's death in 1727, but in the abnormal mental state in which he found himself about that time the possibility of an unconscious parallel working itself out cannot wholly be dismissed. At all events, whether such a connexion existed or not, the similarity in the two life-stories is sufficiently interesting to deserve comment." Stukeley, in his *Memoirs of Sir Isaac Newton's Life*, ed. A. Hastings White (London: Taylor and Francis, 1936), pp. 17-8, writes that: "In Christmas 1725, upon a visit I made him, we had some discourse about Solomons temple: a matter I had study'd with attention, and made very many drawings about it, which I had communicated to my Lord Pembroke (Thomas), to Mr Folkes and more of my friends. I found Sir Isaac had made some drawings of it, and had consider'd the thing. Indeed he had study'd every thing. We did not enter into any particular detail, but we both agreed in this, that the architecture was not like any design or descriptions yet publick. No authors have an adequate notion of antient architecture. Sir Isaac rightly judged it was older than any other of the great temples mention'd in history; and was indeed the original model which they followed. He added that Sesostris in Rehoboams time, took the workmen from Jerusalem, who built his Egyptian temples, in imitation of it, one in every Nomos, and that from thence the Greeks borrow'd their architecture, as they had the deal of thir [sic] religious rites, thir [sic] sculpture and other arts." Westfall, *Never at Rest*, p. 346, observes that Newton read Ezekiel in Hebrew in order to draw up plans for the exact shape, in cubits, of the Temple. On p. 347, Westfall reproduces Newton's

The only legitimate reason, from this Whig perspective, for pursuing an antiquarian study of ancient stone monuments is because of a shared "delight in draughtsmanship" or, possibly, owing to a mild case of "enthusiasm" arising from a "breakdown."

C. Newton, the Egyptians, and the Stonehenge of the Antiquarians

Isaac Newton is as clear in his manuscripts as William Stukeley is in his published works about the nature of, and motivation for, his own interest in the ancient monuments of Britain. As we have seen, Newton's conception of progress involves showing how a truly learned Modern can extend the very boundaries of modern scientific and religious knowledge while illustrating that the learned of antiquity also possessed the very same knowledge. For Newton, progress is a return to the wisdom of the Ancients. Within this perspective, a Modern adept such as Newton ransacks ALL of antiquity for evidence to support his primary thesis that the Ancients possessed the same scientific and religious wisdom as the Moderns.

In his manuscripts from the 1680's, especially the *Theologiae Gentilis Origines Philosophicae* and the later English variants of this piece, Newton argues that the Ancient Egyptians, the descendants of Noah's grandson Mizraim, developed a rational monotheism based upon astronomical observations of the heavens from which they learned that the earth revolves around the central fire of the sun. Newton maintains that the core monotheism of this Ancient Egyptian religion is esoterically symbolized by Egyptian priests in the design of their temples around a central hearth and fire, or Prytanea, a habit retained by the Greeks and Romans. In Greece, each state, city, or village possessed its own central hearth and sacred fire symbolizing the unity and vitality of the community. For the Romans, this fire was kept continuously alight in the Temple of Vesta first, perhaps, by the daughters of the King and later by the Vestal virgins.[34] The point is that these later Prytanea are vestiges of the original Prytanea of the Egyptians for whom this central fire symbolizes the astronomical truth of heliocentrism and the core theological truth of monotheism.

Newton goes on to elaborate his theory of the corruption of this primal monotheism in a manner which we might expect from the author of the *Irenicum*. The "gods" of pagan antiquity Newton identifies with Noah and Noah's descendants. Noah and his sons are first idolized by their people as gods and, ultimately, identified with, especially, planets. The genealogy of

---

diagram of the temple (now in the Babson College Library, Babson MS 434) to which Stukeley refers above.

[34] *The Encyclopedia Britannica*, eleventh edition (New York: Encyclopedia Britannica, Inc., 1910-1), 29 vols., 21:533, s.v. "PRYTANEUM and PRYTANIS."

the gods of the gentiles becomes for Newton, as for other 17th-century Euhemerists, the political history of the kingdoms of the Egyptians, Babylonians, and Assyrians.[35] Newton argues that Noah is ultimately deified as the god Saturn. The most important of his three sons, for the development of idolatrous polytheism, is the outcast Ham who becomes the original of the god (and the planet) Jupiter. Ham's four sons—Cush, Mizraim, Phut, and Canaan—become deified as Hercules, Osiris, Anateus, and Busiris respectively. Osiris' three children include the bastard son Thoth (known much later to the Greeks as Mercury.) Thoth succeeds his father Osiris (originally Mizraim) as the ruler of Egypt and becomes the founder of truly idolatrous gentile theology by beginning the deification of his forebears through the process of naming planets after them (Saturn after Noah, Jupiter after Ham.) At first the process seems innocent but by the time of the Egyptian conquest of her neighbors, including the Jews, the original primal monotheism of the Egyptians has been corrupted into a thorough-going polytheism which the Jews are quick to emphasize as the central feature of the religion of the Egyptians as they observe it. Thoth also begins to combine astronomical learning, especially the heliocentric theory, with other religious rituals all of which are esoterically preserved in the hieroglyphic symbols of the Egyptian priestly class. Ultimately, the esoteric meaning of the hieroglyphs is lost and they are taken to refer literally to a pantheon of gods.

Newton traces a similar pattern of polytheistic and idolatrous accretions to an original rational monotheism in his account of the historical origins of the Babylonian and Assyrian dynasties and religion. Just as Ham's son Osiris (Mizraim) is the founder of the Egyptian dynasty and, through his offspring, Thoth, of the later, corrupt, idolatrous Egyptian theology, Ham's son Cush (later known as Hercules), a mighty warrior, conquers Babylonia. Cush's son Nimrod, later conquers Assyria. In both these empires, a similar idolatrous deification of their own kings takes place.[36] Newton thus traces all ancient political and religious history back to Noah and Noah's offspring and simultaneously gives an historical account of the rise of polytheism and idolatry in these gentile nations as the result of the posthumous deification of their leaders and heroes, a polytheistic process which thoroughly corrupts the core monotheistic truth symbolized by the Prytanea in the original religion of Noah. The Bible shows that wherever the family of Noah went the religion of the central fire accompanied them until it was idolatrously

[35] About 300 B.C., Euhemerus of Messene wrote *The Sacred History* in which he theorizes that the traditional deities are merely earthly rulers first deified and then worshipped by their subjects. See Jean Seznec, *The Survival of the Pagan Gods. The Mythological Tradition and Its Place in Renaissance Humanism and Art*, trans. Barbara F. Sessions (New York: Harper Torchbooks, 1953), pp. 11-3.

[36] Richard S. Westfall, "Isaac Newton's *Theologiae Gentilis Orgines Philosophicae*," pp. 19-21. See note 11 infra.

corrupted in places such as Egypt, Babylonia, and Assyria in the process described above.[37]

The central fires in the temples of the Egyptians symbolize this original monotheism even when that core religious truth eventually becomes obscured by idolatrous accretion. Moses, according to Newton, merely retains the remnants of the Ancient religion of the first Egyptians, the direct descendants of Noah, while he "rejected only what belonged to ye worship of their fals Gods," i.e., the sun, the planets, the elements, Jupiter Hammon, Osiris, Isis, Orus, etc.[38] Still, for the wise of all ages, the Ancient temple architecture of the first Egyptian religion, the religion of Noah and his sons, visibly symbolizes the one God. Newton, the Modern man wise in the scientific and religious wisdom of the Ancients, thus seeks to trace this original monotheism (and heliocentrism) in other ancient temples. He finds such traces, for example, in the "high places" of Canaan and Syria "before the days of Moses." Newton argues, in texts first shown to me by Rob Iliffe, that the "high places" mentioned in the Bible are really the locations of temples devoted to the sacred central fire of the original monotheism:

> So when we find several altars in Israel as in Samuels days when there was one at Mizpeh, another at Gilgal, another at Bethlehem, another at Ramah, wch was built by Samuel, another in Gibsah ['wch was' added] built by Saul [continued till Solomon's reign & called a great high place' added]: it is to be understood that these altars so long as a fire was kept burning on them ['wch seems to have been till the captivity of the ten tribes)' added] were of ye same nature wth Prytanea of other nations.

Newton finds numerous references to these temples consecrated to the sacred central fire throughout the texts of the Ancients. The evidence for his final conclusions that this original monotheism, prior to idolatrous corruption, is "the most ancient religion of all" and that no later religion "became as general as this" is not founded solely in the writings of the Ancients. With the Antiquarians of his day, Newton turns to the physical evidence of the past and cites reports of Prytanea from places as far away as India and China as well as from the nearby ruin on Salisbury plain. Newton writes that:

> In England neare Salisbury there is a piece of antiquity called Stonehenge ['wch seems to be an ancient Prytaneum. For it is' added] an area compassed circularly wth two rows of very great stones wth passages on all sides for people to go in and out at. Tis said there are some pieces of antiquity of ye same form & structure in Denmark... In Ireland one of these fires was

[37] Yahuda MS 41 fol. 4r. This text is cited by Rob Iliffe in his paper "The Egyptian Basis of Stonehenge: Newton, Natural Philosophy and the Origins of Idolatry" which he communicated to me privately.

[38] *Ibid.*, fol. 5r. This text is cited by Rob Iliffe in his paper "The Egyptian Basis of Stonehenge."

> conserved by ye Moncks of Kildare under the name of Brigels fire till of late years & ye Caenobium was called the house of fire.[39]

When Stukeley remarks in his Dedication to his *Palaeographia sacra: or, Discourses on Monuments of Antiquity That relate to Sacred History* (London, 1736) that the "author of all" ... "has sown sparks of an immortal fire" throughout his works, it seems at least possible that he may well have Newton's suggestions in mind. Stukeley elaborates on the implications of Newton's tracing of Stonehenge's design back to the vestal temples of the original monotheism of Noah and his sons and argues that the "Druids" who built Stonehenge were originally Phoenicians who arrived in Britain soon after Noah's flood with knowledge of the Noachide religion.[40]

4. CONCLUSION

Newton, in his manuscripts, shows a vital concern with the wisdom—religious and scientific truth taken together—of the Ancients. Progress, even what is considered to be the most significant modern "scientific" progress in Newton's own writings, he conceives as instantiating this Ancient wisdom. We might still ask what difference it actually makes for Newton's science if, in the privacy of his unpublished manuscripts, Newton is an extreme proponent of the Ancients even while he emphasizes that some few Moderns are able to see what they saw? Is it not true that the *Principia* is still the *Principia* in spite of whatever private theological views Newton may have held? The short answer is that, for us, Newton's theology need not necessarily be related to his science in any way. The longer, less anachronistically refracting, more historically exact answer is that, for Newton, his theory of God's Absolute Dominion profoundly affects his natural philosophy just as it affects his particular theological and political doctrines. If it is our goal to understand Newton in his own terms (in so far as such a goal is even approachable), then we must confront the interaction in his thought—as evidenced in his manuscript writings—between his metaphysical conception of God with what human beings may claim to know and with how human beings can make such knowledge claims about the creation. Of Pythagoras, F. M. Cornford has written that "The vision of philosophic genius is a unitary vision. Such a man does not keep his thought

[39] This text is from Yahuda Ms. 17.3 fol. 10r. This text is cited by Rob Iliffe in his paper "The Egyptian Basis of Stonehenge."

[40] Unlike Newton, however, Stukeley discovers within the Druidical religion "a knowledge of plurality of persons in the Deity" and so here the biographer and his famous subject part company. Cited in Stuart Piggot, *The Druids* (London: Thames and Hudson, 1975), p. 151.

in two separate compartments, one for weekdays, the other for Sundays."[41] The same vision holds true for Newton.

[41] F. M. Cornford, *Before and After Socrates* (Cambridge: Cambridge University Press, 1968), p. 66.

ESSAY 13

The Logic of Millennial Thought: Sir Isaac Newton Among His Contemporaries

REINER SMOLINSKI
Georgia State University

Sir Isaac Newton is commonly praised for his originality in developing a systematic key (or rule) for unlocking the mysteries of prophetic language and the hermetic symbolism of the Bible's prophetic style.[1] In collating the Book of Nature with the Book of Scripture, Newton developed a rigorous taxonomy that allowed him to juxtapose "natural objects" in the Bible with "political events" in human history as a means of decoding prophetic language in a coherent manner.[2] Because of his shared distrust of multiple or allegorical interpretations, Newton's "Rules for interpreting the words and language in Scripture" is characterized by his insistence on a single, obvious, or literal sense of the prophecies, yet allows for a kind of "third or middle way" to govern allegorical or nonliteral interpretations when the plain or literal sense contradicts "reason."[3] With so much well-documented

[1] See Isaac Newton, *Observations Upon the Prophecies of Daniel, and the Apocalypse of St. John* (London, 1733), pp.16-23 and pp. 235-53; Frank E. Manuel, *The Religion of Isaac Newton*, Appendix A, "Fragments from a Treatise on Revelation," (Oxford: At the Clarendon Press, 1974), pp. 107-25; and, for an explanation of Newton's euhemeristic method, Manuel, *Religion*, pp. 94-8.

[2] For related articles on these issues, see Richard H. Popkin, "Newton as a Bible Scholar," in *Essays on the Context, Nature, and Influence of Isaac Newton's Theology*, James E. Force and Richard H. Popkin, eds. (Dordrecht: Kluwer Academic Publishers, 1990), pp. 103-18; Sarah Hutton, "More, Newton, and the Language of Biblical Prophecy," in *The Books of Nature and Scripture: Recent Essays on Natural Philosophy, Theology, and Biblical Criticism in the Netherlands of Spinoza's Time and the British Isles of Newton's Time*, James E. Force and Richard Popkin, eds. (Dordrecht: Kluwer Academic Publishers, 1994), pp. 39-54; Robert Iliffe, "'Making a Shew': Apocalyptic Hermeneutics and the Sociology of Christian Idolatry in the Work of Isaac Newton and Henry More," *Ibid.*, pp. 55-88; and Matania Z. Kochavi, "One Prophet Interprets Another: Sir Isaac Newton and Daniel," *Ibid.*, pp. 89-104.

[3] Newton, "Rules for interpreting the words and language in Scripture," in Manuel, *Religion*, pp. 116-21; William Whiston, "Postulata," in *A New Theory of the Earth...With a large Introductory Discourse concerning the Genuine Nature, Stile, and Extent of the "Mosaick" History of the Creation* (1696), 5th ed. (London, 1737), p. 95. In the notes which follow, I cite the "Introductory Discourse" when quoting from this separately paginated, 95-page "Introductory Discourse" to the *New Theory*. When quoting from

J.E. Force and R.H. Popkin (eds.), Newton and Religion, 259–289.
© 1999 *Kluwer Academic Publishers.*

testimony of Newton's resistance to allegory, it becomes all the more important to examine particular instances of his millennialism where Sir Isaac departs from his own principles and opts for allegorical and mystical readings rather than his avowed literalism.[4] To be brief, I am postulating that in addition to the above-mentioned taxonomies, Newton's decision-making process is also governed by significant *a priori* assumptions that determine the interior logic of his millenarian system. In the subsequent discussion, I shall focus on some of the most hotly debated issues shared by virtually all millennialists of the period and show when and why Newton departed from Joseph Mede (1586-1638)—the principal godfather of English apocalypticism. Moreover, I shall elucidate Newton in the light of his confreres of the Royal Society, especially John Ray (1628-1705), Thomas Burnet (c. 1635-1715), and Cotton Mather (1663-1728), all of whose millenarian hermeneutics were shaped by many of the same ideological and cosmological contingencies that determined Newton's own.[5]

Before we can position Isaac Newton within the context of millenarian theories of the time, it is important to outline the main features of Joseph Mede's premillennialist system that irrevocably shaped those of his disciples.[6] According to the Talmudic tradition of Elias, which Mede followed, the chronometry of the world would last for six millennia only.[7] Theologians in the seventeenth century, perhaps more so than in any other century, were concerned with establishing an exact timeline to determine the Second Coming of Christ.[8] To Mede, one of the most reliable, and indeed prerequisite, signs of the approaching millennium would be the return of the Jewish nation to Palestine and its concomitant conversion just before

---

the body of the *New Theory*, I cite that title; James E. Force, "The God of Abraham and Isaac (Newton)," in Force and Popkin, eds., *Books of Nature*, pp. 179-200.

[4] For helpful background, see Thomas H. Luxon, *Literal Figures: Puritan Allegory and the Reformation Crisis in Representation* (Chicago: University of Chicago Press, 1995), pp. 34-76.

[5] See Margaret C. Jacob's helpful discussion of millenarian theories in Restoration England, *The Newtonians and the English Revolution: 1689-1720* (Ithaca, NY: Cornell University Press, 1976), pp. 100-42 and Rose-Mary Sargent, *The Diffident Naturalist: Robert Boyle and the Philosophy of Experiment* (Chicago: University of Chicago Press, 1995), pp. 109-28.

[6] See C. A. Patrides and Joseph Wittreich, eds. *The Apocalypse in English Renaissance Thought and Literature* (Ithaca: Cornell University Press, 1984).

[7] Joseph Mede, *The Works of the Pious and Profoundly Learned Joseph Mede*, ed. John Worthington, 4th ed. (London, 1677), p. 776. Thomas Burnet makes the same point in *The Sacred Theory of the Earth*, ed. by Basil Willey (Carbondale: Southern Illinois University Press, 1965), pp. 334-5. (The first two Parts of Burnet's *Telluris Theoria Sacra* were published in 1681; King Charles II, greatly impressed, urged Burnet to prepare an English translation which was published in 1684 as *The Sacret Theory of the Earth*. For the second Latin edition of 1689, two more Parts were added. All four Parts were published together in an English translation in 1690). Cotton Mather adopts the same tradition in *The Threefold Paradise of Cotton Mather*, ed. by Reiner Smolinski (Athens: University of Georgia Press, 1995), p. 325.

[8] See Smolinski, "When Shall These Things Be?," in *Threefold Paradise*, pp. 60-78, and Cotton Mather, *Ibid.*, pp. 319-47.

Christ's appearance in the clouds of heaven as a prelude to the destruction of Antichrist by a conflagration issuing out of heaven. At the beginning of the millennium would occur the binding of Satan for a thousand years and the First Resurrection of the martyrs and witnesses as co-regents with Christ in heaven over the Saved Nations huddling around a restored Jerusalem, to be taught and ruled with an iron rod by the Raised Saints. Toward the end of the millennium, Satan and his minions (Gog and Magog) would be released and once more attack the New Jerusalem, but fire issuing from heaven would destroy Christ's adversaries. This cataclysmic event would then be followed by the raising of the dead (Second or General Resurrection) at the Day of Judgment—the righteous to life everlasting, the wicked to eternal death.[9]

1

Against this basic backdrop, we can examine how the blueprints of Thomas Burnet, John Ray, Cotton Mather, and Sir Isaac Newton hold up against the model of Joseph Mede. Closely linked to the issue of Christ's return is the nature of the people principally concerned in the events of the Second Coming especially during the conflagration that would obliterate God's adversaries. Again, Joseph Mede sets the stage for virtually all seventeenth- and early eighteenth-century millennialists in their efforts to construct a logical system. Most interesting in this context is the condition of the "Saved Nations" alive on earth when a literal flood of fire (2 Peter 3:10) would destroy Antichrist's dominion. The English millenarian Dr. Samuel Meddus wrestled with much the same knotty question of eschatology in his endeavor to straighten out the kinks of his millenarian logic. In his 1629 correspondence with Dr. William Twisse on the condition of the Saved Nations at the Conflagration, Meddus put the issue squarely: "Quaere 10... *But if there shall be a* 1000 *years reign on* earth, *what need they* [Saved Nations] *be* caught up *in the* air? *and how* ever *be with the* Lord *from thenceforth, if they and their posterity after them continue for the space of a* 1000 *years subject to MORTALITY*?"[10] Even a cursory study of the eschatological literature of the period reveals how questions of this nature were circulated among the cabal of fellow millenarians, inviting them to debate the issue. The volatile nature of such studies—the result of millenarian efforts to minimize the logical pitfalls of their systems—is highlighted by Mede's response to Meddus' query, which Twisse had passed on to his master for further elucidation. Offering several possibilities,

[9] Mede describes most of these issues in *Works*, pp. 603-5, and 773-7.

[10] *Ibid.* p. 775. Capitalization added.

Mede insisted that the First Resurrection of the immortalized Saints at the beginning of the millennium need not necessarily occur concomitantly with the Rapture of the still mortal Saved Nations but "may admit a great distance of time" even as much as a thousand or more years. In fact, Tertullian, Piscator, and even Beza had argued that the Children of the Resurrection need not be raised

> *all at once* or *altogether*; but the Martyrs *first*, in the *First resurrection*; then (after an appointed time) the rest of the dead in the *Last resurrection; afterward*; when the Resurrection shall be thus compleat; those which remain alive at Christ's coming shall together with those which are risen be *caught up into the clouds, to meet the Lord in the Air*, and from thenceforth be eternally with him. And so the reason why those which Christ found alive at his coming were not instantly translated, should be in part; that they might not prevent the dead, but be consummate with them.[11]

The important distinction between the two groups of Saints bears repeating, for Mede's Saints of the First Resurrection—the Raised Saints or dead martyrs and witnesses—would attain their immortality at the beginning of the millennium and be translated into heaven to serve as co-rulers with Christ. The Saved Nations, on the other hand, were Saints of a different caste. They would be the faithful, alive on earth, as the conflagration incinerates the earth. Their remaining mortality, however, would render them extremely vulnerable to the flood of fire coming down from heaven. As Samuel Meddus put the query: What is the "*Communion between the Saints raised from their graves, and the people then living and remaining on the earth, called the* nations that are saved, *that is, from the fire, whereby the* earth *and* the works thereof *shall be burned*, 2 Pet. 3.10"?[12] These Saved Nations created something of a logistical problem, for their physical presence on earth during the conflagration required a spatial limitation of its initial desolation to allow these mortals to be sheltered from such a cataclysm. For instance, Cotton Mather's answer was derived from Joseph Mede, who argued "that the *Conflagration* will be at first *partial*, and *liesurely* and *progressive*: And that the first Efforts and Effects of it, are like to be felt by *Italy*," the seat of Antichrist.[13] Mede had suggested as much

[11] *Ibid.*

[12] *Ibid.*, p. 773.

[13] Mather, *Christianus per Ignem* (Boston, 1702), p. 69; Mede, *Paraleipomena. Remaines On some Passages in the Revelation* (London, 1640), p. 36, and *Works*, pp. 773-6, and 809-10. For the various positions on a partial or global conflagration, see Cotton Mather's *The Present State of New England* (Boston, 1690), pp. 34-7; *Magnalia Christi Americana* (London, 1702), Bk. 1, Chap. 1, pp. 2, and 4-5; "'Problema Theologicum' [1703]: An Authoritative Edition," ed. by Jeffrey Scott Mares, *Proceedings of the American Antiquarian Society* 104, Pt. 2 (1994), pp. 333-440; *The Diary of Cotton Mather*, ed. by Worthington C. Ford. Collections of the Massachusetts Historical Society, 7th series, Vols. 7-8 (Boston, 1911-12) 2:740, and 748; *Threefold Paradise*, pp. 315-6; Increase Mather, *A Dissertation Concerning*

when he limited the initial conflagration to "exactly M.DC furlongs, or 200 Italian miles" of the papal territories, yet left the ancient terrain "occupyed by the *Babylonian*, and *Persian*, and *Graecian* Kingdomes ... untouched."[14] Such an interpretation of the last things seemed quite logical (and necessary) because the timing and character of the conflagration, like interstices in a house of cards, strictly depended on the nature of the Saved Nations and vice versa. Displace one card and the whole house would come crumbling down.

Isaac Newton, too, wondered about the remaining mortality of the Saved Nations and how they would survive the Petrine conflagration since it would be incumbent upon them to replenish the millennial New Earth with their offspring. At Christ's Second Coming, Newton speculated, there are "mortals remaining on earth," who if found in the Scroll of Life are "saved by being either caught up into the air to be with the Lord or left below on earth in the Kingdome of mortals which he thenceforth rules with a rod of iron and that the rest are adjudged to death and cast into the Lake of fire."[15] Newton's bifurcated answer, too, suggests that he wrestled with this issue, yet his diffidence in resorting either to miracles (the Rapture) or natural occurrences in solving the logistics of sheltering the Saved Nations during the inundation by fire is revealing: "And whilst the Apostle Peter tells us that none but the wicked shall suffer" by this *diluvium ignis*, "I cannot take it for a conflagration of any considerable part of this <habitable globe> globe whereby the rest of the habitable world may be annoyed. And if the world natural be not burnt up there is no ground for such a renovation as they suppose."[16] Up to this point, then, Newton closely followed Mede's literalist position because the underlying hermeneutical principles did not particularly violate any of Newton's firmly-held scientific laws. However, the Petrine prophecy of the Earth's holocaust (2 Peter 3:10) was a challenge of an altogether different caliber.

Indeed, the debate over the spurious authorship of the Second Epistle of Peter had caused Origen and Eusebius to voice their dissatisfaction long before the issue resurfaced in the seventeenth century. In Newton's day, the

---

the Future Conversion of the Jewish Nation (London, 1709), pp. 29, and 31. Thomas Burnet covers the same issues in *Sacred Theory*, pp. 37, 51, 66-7, 83-4, 262, 289-91, 385-7, and 338; John Ray, *Three Physico-Theological Discourses*, 3rd edition (London, 1713), pp. 303-5; and Whiston, *New Theory*, pp. 285-91, and 442-7.

14 *Works*, pp. 593, 809-13; see also Cotton Mather, "Problema," pp. 398-401, and 421.

15 Yahuda MS 6, fol. 16r, at the Jewish National and University Library, Jerusalem. Quoted by permission. Cited in Manuel, *Religion*, Appendix B, "Of the <world to come,> Day of Judgment and World to come," p. 132. See also Yahuda MS 9.2, fol. 138r, at the Jewish National and University Library, Jerusalem. Cited in Manuel, *Religion*, p. 100.

16 Yahuda MS 6, fol. 17r, at the Jewish National and University Library, Jerusalem. Quoted by permission. Cited in Manuel, *Religion*, p. 133.

debate on its authorship had been reopened by the Dutch jurist and theologian, Hugo Grotius (1583-1645), whose *Annotationes in Vetus et Novum Testamentum* (1642) marked a new departure in biblical hermeneutics. Grotius insisted that Peter could not have authored the Second Epistle because he had died during the reign of Emperor Nero (54-68 C. E). The apocalyptic doom pronounced on Jerusalem (2 Pet. 3), Grotius argued, therefore, must have been written *after* the Roman destruction of Jerusalem; Grotius' candidate for author was Simeon, Bishop of Jerusalem, who lived until Trajan's time (c 115 C.E.).[17] This emphasis on historical contextualization of the biblical event embraced literalist exegesis and preterite fulfillment while stripping prophecy of its futurist abrogation. Although disagreeing with this conclusion, Henry Hammond (1605-1660), a pioneer of modern biblical criticism, took the Grotian method of historical contextualization to its inevitable conclusion. Internal evidence (2 Pet. 3:7) suggested to Hammond that Peter's doomsday prophecy "belong[s] all to that judgment of the *Jews* and not to the day of universal doom or destruction of the whole world."[18] The Petrine conflagration, he argued, applied in a historical and literal sense only to Peter's own time; any futurist application of the fire dissolving the heavens and the earth violated the historical context of the prophecy and had to be understood in an allegorical or anagogic sense. Preterists like Hammond historicized Peter's prophecies and allegorized his eschatology, thus subverting the very foundation on which much of the literalists' expectation of the future conflagration depended.[19]

With such formidable guns pointing at the bedrock of prophetic interpretation, old-guard theologians rushed to their exegetical trenches. Any serious physico-theologian could no longer adopt unflinchingly Mede's literalist conflagration without scrutinizing its theological and cosmological underpinnings. Even the most orthodox millenarians during the Restoration period turned to tradition, trying to corroborate biblical predictions about the *diluvium ignis* with external proof from classical and pagan sources;

[17] Hugo Grotius, *Annotationes in Vetus et Novum Testamentum Juxta Editionem Amstelaedamensem, MDCLXXIX* [1642], 2 vols. (London, 1727), 2:390-4.

[18] Henry Hammond, *A Paraphrase and Annotations Upon all the Books of the New Testament* (1653), in *The Works of the Rev. and Learned Henry Hammond*, 4 vols., 7th Edition Corrected and Enlarged (London, 1702), 3:719-20; see also 3:728-30.

[19] In *Paraphrase*, 3:101-8, Hammond applied the same historical contextualization to Matt. 24 and limited the prophecy to the destruction of Jerusalem in 70 C.E. How volatile the issue was can be seen in John Lightfoot's adoption of Hammond's method, in *The Whole Works of the Rev. John Lightfoot, D.D.*, ed. by John Rogers Pitman, 13 vols. (London, 1822-5), 7:78-9, 112-27; 9:328, and 354; 11:303-7, and in the diatribes against the Grotian disciples, in Thomas Burnet, *A Treatise Concerning the State of the Departed Souls Before, and At, and After the Resurrection* (1681), 2nd ed. (London, 1730), pp. 128-44; in his *Sacred Theory*, pp. 295-6, and 306-21; in Ray, *Three Physico-Theological Discourses*, pp. 300-25, 388-93, and in Thomas Sherlock, *Discourses on the Use and Intent of Prophecy* (London, 1726).

ultimately, they turned to nature as a referee. If the conflagration was a universally held belief, Cotton Mather argued, for instance, then extrabiblical evidence might reinforce the Bible's literalism of this fiery destruction.[20] Millennialists, however, were a contentious lot of interpreters and could not agree among themselves about either the extent or the time of its occurrence. Nor was there unanimous agreement on whether the conflagration spoken of in the Petrine prophecies, in Isaiah, in Joel, in Revelation, and elsewhere, was indeed literal at all. Most of the ancient Church Fathers—among them Irenaeus, Eusebius, Origen, Jerome, Augustine, and Gregory the Great—asserted either a partial but not global conflagration on the one hand, or, on the other, a renovation or cleansing of the outward appearance but not substance of the earth.[21] Such disagreement among theological and classical authorities was greatly disconcerting; for how was one to decipher God's cosmic plan if even Church authorities sharply disagreed among themselves? For instance, Newton's colleague, the English nonconformist, biologist, and botanist, John Ray, did what virtually all of his fellow millenarians were doing at that time. In his famous *Three Physico-Theological Discourses* (3rd ed., London, 1713), he collated the opinions of the Church Fathers with those of Stoic philosophers in Greek and Roman history but turned to empirical verification in the constitution of Nature to determine the issue.

Thus John Ray, and many like minds, wondered how to ascertain the intent of God's prophetic spirit. Ray wondered whether a literal conflagration was feasible within the realm of scientific probability or whether supernatural agency would have to be invoked to offset Nature's laws and so to safeguard the literal accomplishment of the prophecies. He, therefore, explored in great detail four potential ways of how the flood of fire might come about:

1. The Waters again overflowing and covering the Earth.

2. The Extinction of the Sun.

3. The Eruption of the *Central Fire* enclosed in the Earth.

4. The Driness and Inflammability of the Earth under the Torrid Zone, and the Eruption of all the *Vulcano's* at once.[22]

For John Ray, a new flood of water was entirely within the realm of natural causality, for rainwater washing down the hills and mountains might

[20] Cotton Mather, *Threefold Paradise*, pp. 194-201; see also Ray, *Three Physico-Theological Discourses*, pp. 326-7 and Thomas Burnet, *Sacred Theory*, pp. 241-58.

[21] See Mather, *Threefold Paradise*, pp. 269-72; Ray, *Three Physico-Theological Discourses*, pp. 320-5, and Thomas Burnet, *Sacred Theory*, pp. 246-53.

[22] Ray, *Three Physico-Theological Discourses*, p. 339.

fill the oceans with such quantities of sedimentary material that a second Noachian deluge was entirely feasible. But the Bible spoke of a flood of fire, not water, to accomplish the dissolution of the earth. Ray also ruled out the natural extinction of the sun as a possible cause of global destruction. The extreme cold accompanying the cooling of the sun, though extinguishing all life on earth, would thus be "quite contrary to a Dissolution by Fire, of which the Apostle speaks." Again, we see how for Ray (as indeed for many of his confreres) natural causes are admissible only if they are in harmony with biblical prophecy. His attitude toward the third postulate was governed by similar concerns. Ray rejected the Cartesian hypothesis "that the Earth was originally a Star, or Globe of Fire, like the Sun, or one of the Fixt Stars, situate in the Center of a *Vortex* continually whirling round with it." Even less pleased was Ray with the Cartesian notion that the Earth's own orbit might deteriorate through a process of cooling until it was "drawn in, and wholly absorpt" by the "vigorous Incroachments of the neighboring *Vortex* of the Sun." Indeed, Ray's cosmology was still overshadowed by his physico-theological endeavor to reconcile scientific advances with the teachings of the Bible. He spoke for most of his peers when he asserted, "But tho' I do reject the *Hypothesis* [of René Descartes]; yet the Being of a Central Fire in the Earth is not, so far as I understand, any way repugnant to Reason or Scripture... . Neither is it repugnant to the History of the Creation in *Genesis*."[23] His concession notwithstanding, Ray hesitated to embrace this hypothesis; he listed it "as a possible" rather "than probable Means of a Conflagration." Likewise, Ray rejected the fourth postulate, the exsiccation of the earth's moisture by the sun's increasing heat, leading to an utter parching and ultimate combustion of the earth: "That there hath not yet been, I appeal to Experience, the Countries lying under the Course of the Sun [equator], being at this Day as fertile as ever they were."[24] That a global drought now or in the future might render the earth flammable seemed illogical to Ray as well, for neither the heat of the sun nor the quantity of water on earth had changed since the beginning of creation. Ray pointed to the universal law of checks and balances in nature that would always offset any disequilibrium and restore the earth's natural harmony:

> And therefore there can be no Desiccation of the Seas, unless by turning all its Water into Vapour, and suspending it in the Air, which to do, what an immense and long-continuing Fire would be requisite? to the Maintenance whereof all the inflammable Materials near the Superficies of the Earth would not afford Fewel enough... . This Fixedness and Intransmutability of Principles secures the Universe from Dissolution by the prevailing of one

[23] *Ibid.*, pp. 374-5, and 377.

[24] *Ibid.*, p. 384.

> Element over another, and turning it into its own Nature; which otherwise it would be in continual Danger of.[25]

Pointing to nature's uncanny ability to return to its intrinsic harmony, Ray also ruled out any gradual deterioration of the earth as a likely cause of bringing about the literal conflagration foretold in the Bible. But since "the Instrumental Efficient of this Dissolution shall be natural" and "be effected by a superficial Fire," it "must be the Work of GOD, extraordinary and miraculous"—unless, of course, we adopt the Grotian method proliferated by Henry Hammond and restrain the Petrine prophecy to the destruction of Jerusalem under Titus.[26] If John Ray's analysis is an indication of how the advancement of the natural sciences complicated the eschatological theories of Restoration and eighteenth-century millenarians, perhaps then we can understand why he was forced in his final analysis to attribute the Petrine conflagration to God's miraculous intervention. But even here is evidence that Ray felt increasingly attracted to the method of historical contextualization. Perhaps John Ray's scientific theory is less sophisticated than those of his contemporaries such as Thomas Burnet and William Whiston; perhaps that is why Ray, more often than his more illustrious colleagues, ultimately deferred to God's wisdom in matters of apocalyptic significance.[27]

Isaac Newton's colleague, Thomas Burnet, FRS, Master of the Charterhouse, and Chaplain to William III, had much to say about such eschatological issues. Burnet's famous *Sacred Theory of the Earth* (1681-90—see Note 7) may well have forced William Whiston and ultimately Sir Isaac himself to scrutinize more closely the scientific underpinnings of their cosmological systems, especially in their efforts to reconcile the literalism of Scripture prophecy with their own physico-theological advances in the natural sciences. Like Mede and Ray before him, Burnet devoted much attention to the issue of how the literal conflagration foretold in the Bible might be effected through natural causes. His collation of Daniel with Thessalonians and the Revelation convinced Burnet that the conflagration there spoken of principally concerned the destruction of Antichrist at Rome,

[25] *Ibid.*, pp. 387-8.

[26] *Ibid.*, pp. 388-9; see also pp. 299-320, and 393.

[27] Perhaps that is why the Puritan clergyman Alexander Nisbet (c. 1623-69) evaded the issue in his *An Exposition of 1 & 2 Peter* (1658), Reprint ed. (Edinburgh: The Banner of Truth, n.d)., pp. 285-6. Nisbet insisted that we should be more concerned with our spiritual salvation "than to be taken up in enquiring and determining whether the visible heavens and the earth, and the rest of the creatures of that kind, shall then be totally and for ever annihilated; or whether there shall be a new edition of them all." Nisbet's opposition resonates similar concerns in the annotations of the *Geneva Bible* (1566, 1599) which, though confirming a literal fire prophesied in 2 Peter 3:10, discouraged further inquiry by asserting that "nothing could be spoken more grauely."

with the inferno gradually spreading to encompass the whole Earth.[28] If that were so, Burnet pondered, then the geological constitution of the Italian peninsula should also corroborate such a reading. And if there were some sort of link between the predicted conflagration and the constitution of Nature, then the verity of the prophecies might also be ascertained through the help of natural science. Burnet was convinced he had found such a seamless juncture between the two books of God:

> Thus much being allow'd from Scripture, let us now return to Nature again; to seek out that part of the Christian World, that from its own constitution is most subject to burning; by the sulphureousness of its Soil, and its fiery Mountains and Caverns. This we shall easily find to be the *Roman Territory*, or the Countrey of *Italy*: which, by all accounts, ancient and modern, is a store-house of fire; as if it was condemn'd to that fate by God and Nature, and to be an Incendiary, as it were, to the rest of the World.[29]

Such a corroboration of the prophecies with the composition of the Italian soil was not at all accidental. Burnet congratulated himself for having found the suture where "both our lines meet in this point," where Scripture and Nature coalesced. The Italian soil substantiated his conclusion that at Christ's return God's fire would first consume "the City of *Rome* and the *Roman* Territory" and then spread further outward.[30]

But what natural evidence did Burnet find so convincing? Untold numbers of volcanic eruptions of Vesuvius, Aetna, Strombole prognosticated that God could easily carry out his plan; earthquakes galore caused by the collapse of subterraneous caverns would raise new mountains near Rome and "vomit out fire" to ignite layers upon layers of sulfurous matter; and lastly, flashes of lightning would kindle the combustible smoke ascending from Italy's brimstone and thus consume whatever lay in its path. Such natural harbingers, recurring since time immemorial, affirmed the Almighty's natural supernaturalism. Unlike Mede who resorted entirely to miraculous causes in explaining the conflagration, Burnet—true to his time—attributed all providential actions to the First Cause while explaining the event itself through natural causation.[31] In this manner, God's fiery destruction of the papal Antichrist might even precede in time the General Conflagration as a forewarning for God's people to ready themselves for the *annus mirabilis*: "And particularly the *Jews*, being Spectators of this Tragedy, and of the miraculous appearance of our Saviour, may see the hand

[28] Burnet, *Sacred Theory*, p. 289. See also 2 Thess. 1:7, and 2:8; Rev. 18:8, and 18; 19:3, and 20; Dan. 7:9, 10-11; and 17:18.

[29] Burnet, *Sacred Theory*, p. 289.

[30] *Ibid.*, p. 289.

[31] *Ibid.*, pp. 289-90.

of God in it, and be convinc'd of the truth and divine authority of the Christian Religion. I say, this supposition would leave room for these and some other prophetick Scenes, which we know not well where to place."[32]

Indeed, Burnet like many of his contemporaries couched Mede's ancient idea of the Jews' miraculous conversion in the scientific language of the Royal Society. Burnet's "Conflagration of the Earth Made Feasible" might thus deprive his allegorizing and historicizing colleagues of the grounds and reasons for abandoning the first principles of biblical hermeneutics.[33] The reason why they had abandoned the time-honored principles of literalist exegesis was that they could not conceive how such a holocaust of the natural world might come about. That is why they made the conflagration "signifie no more than political changes of Governments and Empires."[34] This is of course the route that Isaac Newton had mapped out in his "Rules for interpreting the words and language in Scripture."[35] As if to take a stab at Sir Isaac's hermeneutical system, Burnet lambasted the disciples of Baruch Spinoza for dismissing the poetic flights of the Hebrew prophets as little more than turgid metaphors that chiefly aimed at instilling piety in the hearer "for devotional motives."[36] Any historical fulfillment of the prophecies was simply a smaller type of their larger abrogation at the Second Coming:

> ...I take the true reason and design of that, to be a typical adumbration of what was intended should literally come to pass in the great and universal destruction of the World; whereof these partial destructions were onely shadows and prefigurations. But to determine this case, Let us take the known and approved rule for interpreting Scripture, *Not to recede from the literal sence without necessity*, or where the nature of the subject will admit of a literal interpretation. Now as to those cases in the old Testament, history and matter of fact do show that they did not come to pass literally, therefore must not be so understood; But as for those that concern the end of the World, as they cannot be determin'd in that way, seeing they are yet *future*; So neither is there any Natural repugnancy or improbability that they should come literally

[32] *Ibid.*, p. 291.

[33] Luxon points out that "English Reformers from William Whitaker (d. 1595) to John Weemse (d. 1636) all felt obliged to announce that the Scriptures were perspicuously and singularly literal and to denounce the alleged duplicities and equivocations of medieval exegetical practice, now associated with "Papists" and "Jesuits" (pp. 9-10). Luxon's argument can be extended at least into the early eighteenth century when William Whiston insisted that "A single and determinate sense of every Prophecy, is the only natural and obvious one; and no more can be admitted without putting a force upon plain words," in *The Accomplishment of Scripture Prophecies* (London, 1708), p. 13.

[34] Burnet, *Sacred Theory*, p. 294.

[35] Reprinted in Manuel, *Religion*, pp. 116-9, and 132-3; see also Newton, *Observations*, Chap. 2.

[36] Baruch Spinoza, *A Theologico-Political Treatise* (1670), trans. R. H. M. Elwes, (New York: Dover Publications, 1951), pp. 95, and 98-119.

to pass: On the contrary, from the intuition of that state of Nature, one would rather conclude the probability or necessity of them.[37]

Surprisingly, Thomas Burnet's avowed defense of literalist exegesis might easily deceive unsuspecting readers, for in reading biblical events as prophetic types adumbrating a future accomplishment, he argued in effect for a double accomplishment: a smaller fulfillment in the past adumbrating a larger one in the not-so-distant future. In doing so, Burnet inadvertently supplied grist for the mills of Grotian historicists, who tried to impede the proliferation of multiple accomplishments by insisting on a single, historical abrogation in the time of the prophet's utterance.[38] Ironically, Burnet rebuked those who "would allegorize" the Petrine conflagration just as much as he censured the preterists of Hammond's caste who "would confine them to the destruction of *Jerusalem*." In fact, his own hermeneutics was heavily indebted to the methodology of both parties. The destruction of the Holy City by the Romans (70 C.E.) was not the sole intent of 2 Peter 3:10, Burnet argued. It was a type and prefiguration of the "universal and final destruction of all God's Enemies," a sign from God that he is able and willing to destroy the whole earth: "The same power and providence that have already wrought the one, will also work the other, in due time, the former being still pledges; as well as prefigurations, of the latter."[39] Truth to tell, Burnet's double application of Scriptural prophecy remained literal throughout, but in doubling the literal accomplishment by reading the historical event as a prophetic type, Burnet achieved in effect a quasi-allegorization of the looked-for conflagration. Perhaps William Whiston's charge was therefore not entirely unjustified when he reproached Burnet for tampering with "the Letter of Scripture" and for arguing that the holy prophets "only secur'd the fundamental and general Verities" while accommodating everything else "to the needs of Men, [rather] than to the reality of Things."[40] Whiston's rather self-serving critique, however, is more a literary apologia for and justification of publishing his own rejoinder in

[37] Burnet, *Sacred Theory*, p. 294.

[38] For instance, John Hammond defended his master, Hugo Grotius, by insisting that Grotius did not exclude the literal fulfillment of OT prophecies in the NT but that, in his famous hermeneutical work, *Annotationes*, Grotius was primarily concerned with "the *first* and *literal interpretation*" of the OT prophets, "where there is one *immediate completion* of each *Prophecy* among the *Jews* of or near *that* time, wherein it was written, another more *remote* and *ultimate* concerning *Christ*, or the times of the *Gospel*." In his *Annotationes*, Hammond continued, Grotius therefore established "most *distinctly* the *first*, or *literal* sense, as that is *terminated* in the *immediate completion* ... because it was most *neglected* by other *interpreters*, who were more *copious* in rendring the *mystical notation*" as it applied to the New Testament, in *A Second Defense of the Learned Hugo Grotius* (1655), Hammond, *Works*, 2:81.

[39] Burnet, *Sacred Theory*, p. 296; see also, pp. 287-88.

[40] Whiston, "Introductory Discourse," pp. 77, and 78-9. The best analysis of Whiston's hermeneutics is James E. Force, *William Whiston: Honest Newtonian* (Cambridge: Cambridge University Press, 1985).

New Theory of the Earth (1696) than a valid criticism of Burnet's defects. Whiston's indictment, therefore must not be applied to all aspects of Burnet's *Sacred Theory*.

For instance, if we consider Whiston's allegation in light of the Newtonian taxonomy that decodes apocalyptic metaphors as political allegory, we see the level of insincerity to which Whiston here stoops. To Isaac Newton the shakings of the heavens and earth, the darkening of the sun and moon, and the falling of the stars stands for "the ceasing of a kingdom, or for the desolation thereof" and conversely "for the return of a dispersed people into a body politic or ecclesiastical."[41] As a physico-theologian, Thomas Burnet resisted the eighteenth-century impulse of turning, as Bernard Mandeville did in his *The Fable of the Bees* (1714, 1723), to political allegory as a legitimate hermeneutic device for deciphering prophetic language. In fact, Burnet's literalism led him to embrace natural causes that are not entirely removed from modern scientific theories on the axial tilt of the earth's poles—theories that even Whiston and Newton looked upon favorably:

> This phrase of *shaking of the powers of heaven*, comprehends, in a good measure, both these Heavens, of the Fix'd Stars and of the Planets; >Tis that change of situation in the Axis of the Earth, which we have formerly mention'd; whereby the Stars will seem to change their places, and the whole Universe to take another posture.... And as the heavens and the earth were, in this sence, once shaken before; namely at the Deluge, when they lost their first situation; so now they will be shaken again, and thereby return to the posture they had before that first concussion. And this I take to be the true literal sence of the Prophet *Haggai*, repeated by St. *Paul, Yet once more I shake not the Earth only, but also heaven....* [Consequently,] We need not now look upon these things as Hyperbolical and Poetical strains, but as barefac'd Prophecies, and things that will literally come to pass as they are predicted.[42]

Thomas Burnet's scientific literalism is perhaps as dexterous (or quaint) as any of his colleagues' systems—not excepting those of Whiston and Newton. But that, ultimately, is not really the issue. For the sake of comparison, let us briefly turn to Whiston's rational causation of the earth's literal holocaust. Whiston, too, felt the pressure of staying within the realm of the possible and probable, lest fanciful conjectures cast aspersion on both the prophecies and the natural sciences.[43] He conjectured that the most likely cause of this cataclysmic event might be found in the same cause that effected the formation of the earth in the first place: "a Comet" or its

[41] Newton, *Observations*, p. 18.

[42] Burnet, *Sacred Theory*, pp. 298-9; see also, pp. 324-5.

[43] Whiston, "Introductory Discourse," p. 65.

sulfurous "Atmosphere" constitued the original "Chaos" from which the earth arose.[44] The creation of the earth from the primordial Chaos just like its dissolution and return into Chaos would be precipitated by the same gravitational forces of a comet, Whiston postulated. But unlike Burnet who was ridiculed for arguing that Noah's flood was caused by waters contained in subterraneous caverns which, upon their collapse, caused the inundation described by Moses, Whiston attributed Noah's deluge to yet another comet passing the earth. And, finally, yet another comet (perhaps the same one) would be responsible for precipitating the conflagration and its cataclysmic effects[45] described in Joel 2:30-31, Matt.24:29, and Luke 21:25-26:

> The *Atmosphere* of the Earth, before the Conflagration begin, will be oppress'd with *Meteors, Exhalations* and *Steams*; and these in so dreadful a manner, in such prodigious quantities, and with such wild confused Motions and Agitations, that the *Sun* and *Moon* will have the most frightful and hideous Countenances, and their ancient Splendor will be entirely obscur'd; the *Stars* will seem to fall from Heaven; and all manner of horrid Representations will terrify the Inhabitants of the Earth.[46]

Careful examination of this passage then reveals that Whiston's literalism here depends to a lesser extent (than does that of Ray and Burnet) on natural phenomena that validate the literalist tenor of millenarian taxonomies, than on Whiston's (pardon the mixed metaphor) deus-ex-machina device of an optical illusion. The darkening of sun and moon, the falling of stars, are therefore literally true only if one allows for a deception of the senses: the Mosaic description of the creation is therefore true only to the extent that Moses described what he thought he saw; and the conflagration described by the Hebrew prophets and by Christ's apostles is literally true only because the exhalations of the approaching comets would pollute the atmosphere to such an extent that an eyewitness might *think* the sun, moon, or stars were literally falling. The clouded visions of the doomsday prophets were therefore accurate representations of what would be true to the *eyes* (but not in fact) of those present at the coming of the Lord. By contrast, Burnet's explication embraces a form of hyperliteralism in which the Petrine conflagration is true to the letter of every word:

[44] *Ibid.*, p. 34.

[45] Interestingly, theories of this nature have regained prominence in explaining the extinction of dinosaurs. For instance, the Nobel laureate Luis Alvarez postulated in 1980 what by now is almost religion: "An Extraterrestrial Cause for the Cretaceous-Tertiary Extinction" of dinosaurs through a meteorite that impacted the Earth, causing a dust storm, darkening the atmosphere, and suppressing photosynthesis that led to the extinction of large reptiles. See Charles Officer and Jake Page, *The Great Dinosaur Extinction Controversy* (New York: Addison-Wesley, 1996).

[46] Whiston, *New Theory*, p. 287.

> St. *Peter*, who is our chief Guide in the doctrine of the Conflagration, says, *The Elements will be melted with fervent heat*; besides burning up the works of the Earth. Then adds, *Seeing all these things shall be dissolv'd, &c*. These terms of *Liquefaction* and *Dissolution* cannot, without violence, be restrained to simple devastation and superficial scorching. Such expressions carry the work a great deal further, even to that full sence which we propose... . Wherefore if you would cast the Earth into a new and better mould, you must first melt it down; and the last Fire, being as a *Refiner's* fire, will make an improvement in it, both as to matter and form.[47]

Burnet's hyperliteralism thus stands in stark contrast to Whiston's explanation of an optical illusion. But Whiston's story is not all told yet, for the earth's inferno was "not to extend to the entire Dissolution or Destruction of the Earth, but only to the Alteration, Melioration, and peculiar Disposition thereof into a new State" to receive the Saints and Martyrs of the First Resurrection "for its Inhabitants." Following the millennial reign of these Raised Saints on earth, Whiston, like Mede before him, expected the Second Resurrection and General Judgment just prior to "the final Consummation of all Things," which purported to him that "the Earth will desert its present Seat and Station in the World, and be no longer found among the Planetary *Chorus.*"[48] Certainly, one cannot reproach Whiston for strictly maintaining the literal sense unless it violated reason or nature. But Whiston's method also forced him to resort to such expedients as extraterrestrial phenomena and to the deception of human senses to stay within the boundaries of his own precepts. While physico-theologians with more knowledge of contemporaneous science than faith in the Bible found themselves drifting more and more toward allegory, lesser minds became dogmatic or resigned themselves to the wisdom of God. If this latter posture was another kind of hermeneutical extremism no longer tenable in Newton's day, Whiston's literalism represents an extremism of a different sort: nature as text and context, a combination of reader-response theory and cosmic causality that is perhaps as startling as it is phenomenal.

So, how does Newton navigate through the maelstrom of hermeneutical currents? Did Newton agree with Whiston's explanation of the conflagration? The answer to this question is perhaps subject to conjecture. Yet Newton's taxonomy of Hebrew metaphors provides a glimpse of the (almost) irreconcilable differences between master and disciple. As Richard Westfall reminds us, *Observations upon the Prophecies* was arguably a "product of his old age" even though his theological notebook, dating from

[47] Burnet, *Sacred Theory*, pp. 287-8.

[48] Whiston, *New Theory*, pp. 289 and 291. See also Burnet, *Sacred Theory*, p. 376.

the 1670s, already contained discussions of the prophecies.[49] Newton's taxonomy in "*Of the Prophetic Language*" reveals the same distaste for what he elsewhere calls a "vulgar and litteral sense" as he does in his "Of the Day of Judgment and World to come," which Manuel dates to the period of the 1680s.[50] In his early work on this issue, Newton clearly rejected the "received opinion" of a literal *diluvium ignis* at Christ's Second Coming, but he does so not without some hesitancy:

> But in the <end of the world> day of judgment there is also a litterall conflagration of the world politique in the lake of fire and to those that are cast into it a conflagration also of the world natural, the heaven and earth where they are being on fire and the elements melting with fervent heat. And whilst the Apostle Peter tells us that none but the wicked shall suffer <by> in this conflagration and that this is a time of refreshing to the Godly I cannot take it for a conflagration of any considerable part of this <habitable world> globe whereby the rest of the habitable world may be annoyed. And if the world natural be not burnt up there is no ground for such a renovation as they suppose.[51]

In the manner of Baruch Spinoza, Newton rejects the "fansies" of the dogmatists and their penchant for verbatim interpretations of the prophets' mystical language. Rather, the dissolution of the world by fire "in [prophetic] language signifies the consumption of Kingdoms by war."[52] Newton seems to accomplish his task by cutting his Gordian knot with the sword of Alexander. What leads Newton to metonymize this inferno as human warfare is that the Saved Nations on earth require some sort of hiding place, an escape that allowed Newton's mortal Saints to be sheltered, while the flood of God's fire would consume only the wicked. It is for this reason that Newton—like Joseph Mede before him—limits the scope of the Petrine conflagration to a restricted terrain of the "habitable world."[53]

Newton's early eschatology still shows definite signs of stress between literal accommodation of the prophecies and the logistics of accommodating the Saved Nations, whose mortality would confine them to the habitable globe. To provide for an escape, Newton (like Mede before him) still vacillated between the Rapture of the Saints in the clouds (1 Thess. 4:16-17) and a conflagration limited in scope to a particular part of the globe. At the Day of Judgment, Newton conjectured, the Saints "are adjudged to life and

[49] Richard S. Westfall, *Never at Rest: A Biography of Isaac Newton* (Cambridge: Cambridge University Press, 1980), p. 319.

[50] Newton, *Observations*, pp. 16-23; Cf. Manuel, *Religion*, p. 99.

[51] Yahuda MS 6, fol. 17r. at the Jewish National and University Library. Quoted by permission. Reprinted in Manuel, *Religion*, p. 133.

[52] *Ibid.*, Cf. Manuel, *Religion*, p. 132.

[53] *Ibid.*, Rpt. in Manuel, *Religion*, p. 133. See also Mede, *Works*, p. 593.

saved by being either caught up into the air to be with the Lord or left below on earth in the Kingdome of mortals which he thenceforth rules with a rod of iron and . . . the rest are adjudged to death and cast into the Lake of fire."[54] Again, Newton's early eschatology emphasizes a developmental and process-oriented millennial New Earth rather than a miraculous transformation of the Saved Nations into immortality. Newton's gradualism, then, is akin to the post-millennialism of his English colleague Daniel Whitby, outlined in *True Millennium* (London, 1703), for whom the present world would gradually ripen toward perfection in a climactic progression at the end of the millennium. If Newton's "Of the Day of Judgment and World to come" (1680s) still exhibits all the idiosyncrasies of an imperfect system, his mature *Observations* evinces a harmonious formula that has solved virtually all of its earlier contradictions. We should therefore not be surprised that Newton confidently declares in his old age, "Fire and meteors refer to both heaven and earth, and signify as follows; burning any thing with fire, is put for the consuming thereof by war; a conflagration of the earth, or turning a country into a lake of fire, for the consumption of a kingdom by war."[55] If these cited passages mirror Newton's true views on the conflagration and on other eschatological matters, then Sir Isaac could not have fully approved of Whiston's conflagration. Indeed, whatever taxonomic underpinnings determined Newton's hermeneutics, the interior logic of his developing system governed his literal or allegorical eschatology just as much.

That the Atlantic proved no barrier to the hermeneutical tempests gathering strength in Europe can be seen in the writings of America's "national gargoyle," the New Englander, Cotton Mather. His incessant calculations of prophetic chronometry repeatedly provoked a retrenchment of his avowed literalism. Yet while Mather willingly compromised on his literalist stance on the restoration of the Jewish nation,[56] he drew his line of battle in front of the camp of the metaphorists who ridiculed the hyperbolical language of the Hebrew prophets: "*Thus a Storm*," allegorists were given to object, "*is often represented in such Pompous Terms, as if the whole Frame of Nature had been convuls'd, & the Universe on the Point of Dissolution.*"[57] Grotius' allegorist disciples missed the whole point by ignoring the prophetic intent of the Scripture, Mather countered. Nor did they understand the typological design of the "Prophetic Spirit," for which

[54] *Ibid.*, fol. 16r. Cf. Manuel, *Religion*, p. 132.

[55] Newton, *Observations*, p. 18. For an almost identical passage see Manuel, *Religion*, p. 132.

[56] See Reiner Smolinski, "*Israel Redivivus*: The Eschatological Limits of Puritan Typology in New England," *New England Quarterly* 63 (1990), pp. 357-95.

[57] Mather, *Threefold Paradise*, p. 192; see also Newton, "Rules for interpreting the words and language in Scripture," in Manuel, *Religion*, p. 118.

"the lesser *Particular Judgments*" were "an *Earnest* as well as a *Figure*, of the *General One*, wherein the *Frame* of Nature shall be dissolved." Allegories of the doomsday prophecies positively misjudged their true significance "because the *Prophetic Spirit* had an Eye to a . . . *Storm*, wherein the *whole Frame of Nature* shall be indeed *Convuls'd*, & our *Universe* undergo a *Dissolution*: An Event upon which the Old Testament never used one Word that was *Hyperbolical*; and it is impossible to use too *Pompous Terms*. The Figures are all short of the Reality."[58] Mather had certainly been sensitized to what Spinoza and later Ray, had discouraged as "lofty and tumid Metaphors, and excessive *Hyperbola's* and Aggravations" of "Oriental Rhetorick." Taking Spinoza's *Tractatus* as his matrix, Ray conceded that such flight of the "Poetick Style--though "somewhat conformable to the Oratory of those Countries"—should not be taken too literally but had to be decoded "as a *Jew* or an *Asiatick* would then have understood them."[59] Yet such concessions, Mather retorted, attenuated the prophetic spirit, which true to God's design, intended a double fulfillment: The smaller event generally accomplished in the historical past of the prophet's own time really signified a second, larger but most of all literal fulfillment in the latter days. This was certainly the case with the Petrine prophecy, Mather judged, which in predicting the immediate fall of Jerusalem actually intended the passing away of heaven and earth at Christ's Second Coming.[60] Again we see how, in redressing the contradictions of his millenarian thought, Mather—like his English colleagues—was forced to adjust his taxonomy to maintain the interior logic of his system. At the same time, Mather safeguards his literalism by merging Cartesian notions of the Earth's fiery magma with the modifications introduced by Ray, Burnet, and ultimately by Whiston, whom Mather admired greatly.

2.

Closely related to the millenarians' debate on the issue of the Petrine conflagration is the condition of the Saints of the First Resurrection and their abode in a celestial or terrestrial New Jerusalem. Again, Mede provides the blueprint for all later eschatologists, yet their systems evince considerable variations on a familiar theme. Their millenarian cosmologies, consisting of a network of elements, vary like the determinant of a mathematical matrix, with the change of each element. Mede defined the

[58] Mather, *Threefold Paradise*, pp. 184, 156, and 192.

[59] Ray, *Three Physico-Theological Discourses*, pp. 316-7. See also Spinoza, *Treatise*, pp. 94-6.

[60] See Burnet, *Treatise*, pp. 128-144, and *Sacred Theory*, p. 296. See also Ray, *Three Physico-Theological Discourses*, pp. 311-2.

nature of the Raised Saints and Saved Nations most clearly in his posthumously published *Paraleipomena* (1640) and incorporated in his *Works*. To Mede, the Saints of Christ's kingdom would consist of "two sorts": "*The deceased Martyrs*" (those who lost their lives for God), who are also called the Saints of the First Resurrection. They "shall resume their Bodies and Reigne in Heaven." The second group of Saints are "of the *Living*" (those who had not worshipped the whore of Babylon). They "shall Reigne on Earth," and this class includes "the Virgin-Christians of the *Gentiles*, (who are the Surrogate *Israel*)" and "*the Nation of the Jewes* then converted to the Faith of *Christ*."[61] Most significant for my purposes here is Mede's insistence on the corporeal nature of both classes of Saints—even those who would reign with Christ in the celestial New Jerusalem. The Saints of the First Resurrection living in the New Heavens, the Saved Nations on the millennial New Earth, even the raised dead of the Second Resurrection, would all be endowed with physical bodies that inevitably had extension. The three-dimensional bodies of the Raised Saints, co-rulers with Christ in Heaven, therefore necessitated a multi-dimensional abode to house the bodies of this privileged class: "Tis true," Mede insisted on a literal reading, that the "Glorified bodies have no need of inheritance temporal for their maintenance and nourishment, but for their mansion and habitation they have need of a place of abode."[62] For although the incorruptible bodies of the immortalized Saints in heaven required no food to sustain their bodies—such carnal pleasure belonged to bodies in the state of corruption—their immortal (albeit corporeal) bodies, however, did require some sort of dwelling, one in turn implying a three-dimensional state. If Mede's lengthy explication bespeaks his uncertainty about the physicality of the Raised Saints, their location, and function, it also reveals how much the interior logic of his system required constant readjusting. It therefore does not come as a surprise that many of Mede's disciples tried to develop more cogent answers.

Burnet, for instance, resolved the issue without necessarily compromising his literalist stance. He placed Mede's Raised Saints on earth, rather than in heaven, endowed them with the same immortalized bodies as Mede did, but unlike Mede had his Raised Saints procreate an holy offspring who would become the inhabitants of the New Earth. Because "all Mankind was destroy'd at the Conflagration: ... there is no resource for them any other way, than by a Resurrection." Consequently, "It may be very rationally presum'd, that Those that are raised in the first Resurrection, are those *Just* that will inhabit the *new Heavens and new Earth*" (Rev.20:4). Their reward, "or this Reign of theirs, is upon *Earth*: upon some Earth, new or old, *not* in

[61] Mede, *Paraleipomena*, p. 24, and *Works*, p. 604.

[62] Mede, *Works*, p. 812.

Heaven. For, besides that we read nothing of their Ascension into Heaven after their Resurrection: There are several marks that shew it must necessarily be understood of a state upon Earth."[63] Burnet's solution implies a drastic reinterpretation of this volatile issue and evinces in this case greater resemblance to Thomas Hobbes' political millennium with its capital in a terrestrial New Jerusalem than to Mede's conjecture of the embodied Saints in Heaven. Burnet, then, retains the physical bodies of Christ's co-rulers, yet instead of placing them in heavenly mansions, Burnet opts (like Hobbes) for a residence on earth.[64] Though raised from the dead, their immortalized bodies would partake of such rewards as procreation and filling the cleansed earth with their noble offspring. At the same time, Burnet completely dismissed Mede's idea of the Saved Nations, whose mortality during Christ's millennium posed something of a problem, an encumbrance that even Newton never quite overcame. By implication, Burnet's millennial New Earth (like that of Hobbes) differed substantially from that of Mede, of the young Mather, and of Newton, for the corporeal nature of the earth's inhabitants also determined the nature of the millennial kingdom.

The physical reunion of body and soul, which both just and unjust would undergo at the First and Second Resurrections, was hardly a unique feature in Burnet's eschatology. It can be traced to Johann Heinrich Alsted's *Diatribe de mille annis Apocalypticis* (1627). Translated into English by William Burton and published as *The Beloved City* (1643), Alsted's work left a lasting impression on later chiliasts.[65] Such proponents did not go unnoticed among the New World millenarians. It is therefore not surprising that Mather, in his own "Problema Theologicum" (1703), as indeed in all of his works on the subject, should argue that the souls of the dead would presently rejoin their immortal bodies: "Some, att the *Beginning* of the *Thousand Years*; others, not untill the *Thousand Years* are *Finished*. But the *Latter* then *Live gain*, as to their *Bodyes*; and therefore so much the *Former* too."[66] Mather never changed his mind on that issue, even though it took him until the end of his life to resolve the inconsistencies that characterized his earlier millenarian works. Two years before his own death, he was convinced that the Raised Saints in heaven will have bodies, whose physical and spiritual splendor defied human understanding:

63 Burnet, *Sacred Theory*, p. 329.

64 Thomas Hobbes, *Leviathan; or, The Matter, Forme and Power of a Commonwealth Ecclesiasticall and Civil* (1651), ed. Michael Oakeshott (New York: Collier Books, 1962), pp. 327-9 and 335-7.

65 Robert G. Clouse, "Johann Heinrich Alsted and English Millennialism," *Harvard Theological Review* 62 (1969), pp. 189-207, and "Rebirth of Millenarianism," in *Puritans, the Millennium and the Future of Israel: Puritan Eschatology 1600-1660*, ed. by Peter Toon (Cambridge: James Clarke & Co., 1970), pp. 115-25.

66 Mather, "Problema Theologicum," p. 406.

> Their *Bodies* will be so *Salted by the Garments of Light*, which GOD will putt upon them, that they will become *Incorruptible* under it. *Luminous Bodies*!... Not ceasing to be *Bodies*, or turned into meer *Spirits*; They will be *Material* still; but highly *Spiritualized* ... *Equal to the Angels*: Doubtless, Able to Move, and Mount, and Fly, as the *Angels* do.[67]

And as the Raised Saints soared with their luminous bodies in the New Jerusalem above, so did Mather's imagination take wing, unshackled by the clay of his Maker but not by the law of gravity. Interestingly, Newton's ideas here are not too far removed from those of his New English confrere. Newton's "Children of the Resurrection" were also endowed with weightlessness, albeit Newton never quite went into detail about how far their spiritual existence had obliterated their material bodies: "And as the Planets remain in their orbs, so may any other bodies subsist at any distance from the earth, and much more may beings, who have a sufficient power of self motion, move whether they will, place themselves where they will, and continue in any regions of the heavens whatever, there to enjoy the society of one another, and by their messengers or Angels to rule the earth and converse with the remotest regions."[68] Clearly, then, literalist exegesis of John's Apocalypse made for fascinating flights of the imagination—when supernatural explications could always be counted on as the last resort. But the story of the Saints in heaven is not all told yet, for the mysterious concept of the New Jerusalem is inscribed into this tangled tale.

The New Jerusalem, as Mather envisioned it, was not to be mistaken for a metaphor, nor was it a mere representation of God's grace residing in the heart of mankind, as some of the allegorist Church Fathers had interpreted the visions of Ezekiel and John (Ezek. chs. 40-47; Rev. chs. 4 and 21-22).[69] No doubt, the description of the City so admirably given in the Scriptures invited allegorization. We should therefore not be surprised to find metaphorical interpretations of the Celestial City even among the most ardent literalists. For instance, the annotators of the *Geneva Bible* (1560, 1599) assigned the concept of the New Heaven and New Earth (Rev. 21:1) to "the future estate of the Church in heauen after the last iudgment" and allegorized the cubical form of the celestial New Jerusalem as the Church's future perfection and eternal harmony: "*A fouresquared figure hath equall sides, and outright corners, and therefore the Grecians call by this name*

[67] Mather, *Threefold Paradise*, p. 255; see also p. 261.

[68] Yahuda MS 9.2, fol. 140r, at the Jewish National and University Library, Jerusalem. Quoted by permission. Cited in Manuel, *Religion*, pp. 101-2.

[69] For a review of the debate, see Mede, *Works*, pp. 815-6, and Daniel Whitby, *Treatise of the True Millennium* (1703), in *A Paraphrase and Commentary on the New Testament*, 2 vols. (Edinburgh, 1761), 2:4-5.

those things that are steady, and of continuance, and perfect."[70] Even the nonconformist Biblical exegete Matthew Henry (1662-1714), whose Presbyterian orthodoxy is still appreciated today, evinces a curious mixture of literal and allegorical comments in his *Exposition of the Old and New Testaments* (1708-10):

> The form of it was very regular and uniform: *It was four-square, the length as large as the breadth.* In the new Jerusalem all shall be equal in purity and perfection. There shall be an absolute uniformity in the church triumphant... The measure of the wall ... which is forty-eight thousand furlongs in the whole compass, or fifteen hundred German miles. Here is room sufficient for all the people of God—*many mansions in their Father's house.*[71]

In contrast, for Thomas Hobbes, the New Jerusalem is a place of salvation on earth: "That the place wherein men are to live eternally, after the resurrection, is the heavens... ["called *coelum emphyreum*"] ... is not easily drawn from any text that I can find." Rather, God's "*throne* is heaven" and man shall not ascend any higher than "God's *footstool* the earth."[72] For Hugo Grotius as well, prior to Hobbes, the New Jerusalem signaled the "perfection and eternal stability of the church"[73]; and to Grotius' greatest defender, Henry Hammond, the First Resurrection of the Saints (Rev. 20:1) applied to the resurrected "Church, not of the particular persons." Likewise, the New Jerusalem descending from Heaven signified to Hammond "not the state of glorified Saints in heaven" but "a flourishing condition" of the pure Church "here on earth." Its four-square form and "all this measuring of *the city* is mystically to be understood, and not literally, there is no doubt; but what the mystery is, will not be obvious to determine."[74]

Thomas Burnet, however, could hardly agree with the ground-breaking annotations of Dr. Henry Hammond, who "loves to contract and cramp the sence of Prophecies" far too much. On the one hand, Burnet vehemently objected to Hammond's allegorical application of the First Resurrection to the "state of the Church in general" rather than "to the Martyrs." Yet, on the other, Burnet rather liked Hammond's allegorization of the cubical New Jerusalem as the flourishing state of the Church on earth—except that Burnet added his particular admixture of literalism:

[70] Marginal annotations on Rev. 21:14-7, *Geneva Bible* (1599).

[71] Matthew Henry, *Commentary on the Whole Bible* (1708-12), 6 vols. (Peabody, Mass: Hendrickson Publishers, 1991), 6:953.

[72] Hobbes, *Leviathan*, pp. 326-8.

[73] Grotius, *Annotationes*, 2:436.

[74] Hammond, *Paraphrase*, 4:833, 837, and 839.

> The Scripture on the contrary, makes mention of the descent of the *New Jerusalem*, *Apoc* 21.2. making the Earth the Theatre of all that affaire. And the Camp of the Saints is upon the Earth, *ver.* 9. And these Saints are the same persons, so far as can be collected from the text, that rise *from the dead, and reign'd with Christ*, and were *Priests to God. ver.* 4, 5, 6. Neither is there any distinction made, that I find, by St. *John*, of two sorts of Saints in the *Millennium*, the one in Heaven, and the other upon Earth.[75]

Unlike the Raised Saints who occupy a celestial City in heaven (Mede, Mather, and Newton), Burnet's Raised Saints, though immortalized, would live in the terrestrial New Jerusalem in the Holy Land: "That the *New Jerusalem*-state is the same with the Millennium, is agreed upon, I think, by all Millenaries, Ancient and Modern.... That *Beloved City* is the *New Jerusalem*, and you see it is the same with the Camp of the Saints." If Burnet's hybrid theory echoes the one espoused by Hobbes, Burnet's literalism, however, is perfectly consistent with the internal logic of his eschatological system.[76]

Cotton Mather was similarly wrestling with the hyperbolic dimensions of the City in John's Revelation. Yet Mather's defense of the simple, literal translation of the prophetic language engendered a hyperliteralism rarely to be found elsewhere. Although he conceded that the cubical form of the New Jerusalem may be a "*Figure*," Mather insisted on a literal reading: "I entreat You, Syrs, turn it not all into meer *Shadow*, and Lett not your *Metaphysicks* operate & sublimate upon it, until you have made it no better than one of Your *Entia Rationis*, & a meer *Non Entity.*" Quite to the contrary: The celestial City described in Ezekiel and Revelation could be nothing less than a "*Material City*"—if it were to accommodate the corporeal bodies of the Raised Saints. "*Spiritualize* the Matter as much as You please," Mather challenged the cabal of English allegorists (Grotius, Hammond, and Lightfoot). "But if you think, a *Visible City*, of a *Cubical* Form is too *Corporeal* a Thing, yett you must allow, That there will be a *Place* of Reception for Bodies."[77]

To mistake the celestial City for a metaphor would violate his literalism employed throughout; for if the New Jerusalem had accommodated the physical body of the resurrected Messiah and if the City were to provide mansions for the re-embodied Saints, how then could the City be mere allegory? Neither was its size nor its precious building material far from tangible reality. No, the New Jerusalem above (Rev. 21) was to Mather as literal as the restored Jerusalem in the New Earth below:

[75] Burnet, *Sacred Theory*, pp. 334 and 356.

[76] *Ibid.*, p. 362; Hobbes, *Leviathan*, Chap. 38.

[77] Mather, *Threefold Paradise*, pp. 246, 244, and 245.

> The *Length & Breadth* and the *Heighth* of the CITY; aequal; each being *Twelve Thousand Furlongs*; or *Fifteen Hundred Miles*, every Way, in each of the Three Dimensions. The *Cubical Arrangement* of the *Mansions*, will not at all obtenebrate any of them, while the *City* is thus filled with the *Glory of the Lord*. No; It will be *the Inheritance of the Saints in Light*.[78]

Granted, the hyperbolical dimensions outlined in John's Apocalypse seemed unfathomable and naturally would give rise to metaphoric interpretations, especially in light of such Newtonian precepts as "What Ancient Tradition asserts of the Constitution of Nature ... is to be allow'd for True, where 'tis fully agreeable to Scripture, Reason, and Philosophy."[79] Although Cotton Mather rendered the latter doctrine in a preterite and allegorical sense in his old age, he always spoke of two separate Jerusalems: the New above and the Old below. During Christ's Sabbatism, "There shall then be *Two Jerusalems*," Mather expounded the issue in 1703: "*Johns* New *Jerusalem*, in which there is *No Temple*, and *Ezekiels* New *Jerusalem*, in which there shall be *a Temple*."[80] Interestingly enough, in this passage Mather did not identify Ezekiel's visionary Temple with John's, as he would do in his *Threefold Paradise* twenty-three years later. For as he saw it in "Problema Theologicum," John's vision applied to the celestial City in the New Heavens, while Ezekiel's applied to the restored Jerusalem in the New Earth. Assuredly, the two cities would be in complete concord, for the City of God in the clouds would send down its winged messengers, the Raised Saints, with instructions to the rulers below.[81]

Modern readers may smile at such hermeneutical niceties of long-forgotten lore. However trivial it may appear to us today, though, Mather's distinction between the two Jerusalems has significant ramifications for modern intellectual historiography concerning the Puritan Errand into the American wilderness.[82] While figural interpretation was closely aligned with Augustine's allegorization of these visions "as the Church of Christ extending over the whole world,"[83] Cotton Mather's literalism adopted as proof Tertullian's mirage described in his *Five Books Against Marcion* [3.24] and placed the New Jerusalem in the New Heavens, but visibly

[78] *Ibid.*, p. 244.

[79] Whiston, "Introductory Discourse," p. 95.

[80] Cotton Mather, "Problema Theologicum," p. 423.

[81] C. Mather, *Threefold Paradise*, pp. 304-6, and 273.

[82] In "*Israel Redivivus*" and in *Threefold Paradise*, pp. 53-9, I argue that New England's millenarians did not locate the capital of Christ's future kingdom in their New English Boston (nor on the American continent) but in the Judean Jerusalem upon the restoration of Israel. In attributing the sources of Manifest Destiny to Puritan New England, intellectual historians from Perry Miller to Sacvan Bercovitch have erroneously superimposed an essentially nineteenth-century ideology onto the seventeenth century.

[83] St. Augustine, *The City of God*, trans. Marcus Dods (New York: Modern Library, 1950), Book. 20, Chap. 11, p. 729; see also Chap. 17.

hovering (as Jonathan Swift's flying island Laputa) over the Judean Jerusalem in the New Earth: "The *City of God* in the *New Heavens*," Mather postulated in 1703, will extend "Fifteen Hundred Miles," and "It will be seated over the Land of *Israel*, which will now again be possessed by the *Israelitish* Nation." He underscored this literalist interpretation again in 1726/27 while writing his definitive treatise on eschatology: "The Situation of it [New Jerusalem], will be in a Part of the *Atmosphere*, which will be nearer to the *Earth*, where the *Nations* are to *Walk* in the *Light of it*, than as yett it is, and it will be conspicuous to the Nations."[84]

No matter how peculiar Mather's exegetical stance may strike modern readers, this hyperliteralist depiction was borne out by Mather's frustration with the onslaught of his allegorizing colleagues who had already succumbed to Arminianism, Arianism, and Hobbesian materialism that denied the existence of the soul: "A *Devil*, is in the apprehension of those Mighty acute Philosophers, no more than a *Quality*, or a *Distemper*.... We shall come to have no *Christ* but a *Light within*, and no *Heaven* but a *Frame of mind*, if the Scriptures must be expounded after the Rules of the modern *Sadducees*. Perhaps the *Scriptures* are *Fables* to that sort of men."[85] With such mounting challenges to hermeneutical orthodoxy, it is not surprising that Mather's defense of literalism became all the more pronounced—even though he was not beyond becoming a Grotian historicist when it suited his purpose.[86]

Had Isaac Newton lived long enough to read Mather's recently published manuscript "Threefold Paradise," he would certainly have found many explications to his liking. In fact, most of their common threads on the issue of the mortal Saved Nations and the Children of the Resurrection can be traced directly to Joseph Mede. Major differences, however, can be found in Newton's explication of the New Jerusalem and the Raised Saints. For Sir Isaac, the City of God

> is a city of mortals assembled from captivity and rules over the nations and continues for ever and how (as in the Apocalypse) the Gentiles come to her light... . [S]he is the Lamb's wife. By her being returned from captivity, her inhabiting the desolate cities and her inheriting the nations and by their making war upon her with weapons formed by the Smith you may know that she is a city of mortals; a city not in a literal sense, but mystically put for the

[84] C. Mather, "Problema Theologicum," p. 423; see also p. 421; and *Threefold Paradise*, p. 245; see also pp. 268-9, 287-9.

[85] C. Mather, *Memorable Providences* (Boston, 1689), Part. III, pp. 14-5, 16-7.

[86] See my discussion "The 'New Hermeneutics' and the Jewish Nation in Cotton Mather's Eschatology," in *Threefold Paradise*, pp. 21-37.

> whole nation of the Jews, the pretious stones and pillars and foundations thereof being the saints and Apostles.[87]

In Newton's explication, then, the New Jerusalem stands mystically for the whole Jewish nation ("the Lamb's wife") and the City's precious stones represent the Saints and Apostles as pillars of the spiritual temple. Even though Newton—like Mather—speaks of a double New Jerusalem, the two millenarians would have parted company on precisely what that doubled City looked like or what it represented. Newton's quasi-literalism gives way to allegorical representations, in which the City's "glorious description" (Rev. 21:11-27) is "a commentary upon this city."[88] On the one hand, then, Newton's New Jerusalem mystically represents the whole Jewish nation returned to her ancient patrimony while typifying God's spiritual temple as described in Revelation. On the other hand, Newton's New Jerusalem—unlike Mather's—is no literal city in Heaven, but rather describes Christ's assembly of Saints, whose spiritual function was typified in the dimensions and layout of Ezekiel's visionary temple:

> This city must be understood to comprehend as well Christ and the children of the resurrection as the race of mortal Jews on earth. It signifies not a material city but the <spiritual> body politique of all those who have dominion over the nation whether they be the saints in heaven or their mortal viceregents on earth and therefore the Apostle Paul in [Heb. 11] ... understands it of the saints in heaven and in Gal. 4.26 calls it Jerusalem which is above. Hence this city is not only long and broad as other cities are but rises high from the earth into heaven. Hence also the dimension of the sides thereof are double to those of the terrestrial Jerusalem described by Ezekiel [Ezek. 40:5 vs. Rev. 21:15].[89]

Clearly, then, Newton's analysis here combines a terrestrial City embodying the Jewish nation and a celestial City whose cubical nature represents Christ's "<spiritual> body politique." In short, Newton here retains a middle way between literalism and metaphor for things confined to the earth, yet employs a mystical or anagogical rendition for things that extend into the heavens. We recall that Burnet solved this issue by confining the Raised Saints and their offspring to the terrestrial New Jerusalem on the New Earth. Cotton Mather's literalism on the other hand required a literal "Castle in the Air" to accommodate the material bodies of the Raised Saints in the heavens. Newton, too, has his "Children of the Resurrection" fly

[87] Yahuda MS 6, ff. 13r and 14r, at the Jewish National and University Library, Jerusalem. Quoted by permission. Cf. Manuel, ***Religion***, pp. 128-9.

[88] Yahuda MS 9.2, fol. 138r, at the Jewish National and University Library, Jerusalem. Quoted by permission. Cf. Manuel, ***Religion***, p. 100.

[89] Yahuda MS 6, fol. 18r, at the Jewish National and University Library, Jerusalem. Quoted by permission. Cf. Manuel, ***Religion***, p. 134.

through the stratosphere of the heavens, even through the solar system, yet Newton does not concern himself whether these fortunate travelers are endowed with physical bodies.[90] In contrast, Mather's "Children of the Resurrection," though airborne, still retained "An Heavy Tendency to the *Earth*, doubtless, they will be so disencumbred from it, that *They shall mount up with Wings as Eagles.*"[91]

Why does Newton refrain from extending his avowed literalism into the heavens above, even though he does clearly envision his "Children of the Resurrection" to have "sufficient power of self motion, move whether they will, place themselves where they will, and continue in any regions of the heavens whatever"?[92] Answers to this question are, of course, open to speculation. But if we examine Newton's position on Christian mortalism, we may perhaps find clues in Newton's belief in psychopannychism.[93] In fact, Newton's explication of Gog and Magog (Satan's minions), who would attack the terrestrial New Jerusalem in a final, albeit ineffectual, battle at the end of the millennium, can be directly linked to his position on the nature of the souls of the dead. Examining Ezekiel chs. 38-9 for evidence, Newton looked for Gog and Magog among the nations: "Persia and Arabia and Afric and the northern nations of Asia and Europe" would briefly rise up against Israel made "invincible by holiness," but God would crush His enemies.[94] For Newton, then, Gog and Magog are rebellious unbelievers among the Saved Nations whose mortality, hence propensity for sinfulness, would still persist during Christ's millennial reign.

Newton's position can be better appreciated if we examine those of his fellow millenarians. Burnet, for one, was somewhat at a loss about how to account for Gog and Magog. He clearly recognized that these adversaries spoken of in Ezekiel and in Revelation could not come from his immortalized Raised Saints who had replenished the earth with their own sinless offspring. Because Burnet had earlier insisted that only God's Raised Saints would be spared from the conflagration at Christ's Second Coming, Burnet had to look for Ezekiel's Gog and Magog elsewhere. They would be "a People distinct and separate from the Saints" distinguished by their different manners, habitations, and origin: "There will be a double race of

[90] Yahuda MS 9.2, ff. 139r and 140r, at the Jewish National and University Library, Jerusalem. Quoted by permission. Cited in Manuel, *Religion*, p. 101.

[91] Mather, *Threefold Paradise*, p. 256.

[92] Yahuda MS 9.2, fol. 140r, at the Jewish National and University Library, Jerusalem. Quoted by permission. Cf. Manuel, *Religion*, pp. 101-2.

[93] For background and evidence, see James E. Force, "God of Abraham and Isaac (Newton)," pp. 179-200 and Philip C. Almond, *Heaven & Hell in Enlightenment England* (Cambridge: Cambridge University Press, 1994), pp. 38-80.

[94] Yahuda MS 6, fol. 18r, see also ff. 13r-15r, at Jewish National and University Library, Jerusalem. Quoted by permission. Cf. Manuel, *Religion*, pp. 134, and 127-9.

Mankind in that *Future Earth...* . The one born from Heaven," the immortal Raised Saints, who are "heirs of the Millennium. The others born of the Earth ... generated from the slime of the ground, and the heat of the Sun, as brute Creatures were at first." Burnet's allusion to Ovid's creation story of Deucalion and Pyrrha as ideational source for Gog and Magog clearly demonstrates how Burnet's system strained to the breaking point.[95] Cotton Mather, on the other hand, vacillated between several rivaling systems. In his early treatise on this issue, Mather still rejected the popular conjecture of his French colleague Pierre Poiret, whose *L'Oeconomie Divine* (Amsterdam, 1687) identified God's adversaries as "the Devils and the Damned" risen from hell at the end of the millennium, and attacking the saints terrestrial and celestial.[96]

In 1703, Poiret's conjecture seemed almost blasphemous to Mather. He could not see how the "Changed Saints" in the terrestrial New Jerusalem, having been filled with the Shechinah of God for a thousand years, could again fall prey to the Devil's machinations. Likewise, Mather could not see how the Raised Saints in the celestial New Jerusalem, who had triumphed in their former lives against Beelzebub, had undergone martyrdom for God, had received their reward in heaven, would now again succumb to Satan's allurements in his last stand: "One would have expected such a Fancy rather in a poem of *Miltons*," Mather sneered in his "Problema Theologicum" (1703), "than in a Treatise of Divinity." Consequently, it seemed more likely to Mather that Satan would prevail for a short while over many of the mortal nations on earth, especially those that had been on the "outskirts" of the City. Twenty-three years later, Mather was now resolved that the Gog and Magog mentioned in Ezekiel 38-9 was not the same as the one in John's Revelation (20:7-12), but removed by a thousand years. Rabbi David De Pomis' *Zemah David* (Venice, 1587) had identified Ezekiel's Gog as Armilus, the Messiah's adversary. This explanation now seemed logical to Mather, who noted that the prophecies of Ezekiel and Revelations applied to two entirely different battles, the former to the destruction of Antichrist at the beginning of the millennium, and the latter to Satan, loosed at the Battle of Armageddon at the end of Christ's millennial reign. Still, the multitude of nations who would join forces with Satan needed to be accounted for. In his *Threefold Paradise*, Mather was convinced that they could come neither from the Saints in Heaven nor from the Changed Saints on earth. Of the many works that Mather had read on the subject, those by Thomas Staynoe, John Smith, and Pierre Poiret now seemed to offer the best answer. They had conjectured (and Mather now agreed) that the numerous nations spoken

[95] Burnet, *Sacred Theory*, p. 374; Ovid, *Metamorphosis*, trans. Rolfe Humphries (Bloomington, IN: Indiana University Press, 1955), Book I, lines 418-36.

[96] Mather, "Problema," pp. 410-5.

of in Revelation were not of the godly Saints, but rather of the "Vast Number of *Humane Ghosts*," the evil souls of mankind, who had been confined in "Dismal *Receptacles*" for nearly "Seven Thousand Years" since the beginning of time and had "even lain in the Torments of Hell" for an entire millennium. Mather was knowledgeable enough in the wonders of the invisible world (so aptly described by his friend Henry More) to understand "that even *at This Day*, the *Ghosts* of the Dead which have *died in their Sins*, may be employed in some *Diabolical Services.'*"[97]

Why did Newton not adopt a similar position and identify Gog and Magog as the wicked souls of the damned? For Newton, such an ethereal army of wicked spirits and damned souls was implausible. Newton's own mortalist belief in psychopannychism did not concern itself either with the location or dimensions of souls as Mather's literalist system required.[98] Again, we see how Newton's a priori assumptions about psychopannychism determined his choice of candidates. Gog and Magog applies to the Devil and his *human* forces ("Persia and Arabia and Afric and the northern nations of Asia and Europe"), who in the final rebellion against the terrestrial New Jerusalem would be "devoured by fire from the throne, that is by war." Again the interior logic of Newton's millenarian system allowed him to stay in the realm of human possibility, and true to Rule # 9 of his "Rules for methodising the Apocalypse," Newton chose "those <interpretations> constructions which without straining reduce things to the greatest simplicity."[99]

Mather's Gog and Magog, however, were armies of wicked spirits and therefore could not be destroyed by human warfare alone. It is for this reason that Mather maintains a literal, yet supernatural fire issuing from heaven. Characteristically, Newton, the psychopannychistic mortalist, rejects any activity of the souls of the dead before their resurrection and incorporation at the day of Universal Judgment. He, therefore, locates his adversaries among terrestrial enemies of Israel and identifies the biblical conflagration as human warfare. In contrast, Burnet, More, and Mather believed that the souls of the wicked would be "imprison'd within their several Atmospheres in our Vortex" and required therefore a literal conflagration—one that would penetrate the atmosphere and destroy the

[97] *Ibid.*, p. 412 and Mather, *Threefold Paradise*, pp. 293-94; see also Mather's collection of ghost stories, pp. 112-51.

[98] For John Calvin's attack on the Anabaptist psychopannychists, see *Psychopannychia; or, A Refutation of the Error that the Soul Sleeps in the Interval Between Death and the Judgment* (1542), in *Tracts and Treatises in Defense of the Reformed Faith*, 5 vols. Trans. Henry Beveridge (Grand Rapids, MI: Wm. B. Eerdmans Publishing Co., 1958), 3:413-90.

[99] Yahuda MS 6, ff. 17r and 13r; Yahuda MS 1, fol. 14r, at the Jewish National and University Library, Jerusalem. Quoted by permission. Reprinted in Manuel, *Religion*, pp. 134, 127, 120.

wicked spirits and souls in league with the Devil. A literal (albeit supernatural) fire could only accomplish such a task.[100]

A juxtaposition of the cosmologies of some of the principal millennialists of the period demonstrates how they position themselves within an ongoing debate allowing us to draw several inferences. Newton's eschatological position is shaped by his mortalistic psychopannychism, which reveals his own ideological war against the Athanasian creed. In his "Paradoxical Questions concerning ye morals & actions of Athanasius & his followers," Newton was far too much concerned with pressing his anti-Trinitarian charges against Athanasius by blaming him for introducing into Christianity the demonology of Egypt and of Greece: "And Athanasius by making Anthony see ye souls of Ammon ascend up to Heaven, laid a foundation for introducing into ye Greek Churches this heathen doctrine of Daemons, together with that Popish one of Purgatory." So, too, Thomas Hobbes' Epicurean materialism that denied the existence of the soul separated from the body aimed at combating Roman Catholic belief in Purgatory and the invocation of the Saints by deriding the "contagion of the demonology of the Greeks."[101] Last but not least, Mather's own hermeneutical agenda was as much determined by his Calvinist defense of an immortal soul and its activity in a separate receptacle (Second Paradise) before resurrection as by his hostility toward Hobbesian materialism. As the shepherd of the largest flock in British North America, Mather is much more concerned with saving souls than with metaphysical conjectures. He therefore needs spiritual ammunition to combat what he calls the atheism of the "modern *Sadducees*" of his age. Hence, to Mather as to Thomas More, and to Thomas Burnet, stories of walking ghosts, poltergeists, deathbed narratives of parishioners with out-of-the-body experiences, and manifestations of witchcraft were precisely the type of tangible and extrabiblical evidence they needed to rebut the Hobbesian Leviathan. Mather's collection of energumen clearly demonstrates his pastoral intent, especially in the case of a young man, whom Mather cured from his uncontrollable temptations of "Atheism & Blasphemy" by confronting him with one of the possessed Goodwin children, three years before the specter of Salem Witchcraft would come to haunt him.[102] In like manner, Newton's hermeneutical taxonomy is shaped

[100] Henry More, *The Immortality of the Soul* (London, 1659), Book. III, Chap. 19, Par. 2; Mather, *Threefold Paradise*, pp. 219-30.

[101] Newton, "Paradoxical Questions concerning ye morals & actions of Athanasius & his followers," Clark Library MS. Quoted by permission. Cited in Force, "God of Abraham and Isaac (Newton)," p. 192. The Hobbes citation is from *Leviathan*, p. 445.

[102] Mather, *Memorable Providences*, Pt. III, pp. 14-5; see also my discussion in "Salem Witchcraft and the Hermeneutical Crisis of the Seventeenth Century: Cotton Mather's Response to Thomas Hobbes and the 'Modern Sadducees'," in *Die Salemer Hexenverfolgungen: Perspektiven, Kontexte, Repräsentationen*, ed. Winfried Herget (Trier: WVT Wissenschaftlicher Verlag, 1994), pp. 143-83.

by similar *a priori* assumptions and ideological concerns. By embracing mortalist psychopannychism, Newton could keep his explanations simple by evading such issues as the extension of the disembodied souls and their postmortem location. Perhaps that is why Newton's attack on the demonology of the Greeks and Egyptians, his rejection of ghosts and goblins and evil spirits of men haunting the living, strikes us as modern and enlightened—so worthy of a Newton. Mather's evidence, in contrast, smacks of gullibility and of the darkest medievalism against which Hobbes had raised such a cry. It needs to be added here that, unlike Newton, Mather emulated the collections of energumen underlying Henry More's *Antidote against Atheisme* (1653) and *The Immortality of the Soul* (1659). For Newton, the psychopannychist, such issues were mute.

The eschatological systems of such diverse millenarians as Joseph Mede, John Ray, Thomas Burnet, Cotton Mather, and Isaac Newton clearly evince the existence of *a priori* assumptions governing the internal logic of their systems. For Newton and his colleagues, decoding the hyperbolical metaphors in prophetic literature, then, does not exclusively depend on taxonomic guidelines but on a systemic harmony that determined whether literal or allegorical accomplishments were given preference. If we compare the confused and often contradictory systems of Newton's early eschatology with his late work, his millenarian system is characterized less by its interior logic than by the consistency with which Newton adheres to his own taxonomy. Further research, however, is necessary to trace the developmental stages of Newton's eschatological theories buried in innumerable folio versions of manuscripts yet to be explored.

INDEX

Aaron 102
Aben Ezra 90
Abraham 145, 195, 196, 246, 251
abstract symbolism 53
Access to supernatural powers 8
Adam 2, 8, 17, 25, 54, 55–57, 60, 62–64, 127–129, 132, 227
Adamic antecedents 8
Adamic
~ inheritance 61, 64
~ innocence 70
Adamic kunst signata 54
Adamic language 51, 60, 62, 66, 67, 69, 70, 133
Adamic legacy 65
Adamic man 47
Adamic meaning 53
Adamic names 56
Adam and Eve 2, 127, 128
Adam's Fall 1, 62, 227
Agrippa, Henry Cornelius 8, 30
~ *De occulta philosophia* 4, 5
Albigensians 206
Alcazar, Ludovicus ab 203, 208, 209, 219
~ *Vestigatio arcani sensus in Apocalypsi* 207
Alchemical experimentation 197, 199
Alchemical historiography 187
Alchemical ideas 15
Alchemical interests 33
Alchemical literature 1
Alchemical materials 9
Alchemical practice 187
Alchemical process 9, 141
Alchemical secrets 9
Alchemical studies 3, 198, 248
Alchemical sublimation 137
Alchemical texts 194, 198
Alchemical theorists 10, 18
Alchemical transformation 10
Alchemical work 12, 18, 194
Alchemical writings 25
Alchemist 198
Alchemy 2, 8–11, 25, 26, 28, 29, 76, 137, 165, 179, 184, 186, 194
Aldridge, A. Owen 239
Aldrovandi, Ulisse 48, 50, 64, 69, 70
Alexander, H.G. 164
~ *The Leibniz-Clarke Correspondence* 164
Alexander of Bremen 215
Alexandrinus 82
allegory 180, 186
Allen, Don Cameron 133
~ *The Legend of Noah: Renaissance Rationalism in Art, Science and Letters* 133
Almond, Philip C. 285
~ *Heaven & Hell in Enlightenment England* 285
Alphabet 8
Alsted, Johann Heinrich 278
~ *Diatriba de mille annis Apocalypticis* 278
~ *The Beloved City* 278
Alvarez, Luis 272
Ammon 110
Anabaptists 11, 286
ancient church 162
Andreae 33
~ *Chemical Wedding* 33
angel(s) 2–5, 7–9, 11–14, 20, 102, 151, 209–211, 214, 218, 279
~ angel Gabriel 10
~ angel Michael 11, 213
~ angel Uriel 5
~ angel Nariel 5
~ conversations 5, 7, 10, 14
angelic messengers 4
angelic patron 9
angelic stone 12, 13
angelic teachers 11

angelology 4, 11
Anglican 154, 161
Anglicanism 159, 161
Anglican Church 28, 150, 152, 153, 157, 162
Anglican Trinitarianism 162
Antichrist 81, 84, 100, 119, 146, 172, 175, 204, 216, 260–262, 267, 268, 286
Anticatholicism 99, 165, 166, 219
antipapist 84
antipopish 84
antitrinitarian(s) 149, 150, 159, 161, 173
antitrinitarian belief 79
Antitrinitarianism 159, 161, 162, 165, 166, 288
Antony 110, 111
Apoc 1 169
Apoc 12 168
Apoc 19 168
Apoc 21 281
apocalypse 2, 9, 39, 94, 105, 122, 146, 165, 166–168, 172, 176, 177, 203–213, 216–220, 233, 241, 246, 279, 281, 283, 287
Apocalypseos 175
apocalyptic 217
apocalyptic account 177
apocalyptic commentary 216
apocalyptic end 126, 128
apocalyptic language 243
apocalyptic metaphors 271
apocalyptic scheme 169
apocalyptic scenario 239
apocalyptic sentiment 6
apocalyptic significance 267
apocalyptic signs
apocalyptic speculation 7, 218
apocalyptic texts 6
apocalyptic tradition 19
apocalyptic treatises 6
apocalyptic rebirth 10
apocalyptic war 247
apocalypticism 7, 11, 165, 260
Apocalypse 36
Apocrypha 91
Apollodorus 186
Apostacies 168, 171, 175–177
Apostates 177
Apostle John 98, 209, 210, 212, 213, 215, 216
Apostle Paul 150, 284
Apostle Peter 263, 264, 274
Apostle(s) 79, 107, 116, 151, 152, 155, 245, 266, 272, 284
Apostles' Creed 152, 153
apostolic 205
Apostolic age 208
Apuleius 186
Aquilecchia, Giovanni 21
~ *Giordano Bruno* 21
Aquinas 19
Arcadius 171
Arian controversy 81, 98
Arian doctrine 22
Arian position 242
~ Antitrinitarian Arians 111
Arianism 242
Arias, Robert John 157
Arianism 148, 154, 155, 161, 166, 178, 283
Arians 159
Aristotelian core 23
Aristotelian(s) 57, 59
Aristotle 19, 186, 222
Ark 210
Armageddon 219, 286
Arminian justification 162
Arminianism 283
Armogathe, Jean Robert 81
Ars notoria 25, 26
ars signata 55
Arsinoë, Bishop Nepos of 203
~ *Refutation of the Allegorists* 203
artifacts 50
Artopeus 213
Asclepius 18, 20
Ashcraft, Richard 38, 159, 160
~ "Latitudinarianism and Toleration: Historical Myth versus Political History" 159, 160
Ashmole, Elias 1, 12, 13
~ "angelic stone" 13
~ "Prolegomena" 12, 13
~ "Widow Jones" 12
~ *Theatrum Chemicum Britannicum* 12

Ashmole MS 580 1
Ashmole MS 1790 1
Ashworth, William 46–52, 60–62, 64, 70–73, 249
~ "Emblematic World View" 49–51, 61, 62, 64, 70
~ "Natural History and the Emblematic World View" 46, 249
Astbury, R. 91
~ "The Renewal of the Licensing Act in 1963 and its Lapse in 1695" 91
Aston, M. 99
~ *England's Iconoclasts*, Vol. 1, *Laws against Images* 99
Astrology 14
Athanasian Creed 153, 154, 159, 288
Athanasians 110
Athanasius 812, 100, 108–111, 122, 142, 148, 204, 288
~ *Life of Anthony* 109, 110
Athenagoras 186
atomism 223
atomist 223
Atoms 36
Atonement 21, 210
Atterbury, Francis 243
~ *Epistolary Correspondence* 243
Aubrey, John 115, 116
Audcal 9
Augustine 19, 20, 24, 25, 186, 216, 265, 282
~ Augustine's doctrine of "legitimate use" 25
~ *The City of God* 282
Augustinian anthropology 19
Augustinian theology 22
Augustinian view 19, 42
Augustus 7
~ *De publica poenitentia... exorta peregrina stella 1572* 7
Autpertus, Ambrosius 213
Author and Benefactor, true 102, 139
Avebury 250, 251
Avicenna 59
Avila, Teresa of 207
Azazel 211
Baal 103
Babel 62–64, 66–68, 70, 71
~ fall of 64
Babell 62
Babels 62
Babson MS 420 141
Babson MS 434 253
Babson MS 437 103
Babylon 214–216
babylonic confusion 64
Bacchanalia 104
Bachmann, E. Theodore 206
~ Word and Sacrament 206
Bacon, Francis 25, 27, 29, 38, 39, 41, 51, 61, 71, 72, 112, 129, 225–229, 249
~ "Great Instauration" 39
~ *New Organum* 226, 249
~ *The Proficience and Advancement of Learning, Divine and Human* 226
Bacon, Roger 25, 42
Bale 218, 219
Ball, Bryan 6, 100, 172
~ *A Great Expectation: Eschatological Thought in English Protestantism to 1660* 6, 100, 172
Barone, Robert William 7
~ "The Reputation of John Dee: A Critical Appraisal" 7
Barnard, John 131
~ *Theologo-Historicus, or the True Life of the Ost Reverand Divine, and*
~ *Excellent Historian Peter Heylyn* 131
Barnes, A. 80, 81
~ *Jean le Clerc (1657–1736) et la République des Lettres* 80, 81
Barnes, Robin Bruce 9, 206, 218
~ *Prophecy and Gnosis: Apocalypticism in the Wake of the Lutheran Reformation* 10, 206, 218
Barnstone, W. 89
~ *The Poetics of Translation. History, Theory, Practice* 89
Basnage de Beauval, Jacques 30, 31
~ *Antiquitez judaïques, or Remarques critiques sur la République de Hébreux* 31

~ *History of the Jews* 30
Bauckham, Richard 6, 172, 208, 213, 214, 218, 219
~ *Tudor Apocalypse: Sixteenth Century Apocalyticism, Millenarianism, and the English Reformation from John Bale to John Fox* 6, 172, 208, 213, 214, 218, 219
Baumer, Anne and Manfred Buttner 154
Beast(s) 119, 178, 208, 209, 212, 213–215, 217, 218
Bechler, Z. 179, 238, 241, 242
~ *Contemporary Newtonian Research* 179, 238, 241, 242
Beer, de E.S. 77, 168, 169
~ *The Correspondence of John Locke* 77, 168
beginning of the world 207
Behn, Aphra 142
~ "The Golden Age" 142
believe 229
Belon 69
Belot 48
Beloved City 281
Benefactor 200
Bentley, Richard 189, 190, 231
Bercovitch, Sacvan 282
Beretta, Marco 197
~ *A History of Non-Printed Science: A Select Catalogue of the Waller Collection* 197
Berosius 186
Bersmanius, G. 184
~ *Metamorphoses* 184
Beza, Theodore 82, 83, 262
Bezae 82
Bianchi, Massimo Luigi 58
~ *Signatura rerum. Segni, Magia e Conoscenze Da Paracelso a Leibniz* 58
Bible(s) 1, 45, 85, 87–89, 92–95, 129, 134, 135, 153–156, 165, 166, 169, 175, 177, 200, 205, 227, 245, 250, 254, 255, 259, 265–267, 273
~ *Geneva Bible* 267, 279, 280
~ King James Bible 100
Bible des Rose-Croix. Traduction et Commentaire des trois premiers écrits rosicruciens 26
Bible's authority 127
biblical allusions 251
biblical antiquity 252
biblical archaeology 184
biblical authors 210
biblical commentaries 190
biblical conflagration 287
biblical criticism 77–81, 83, 90, 91, 93–96, 264
biblical cultures 72
biblical event 264, 269
biblical exegete 280
biblical figures 73, 184
biblical government 157
biblical hermeneutics 263, 269
biblical history 132, 184
biblical inspiration 96
biblical knowledge 76
biblical narratives 47, 73
biblical polities 155, 157
biblical prophecies 35, 36, 121, 122, 166, 168, 175, 266
biblical predictions 264
biblical scholarship 14, 82, 129, 245
biblical sources 42, 72
biblical studies 248
biblical text 35, 96, 226
biblical truth 154
Birch, Thomas 238
~ *The History of the Royal Society* 238
Biringuccio, Vannoccio 9
~ *The Pyrotechnia of Vanoccio Biringuccio* 9
Bishop of Rome 148
Blackhall, Ofsping 92
Blasphemy Act of 1698 161
Bloom, Allen 237, 238
~ The Closing of the American Mind 237
Bloom, Harold 199
~ *Kabbalah and Criticism* 199
Blumenberg, Hans 21
~ *The Legitimacy of the Modern Age* 21
Boccaccio 50
Bochart, Samuel 183, 186, 239, 240
Bodin 218
Bodleian MS 171
Bodleian MS Locke 85, 86

Bono, James 45, 52, 53, 59, 60, 64, 66, 72, 74
~ "Harvey versus Fernel" 53
~ "Medical Spirits and the Medieval Language of Life" 53
~ "Reform and the Languages of Renaissance Theoretical Medicine: Harvey versus Fernel" 52
~ *The Word of God and the Languages of Man: Interpreting Nature in Early Modern Science and Medicine* 52, 53, 59, 64, 66, 72, 74
Book of Common Prayer 153, 160
Book of Nature 8, 10, 45–47, 49, 51–53, 57, 59, 60, 63, 71–76, 135, 228, 242, 245, 259
Book of Scripture 259
Bossuet 208
Botts, H. 94
Bousset, Wilhelm 203–205, 207–209
~ *Die Offenbarung Johannis* 203
Bouwsma, William 19, 20
~ "The Two Faces of Humanism: Stoicism and Augustinianism in Renaissance Thought" 20
Boyle, Robert 13, 14, 27, 71, 73, 79, 86, 92, 112, 135, 189, 238
~ *Some Consideration Touching the Style of the Holy Scripture* 135
Boylean language 113
Brandt, Reinhardt 234
Brass age 126
Brightman 213, 218
Brixiensis, Philostrius 186
Brown, Norman O. 135
~ *Life Against Death: The Psychoanalytic Meaning of History* 135
Brown, Peter 19
~ *Augustine of Hippo* 19
Bruno, Giordano 21, 22, 24, 25
Burnet, Gilbert, Bishop of Salisbury 80–83, 118, 162, 284
Burnet, Thomas 239, 260, 261, 263–265, 267–273, 276–278, 280, 281, 285–289
~ "Conflagration of the Earth Made Feasible" 269
~ *Sacred Theory of the Earth* 260, 263–265, 267, 269–271, 273, 276, 278, 281, 286
~ *Telluris Theoria Sacra* 260
Burnet, William 264, 276
~ *A Treatise Concerning the State of the Departed Souls Before, and At, and After the Resurrection* 264, 276
Burns, Norman T. 245
~ *Christian Mortalism from Tyndale to Milton* 245
Burton, William 278
Burtt, E.A. 33
~ *The Metaphysical Foundations of Modern Physical Science* 33
Busch, Georg 7
~ *Die andere Beschreibung von dem Cometen welcher in 1752 war erschienen* 7
Buslidianus 82
Butt, John 220
Buxtorf, Johann 156, 239
Byzantius, Stephanus 186

Cabbala 4, 8, 13, 30–32, 200
~ Christian cabbalists 8, 30, 32
~ Jewish cabbala 200
~ Of nature 10
Cabbalist(s) 162
Calvin, John 205, 218, 286
~ *Psychopannychia; or, A Refutation of the Error that the Soul Sleeps in the Interval Between Death and the Judgment* 287
Calvinism(s) 29, 219, 288
Calvinist Church 28
Calvinist hostility 232
Calvinist reading 74
Calvinist thought 135
Camden, Carroll 7
~ "The Wonderful Yeere" 7
Camden, William 249
Canticles 89
Capp, B.S. 217, 218
~ *The Fifth Monarchy Men: A Study in Seventeenth Century English Millenarianism* 217, 218
Cardinal Virtues 101, 139
Cardwell, Edward 162

~ *Carion's Chronicle* 206
Carlstadt 205
Cartes 225
Cartesian hypothesis 266
Cartesian notions 276
Cartesianism 35, 40, 229
Cartesian(s) 162, 239
Carthusian, Dennis the 206
Casaubon, Meric 1, 11, 12, 14
~ *A True and Faithful Relation* 1
Castillejo, David 135, 146, 179, 180
~ *The Expanding Force in Newton's Cosmos* 135, 146, 179
Catholic(s) 90, 96, 98, 99, 105–107, 111, 112, 117, 119, 142, 148, 206
Catholic belief 288
Catholic countries 26
Catholic Church 20, 28, 98–100, 107, 108, 117, 119
Catholic humanism 88
Catholic polemicist 84
Catholic practices 99
Catholic prohibitions 95
Catholic version of the Old Testament 88
Catholicism 100, 101, 105–107, 112, 117, 119, 122, 148
~ Roman Catholicism 101
Catholicks 98, 105, 106, 245
Cavendish, Margaret 13, 14
~ *The Blazing World and Other Writings* 13
~ *The Description of a New World, Called the Blazing World* 13
Cavendish Laboratory 227
Céard, Jean 66–70
~ "De Babel à la Pentecôte: la transformation du mythe de la confusion des langues au XVIe siècle" 66–70
~ *La Nature et les Prodiges: l'insolité au seizième siècle* 66
Celestial City 279
Cerinthus 203
Certeau, Michael de 90, 92
~ "L'idée de Traduction" 92
~ *The Practice of Everyday Life* 96
Chaldaic orthography 87
Chaldean paraphrases 88
Chaldean retell 126
Champion, Justin A.I. 77, 155, 158, 205
~ "'Acceptable to inquisitive men': Some Simonian Contexts for Newton's Biblical Criticism, 1680–1692" 205
~ "Pere Richard Simon and English Biblical Criticism, 1680–1696" 85
~ *The Pillars of Priestcraft Shaken: The Church of England and its Enemies, 1660–1730* 155, 158
Chemosh, Molech, Asterosh 102
children of the Resurrection 246, 247, 262, 279, 283–285
Christ 7, 11, 20, 42, 57, 59, 97, 98, 100, 103, 105, 106, 108, 110, 116, 147, 151, 155, 166–169, 171, 174–178, 200, 246, 247, 260, 262, 268, 270, 278, 281–286
Christians, gentile 155
Christianson, P. 100, 172, 217–219
~ *Reformers and Babylon: English Apocalyptic Visions from the Reformation to the Eve of the Civil War* 100, 172, 217–219
Christo 175
Church 27–29, 100, 106, 107, 110, 118, 119, 14[illegible] 145–149, 153–155, 166, 168, 172, 177, 199, 200, 206, 212, 215, 216, 227, 232, 279, 280
~ adulterate church 212, 216, 265
~ corrupt Roman church 216
~ false church 212
~ hidden church 206
~ true church 212, 213, 216
Church council 153
Church corruption 148, 152
Church Fathers 22, 32, 33, 43, 106, 263, 279
Church of Christ 282
Church of England 149, 150, 152–154, 158, 160 161, 164, 251
Church of Rome 172, 232
Churchmen 81
Chus 104
Cicero 36, 186, 192
Ciliberto, Michele 21
~ *Giordano Bruno* 21

Civil War 124, 149, 155
Claramontanus 82
Clark Library MS 109–111, 117, 288
~ *The Original of Religions* 109, 122, 195
Clarke, Samuel 160, 162, 164, 245
Clement of Alexandria 28, 186, 188, 190
~ *The Stromata, or Miscellanies* 188
Clifton, R. 99
Clouse, Robert C. 278
~ "Johan Heinrich Alsted and English Millennialism" 278
~ "Rebirth of Millennialism" 278
Clucas, Stephen 25, 26
~ "Wondrous Force and Operation: Renaissance Praxiology reassessed" 26
Clulee, Nicholas 25
~ *John Dee's Natural Philosophy: Between Science and Religion* 25
codicological and textual problems 91
Cohen, I. Bernard 26, 140
~ *The Newtonian Revolution* 140
Cohn, Norman 6, 33, 34
~ *The Pursuit of the Millennium: Revolutionary Millenarians and Mystical Anarchists of the Middle Ages* 6
Colors 113
Comenius 27, 29
Comet of 1577 1
Comet of 1680 2
Comprehension Act 159
Conduitt, Catherine 146
Conduitt, John 1, 157
Confession(s) 24, 30
Constantinople Creed 153
Contamination 147
Cook, Margaret 13
Copenhaver, Brian P. 8, 33
~ "Hermes Trismegistus, Proclus, and the Question of a Philosophy of Magic in the Renaissance" 18
~ *Hermetica: The Greek "Corpus Hermetica" and the Latin "Aesclapius"* 13
~ "Jewish Theologies of Space in the Scientific Revolution: Henry More, Joseph Raphson, Isaac Newton and their Predecessors" 32
~ "Lefevre d'Etaples, Symporien Champier, and the Secret Names of God" 8
Copernicianism 21
Copernicus 26, 186
Corinthians I 226
Cornford, F.M. 256
~ *Before and After Socrates* 256
Corrupted ecclesiastical polity 146
Corrupters of Christianity 116
Corruption 10, 60, 122, 126–128, 130, 133, 134, 136, 142, 145, 147, 148, 152, 162, 183, 200, 254, 255
~ History of 132
corruption of interpretation 96
corruption of a primitve monotheistic religion 121
corruption of the physical world 140
corruptions of scripture 168
corruption of true worship 100
Cosmic restitution 9
Cotes 224
Coudert, Allison P. 17, 29, 40, 162–164
~ "Henry More, the Kabbalah and the Quakers" 38
~ *Leibniz and the Kabbalah* 29, 40, 163, 164
Council of Ephesus 153
Council of Nice 108
Council of Sardica 108
Council of Trent 89, 95
Cranach 205
Creation 54, 58, 59, 65, 69, 73, 106, 133, 138, 227–229, 232, 234, 238, 245, 256, 266, 271
Creator, the 31, 32, 39, 57, 63, 132, 165, 174, 175, 190, 230
Creator and Father 31
Creed 108
Criticks 94
Crocker, Robert 38
~ "Mysticism and enthusiasm in Henry More" 38
Croll's Calvinist Paracelsianism 74
Crown 124

Cryptographer 14
Cryptography 14
Cudworth 33
~ *Principium Hylarchicum* 33
Cult of Bethels 105
Cultural anthropology 123
Curiosity 24
Curry, Patrick 1
~ *Curry's Astrology, Science and Society, Historical Essays* 1
~ "Revision of Science and Magic" 18
Cust, R. and A. Hughes 100
Cyprian 130
Cyprian
~ authorities 82, 129
Cyril 186

Dan, Joseph 28
Daniel 35, 36, 122, 135, 146, 165, 167, 190, 205, 218, 237, 241, 267
Daniels's Mystery of Iniquity 148
Daniel 7-168, 169, 268
Daniel 8:10-11 147
Daniel 10-268
Daniel 12-165, 237, 247
Daniel 17-268
Dante 206
Darwin 238
Day of Judgement 245, 246, 247, 261, 263, 274
Debrock, G. 35, 36
De Dieu 209
Dee, John 2–14
~ *Actions with Spirits* 1, 3, 4, 6, 8–11
~ "Celestial Hierarchy" 4, 5
~ Dee's angel conversations 1, 4, 5, 9, 11–15
~ Dee's angel diaries 11, 12, 14
~ Dee's apocalyptic interests 10
~ *Mona hieroglyphica* 10
~ *De Mystica Theologica* 4
~ "Divine Names" 5
~ Personal diary 7
~ *True and Faithful Relation* 8, 9, 11
Dei 191
Deism 145, 154, 155, 164
Deist(s) 154, 155, 158, 159
Deity 5, 15, 98, 133, 135, 138, 173, 174, 228, 230, 231, 234, 255
Demetz, Peter 24
demythologizing 202
Descartes, René 14, 26, 35, 49, 71, 225, 228, 229, 230, 234, 242, 243
~ *Principles of Philosophy* 228, 234
Deuill 134
Deus Pantrocrator 174, 178
Deut XXXI.19 86
Deuteronomy 211
Devil 105, 107, 110, 134, 152, 204, 283, 286–288
devices 52
Devine 63
Digby, Kenelm 106
diluvium ignis 263, 264, 274
Diocletian 220
Diodorus 186, 192
Dionysius the Areopagite's 4 , 5
~ *De Mystica Theologica* 4
~ *Opera* 4, 5
~ *The Celestial Hierarchy* 4
~ *The Divine Names* 4
Dionysius Halicarnassus 196
Dionysius of Alexandria 203–205, 207, 216
~ *On Promises* 203
Dispensations 117, 118
divine 45, 52
divine insight 56
divine intention 128, 136
divine justice 122
divine knowledge 63
divine language 10, 51, 65, 69
Divine Mens 75
divine mind 61
divine omnipotence 72
divine plan 55
divine punishment 68
divine revelation 6
divine retribution 130
divine text of nature 63
divine truth 73
divine wisdom 136, 137
divinity 138

Dlafod 9
Dobbs, Betty Jo Teeter 1, 9, 33, 36, 45, 46, 75, 76, 121, 137, 179, 193, 237, 248
~ *Alchemical Death and Resurrection: The Significance of Alchemy in the Age of Newton* 9
~ "Alchemy and Newton's Principle of Gravitation" 36
~ "Newton as Final Cause and First Mover" 237
~ *The Foundations of Newton's Alchemy: or, the Hunting of the Greene Lyone* 33
~ *The Janus Faces of Genius: The Role of Alchemy in Newton's Thought* 1, 2, 15, 46, 76, 121, 137, 200, 248
Dobbs, B.J.T. and Margaret C. Jacob 121
~ *Newton and the Culture of Newtonianism* 121
dragon 209, 212–215, 246
Drew, David 249, 250
Dronke, Peter 186
~ *Fabula* 186
Dugdale, William 249
Dürer 205
Duret, Claude 67
~ *Histoire admirable des plantes et herbes* 67
~ *Thresor de l'histoire des langues de cest univers (1613)* 67
duty of criticism 96

Eamon, William 23
~ *Science and the Secrets of Nature: Books of Secrets in Medieval and Early Modern Culture* 23, 24
~ "Technology as Magic in the late Middle Ages and the Renaissance" 23
early church 156, 162
Earthquake 7
Ebionites 152, 155, 158
Eccles 9.5, 10 247
Ecclesiastes 89
Ecclesiastical Commission 117, 118
Ecclesiasticus 37
Edenic abundance 140
Edenic age 132
Edenic creatures 47
Edenic grace 131
Edenic innocence 127
Edenic past 41
Edenic state 130
Edwards, Thomas 83
~ *Gangraena* 83
Egypt 30
Eikon Basilike 91
Einstein, Albert 180, 238
Eire, C.M.N. 100
~ *War Against the Idols: The Reformation of Worship from Erasmus to Calvin* 100
Elias 260
Elizabeth 14
Ellenius, Allan 53
Ellys, Sir Richard 250
emanation 162, 163
emanational cosmogony 163
emanational heresy 162
emanational kabbalistic metaphysics 162
emanationist 162
emblematic learning 62
emblematic natural history 49, 51, 59, 64, 74, 75
emblematic reading 75
emblematic traditions 50
emblematic world view 46–52, 60–62, 64–66, 68–70, 72–76
emblematical way of writing 74, 75
emblems 48, 52, 63
Emerson, Roger L. 159
~ "Latitudinarianism and the English Deists" 159
Emeth 5
End of the world 1, 207
English Reformation 218
Enlightenment 27–29
Ennius 186
Ephesus 171
Epicurian sources 21
Epiphanius 32, 33
~ *Opera omnnia* 32
Episcopalianism 156
Epistles 169, 171

Erasmus 27, 29, 82, 91, 169, 204
~ *Annotationes* 82, 204
Erastian 156, 160
Erastianism 156, 158
Eratosthenes 186
Eriksson, Gunnar 197
~ "Historia och naturalhistoria hos Olof Rudbeck" 197
~ "Olof Rudbeck d. ä" 197
~ "Rudbeck och Linné: Enmansinstitutioner" 197
~ *The Atlantic Vision: Olaus Rudbeck and Baroque Science* 197
Eros, John Francis 51
~ "Diachronic Linguistics in Seventeenth-Century England,
~ Special Attention to the Theories of Meric Casaubon" 51
eschatological ideas 15
eschatological signs 2, 6, 7
eschatological treaties 6
eschatology 2, 7, 8
Esdras 87
Eucharist 106
Euhemerists 253
Euhemerus of Messene 254
~ *The Sacred History* 254
euphorus 186
Euphrates 215
Eusebius 33, 186, 192, 203, 263, 265
~ *Ecclesiastical History* 203
Evangelist, John the 204, 205
Eve 127, 129, 132
Every, George 159
~ *The High Church Party 1688–1718* 159
exegesis 10

Exod XVII.14 86
Exodus 213, 215
extrabiblical evidence 265, 288
Ezek 38-39 285, 286
Ezek 40:5 284
Ezekiel 213, 252, 279, 281, 282, 284, 285

faith, true 245
faithful remnant 215
Falkland 106
Fall (of mankind) 24, 53, 54, 58, 60, 64, 71, 76, 122–124, 127, 128, 130–133, 138, 142, 227, 250
Fall(en) (of) nature 55, 127, 129, 139
fallen world 128, 135
false goddess 139
false Gods 101, 104, 139, 183, 192, 200, 255
false interpretation 81
false Prophet 107
falsity of the text 83
Fama 30
Fasti 184
Father 98, 173, 174, 176
Faur, José 31, 35
~ "Newton, Maimonides, and Esoteric Knowledge" 31, 35
Faust Legend 24
Fauvel, John, Raymond Flood, Michael Shortland and Robin Wilson 239
~ *Let Newton Be! A New Perpsective on his Life and Works* 239, 248
Feast of Temples (or Tabernacles) 210
Ferguson, Arthur 126, 127
~ *Clio Unbound: Perception of the Social and Cultural Past in Renaissance England* 126
~ *Utter Antiquity: Perceptions of Prehistory in Renaissance England* 127
Ferguson, R. 107
~ *No Protestant-Plot, or, the present pretended Conspiracy of Protestants against the King and Government; discovered to be a Conspiracy of the Papists against the King and his Protestant-subjects* 107
Fernel, Jean 51, 52, 59
Festugière, A.J. 19
~ *La Rèvèlation d'Hermes Trismégiste* 19
Feuillet, André 210
~ *L'apocalypse: L'État de la question* 210, 211
Ficino, Marsilio 4, 20–24, 28
~ *Corpus Hermeticum* 28
~ *Index eorum, quae hoc in libro*

habentur. Iamblichus, de mysterils Aegyptiorum... 4
Fifth Monarchists 217
Filioque 153
Filmer, Robert 127
final coming of Christ 200
Findlen, Paula 51, 64, 65
~ "Jokes of Nature and Jokes of Knowledge: The Playfulness of Scientific Discourse in Early Modern Europe" 65
~ "Museums, Collecting and Scientific Culture in Early Modern Italy" 51
~ "The Economy of Scientific Exchange in Early Modern Italy" 51
~ "The Museum: Its Classical Etymology and Renaissance Genealogy" 51
Finocchiaro, M.A. 223
~ "Newton's Third Rule of Philosophizing: A Role for Logic in Historiography" 223
Fiore, Joachim, of 203
first author 173
first coming of Christ 167
First Epistle 204
first law of motion 224
first reformation 206
Firth, Katharine R. 6, 100, 172, 216, 218, 219
~ *The Apocalyptic Tradition in Reformation Britain 1530–1645* 6, 100, 172, 216, 218, 219
Flood 9, 133, 136
Florentine Platonists 20
Ffolkes, Mr 252
Fontenelle, Bernard 239, 243
Force, James E. 35, 36, 73–75, 77, 79, 85, 149, 150, 234, 237, 241, 244, 246–248, 259, 270, 285, 288
~ "Biblical Interpretation, Newton and English Deism" 154, 237, 241
~ "Hume and the Relation of Science to Religion Among Certain Members of the Royal Society" 234
~ "Millenialism" 246
~ "Newton and Deism" 154
~ "Newton's God of Dominion: The Unity of Newton's Theological, Scientific and Political Thought" 174, 244, 248
~ "Newton, The Lord God of Israel and Knowledge of Nature" 35, 150, 241
~ "Samuel Clarke's Four Categories of Deism" 241
~ "Science, Deism and William Whiston's 'Third Way'" 241
~ "Sir Isaac Newton, `Gentleman of Wide Swallow'? Newton and the Latitudinarians" 159
~ "The God of Abraham and Isaac Newton" 247, 259, 285, 288
~ *William Whiston: Honest Newtonian* 79, 270
Force, James E. and Richard H. Popkin 30, 73, 81, 121, 159, 165, 174, 186, 204, 244, 245, 259
~ *Essays on the Context, Nature and Influence of Isaac Newton's Theology* 73, 81, 121, 159, 165, 174, 186, 204, 244, 259
~ *Narcissus and the Invention of Personal History* 186
~ "Samuel Clarke's Four Categories of Deism, Isaac Newton, and the Bible" 74
~ *The Books of Nature and Scripture: Recent Essays on Natural Philosophy, Theology, And Biblical Criticism in the Netherlands of Spinoza's Time and British Isle of Newton's Time* 30, 36, 40, 73, 81, 100, 179, 203, 247, 259, 260
Ford, Worthington C. 262
Foucault, Michel 53
~ *The Order of Things: An Archaeology of the Human Sciences* 53
Foxcroft, Ezekiel 33
Foxe 206, 213
~ *Eicasmi* 213
Francis, Alban, Father 117
Francis of Verulam 38
Frazer, James 91
Freedman, L. 159
Freedom 20
Froom, LeRoy Edwin 146
~ *The Prophetic Faith of Our Fathers*

146
Fulfillment of Biblical prophecies 11
Fulke, William 214
Funkenstein, A. 174
~ *Theology and the Scientific Imagination from the Middle Ages to the Seventeenth Century Theology and the Scientific Imagination from the Middle Ages to the Seventeenth Century* 174
Furley, Benjamin 84
Furley's son 85

Gabriel, the angel 10
Gal 4.26, 284
Galen 19, 59
Galenists 57, 59
Galileo 26, 29, 71, 72, 104, 186, 190
Galileus 190
Galuah 12
Garden of Eden 55, 127, 128, 132, 227, 246
Garden of Paradise 62, 127, 132
Garin 19
Gassendi 229
Gatti, Hilary 21, 22, 24
~ *The Renaissance Drama of Knowledge: Giordano Bruno in England* 21, 24
Gematria 10, 11
Gemma, Cornelius and Guillaume Postel 7
~ *De peregrina stella 1572* 7
General conflagration 268
Genesis 8, 19, 25, 35, 127, 129, 142, 266
Gentiles 277
Gesner, Conrad 48–50, 64, 66, 67, 69, 70
~ *Historica animalium* 66
~ *Mithridates: De Differentiis Linguarum* 66
Giamatti, A. Bartlett 24, 127
~ "Proteus Unbound: Some Versions of the Sea God in the Renaissance" 24
~ *The Earthly Paradise and the Renaissance Epic* 127
Gimpel, Jean 23
~ *The Medieval Machine: The Industrial Revolution of the Middle Ages* 23
Ginzburg, Carlo 24
~ "High and Low: The Theme of Forbidden Knowledge in the Sixteenth and Seventeenth Centuries" 24
Giorgi, Francesco 4
~ *De harmonia mundi* 4
Gjertsen, Derek 241
~ *The Newton Handbook* 241
Glacken, Clarence 130
~ *Traces on the Rhodian Shore: Nature and Culture in Western Thought from Ancient Times to the End of the Eighteenth Century* 130
Glanvill, Joseph 40, 221, 228–231
Gnostic doctrines 31
Gnostic heresies 32
Gnostic texts 28
Gnosticism 28–30, 32, 33, 36, 40–42, 163
~ criticism of gnosticism 30
~ Jewish Gnosticism 28
Gnosticks 30, 190
Gnostic/Kabbalistic emanation theory 163
Gnostics 28–30, 32, 33, 43, 147, 152, 162
God 2, 3, 5, 8, 20, 22, 28, 31–33, 34, 36–42, 45–47, 51–56, 58–65, 68, 71, 73–76, 79, 80, 97, 101–106, 118, 122, 126, 128, 129, 132–140, 142–145, 147, 151, 163, 165–169, 174–178, 190, 191, 196, 200, 204, 210, 211, 217, 222, 224, 227, 229–234, 238, 241, 244, 245, 250, 251, 255, 256, 261, 265, 267–270, 273, 274, 276, 277, 279, 280, 283–285
~ God's "book of nature", *see* book of nature
~ reflection of God 61
God Almighty 250, 268
God Himself 59, 63
God Jehova 104
God's image 42
God's Word 57, 76
Goldammer, Kurt 56
~ "La conception paracelsienne de l'homme entre la tradition théologique, la mythologie et la science de la nature" 56
Golden, S.A. 79, 83
~ *Jean Le Clerc* 79, 83
Golden age 123, 126–128, 130, 132, 133, 140, 141, 143

Golding, Andrew 6
~ *See also* Tuveson E.L.
Goldish, Matt 30, 32, 145, 177, 179
~ "Newton and the Kabbalah" 30, 31, 179
~ "Newton's Of the Church: Its Contexts and Implications" 177
Goldstone, Jack 122–125
~ *Revolution and Rebellion in the Early Modern World* 122–124
Gooding, D., T. Pinch and S. Schaffer 114
Goodman, Godfrey 123, 129–131, 134, 140, 143
~ *The Fall of Man, or the Corruption of Nature* 130
Goodman, Lenn E. 31
Goodman, Nelson 221
~ *Fact, Fiction and Forecast* 221
Gorceix, Bernard 26
Gordeian knot 274
Goropius Becanus, Jan 51, 249
Gospel 204, 205, 270
gospel harmonies 94
Gospel of Luke 91
Gospels 91, 167
Goux, Jean-Joseph 137
~ *Symbolic Economies after Marx and Freud* 137
Grace for salvation 20
Grafton, Anthony 126
~ "Joseph Scaliger and Historical Chronology: The Rise and Fall of a Discipline" 126
Graminaei, Theodorus 7
~ *Erklerung oder Auszlegung eines Cometen* 7
Graven images 102, 136
Gravitational forces 39
Gravity 36, 40, 41
Great Persecution 220
Greater Catechism 153
Greene, Thomas 24
Gregory, David 36, 103, 104
Gregory the Great 265
Greig, Martin 162
~ "The Reasonableness of Christianity? Gilbert Burnet and the Trinitarian Controversy of the 1690's" 162
Griffin, Martin I.J. 159–162
~ *Latitudinarianism in the Seventeenth-Century Church of England* 150, 160
Grindal 107
Grotian disciples 264
Grotian historicists 269, 283
Grotian method 267
Grotius, Hugo 28, 29, 98, 157, 158, 208, 209, 219, 263, 264, 270, 275, 280, 281
~ *Annotationes in Vetus et Novum Testamentum* 263, 264, 270, 280
~ *De iure belli ac pacis* 158
Grove, Richard H. 129
~ *Green Imperialism: Colonial Expansion, Tropical Island Edens, and the*
~ *Origins of Environmentalism 1600–1860* 129
Guinsberg, Arlene 38
~ "Henry More, Thomas Vaughan and the late renaissance magical tradition" 38

Hajek, Tadeas 7
~ *Dialexis de nova stella* 7
Hakewill, George 130
~ *An Apologie, or Declaration of the Power and Providence of God in the Government of the World* 130
Hale, Matthew 127, 133
~ *The Primitive Origination of Mankind, Considered and Examined According to the Light of Nature* 127, 133
Hall, A. Rupert and Marie Boas Hall 163
~ *Unpublished Scientific Papers of Isaac Newton* 163
Haller 156
~ *Liberty and Reformation* 156
Halley 193
Hammond, Henry 208, 219, 263, 264, 267, 269, 280, 281
~ *A Paraphrase and Annotations Upon All the Books of the New Testament* 264, 280
Hammond, John 270
~ *A Second Defense of the Learned Hugo*

Grotius 270
Hampden, John 85, 95
Harkness, Deborah E. 1, 3, 4
~ "Shows in the Showstone: A Theater of Alchemy and Apocalypse in the Angel Conversations of John Dee" 10
~ "The Scientific Reformation: John Dee and the Restitution of Nature" 1, 3, 4, 8
Harms, Wolfgang 53
~ "Bedeutung als Teil der Sache in zoologischen Standardwerken der frühen Neuzeit 53
~ "On Natural History and Emblematics in the Sixteenth Century" 53
Harrington, James 156, 157
Harris, Marvin 125
~ *Cannibals and Kings: The Origines of Cultures* 125
Harrison, John 3, 32, 33, 157, 158, 184
~ *The Library of Isaac Newton* 3, 32, 157, 158, 184
Harrison, F.J. and P. Laslett 85
~ *The Library of John Locke* 85
Harrison, Thomas P. 7, 33
~ *Studies in Honor of De Witt T. Starnes* 7
Hartlib 27, 29
Harvey, Gabriel 7
Haynes 150
He 59, 174
Hebraic 195
Hebraists 239, 240
Hebrew 8, 54, 66, 67, 87, 88, 93, 95, 147, 208, 241, 252
Hebrew 5 150
Hebrew 11 284
Hebrew manuscripts 89
Hebrew metaphors 273
Hebrew prophets 209, 241, 269, 272, 275
Hebrew ritual 203
Hebrews 35, 185, 190, 205
Heer, Friedrich 27
~ *Die Dritte Kraft: Der europäische Humanismus zwischen den Fronten des konfessionellen Zeitalters* 27
Hellegers, Desiree 129
~ *Handmaid to Divinity: Science, Politics, and the Gendered Poetics of Resistance from John Donne and Francis Bacon to Anne Finch* 129
Helvetius, Christian 183
~ *Theatrum Historicum* 183
Henry, John 14, 40
~ "'Pray Do Not Ascribe that Notion to Me': God and Newton's Gravity" 40
~ "Robert Hooke, the Incongruous Mechanist" 14
Henry, Matthew 280
~ *Commentary on the Whole Bible* 280
~ *Exposition of the Old and New Testaments* 280
Herget, Winfried 288
hermeneutics 180, 186
hermeneutics of nature 53
Hermes 13
Hermes Trismegistus 38
Hermetic art 26
Hermetic magic 25
Hermetic soliloquy 38
Hermetic texts 28, 29
Hermetic vision 38
Hermeticism 18, 19, 21, 22, 24, 26, 36, 42
~ Hermetic image of man 24
Herodotus 186
Herr, Friedrich 27
~ *Die Dritte Kraft* 27
Herzog, Isaac 156
~ "John Selden and Jewish Law" 156
Hesiod 126, 186
Hesse, Mary B. 18
~ "Hermeticism and Historiography: An Apology for the Internal History of Science" 18
Hesychius 186
Heylyn, Peter 123, 131, 143
~ *Cosmographie in Foure Books, Contayning the Chorographie & Historie of the Whole World, and All the Principal Kingdomes, Provinces, Seas and Isles, Thereof* 131, 143
Hibbard, C.C. 100
~ *Harles Land and the Popish Plot* 100

Hickman, Bartholomeuw 3
hieroglyphic emblem 189
hieroglyphic symbols 254
hieroglyphics 48, 52, 73, 188, 254
Hill, Christopher 100, 129, 135, 172
~ *Antichrist in Seventeenth Century England* 100, 146, 172
~ *The English Bible and the Seventeenth-Century Revolution* 129, 135
Hill Green, Thomas and Thomas Hodge Grose 240
~ *The Philosophical Works* 240
Histories of ancient empires 121
Hobbes, Thomas 89, 156, 278, 280, 281, 288, 289
~ *Leviathan* 89, 156, 280, 281, 288
Hobbesian materialism 283, 288
Holland, A.J. 18
Holy Books 245
Holy Ghost 98
Holy Spirit 204, 205
Homer 186
Honorius 171
Hooke, Robert 13, 14, 112, 115, 116, 242, 243
~ *The Posthumous Works* 13
Hooker, Richard 149, 157
Horn, Georg 127
~ *Arcae Noae sive historia imperiorum et regnorum a condito orbe ad nostra temorum* 127
Huggard, E.M. 31
~ *Theodicy* 31
Huguenot successor 157
Humanists 28
Hume, David 221, 222, 225, 229, 233, 239, 240
~ "Of the Rise and Progress of the Arts and Sciences" 239
~ *The History of England. From the Invasion of Julius Caesar to the Revolution in 1688* 240
Humean world 222
Hunter, Michael 13, 77
~ "Alchemy, Magic and Moralism in the Thought of Robert Boyle" 13, 14
Hunter, Michael and Simon Schaffer
~ *Robert Hooke: New Studies*
Huppert, George 53
~ "'Divinatio et Eruditio': Thoughts on Foucault" 53
Hutin, Serge 32
~ "Note sur la création chez trois kabbalistes chrétiens anglais: Robert Fludd, Henry More et Isaac Newton" 32
Hutton, Sarah 36, 37, 73, 74, 165, 172, 173, 203, 216, 259
~ *Henry More (1614–1687) Tercentenary Studies* 38
~ "More, Newton and the Language of Biblical Prophecy" 36, 73, 74, 165, 172, 173, 203, 216, 259
Hyginus 186

I Kings 4.33 102I Kings 11.7 102
I Kings 11.9 102
I John 5:7 78, 80–84, 91, 98, 168
1 Timothy 1.13 245
I Timothy 3:16 78, 80, 168
iconographic lexicons 184
iconographic opening 196
iconographic work 184
iconography 189
Idel, Moshe 30, 42, 199, 200
~ *Kabbalah: New Perspectives* 30, 199, 200
~ "Rabbinism versus Kabbalism: On G. Scholem's Phenomenology of Judaism" 42
idol 213
Idolaters 103, 104, 107
idolatrous accretions 254, 255
idolatrous corruption 255
idolatrous ideas 73
idolatrous polytheism 254
idolatrous position 149
idolatrous practices 102, 103
idolatrous religion 107
Idolatria 102
idolatry 101–108, 119, 123, 133, 134, 136, 138, 139, 141, 142, 152, 192, 212, 248, 251, 254, 255

Idoll 134
idols 102, 103
Iliffe, Robert 77, 80, 97, 100, 111, 121, 165, 194, 209, 210, 218, 219, 255, 259
~ "'Making a shew': apocalyptic hermeneutics and the sociology of Christian idolatry in the work of Isaac Newton and Henry More" 100, 165, 209, 210, 218, 219, 259
~ "Isaac Newton: Lucatello Professor of Mathematics" 111
~ "The Idols of the Temple: Isaac Newton and the Private Life of Anti-Idolatry" 121
~ "The Egyptian Basis of Stonehenge: Newton, Natural philosophy and the Origins of Idolatry" 255
Images 103, 104
Images of gold and wood 102
Impey, Oliver and Arthur MacGregor 51
~ *The Origins of Museums: The Cabinet of Curiosities in Sixteenth- and Seventeenth-Century Europe* 51
Incantations 105
incompetent scribes 82
infallibility 106
Intervention 21
Invocation of Saints, Ghosts or Daemons 109
Irenaeus 32, 265
~ *Adversus hereses* 32
Iron age 126
Isagogius 186
Isa 60 246
Isaiah 213, 265
Israelite ceremonies 215
Israelite cults 215
Israelite festivals 210
Israelites 102, 103, 210, 211, 213
Italy 262, 268

Jacob, James R. 156, 158
~ *Henry Stubbe: Radical Protestantism and the Early Enlightenment* 156, 158
Jacob, Margaret C. 159, 260
~ *The Newtonians and the English Revolution, 1689–1720* 159, 260
Jacobs, Charles M. 206
Jagt, Arjo van der, *see* Van der Jagt
James 205
Jeffrey, Judge 117, 118
Jerome 24, 81, 82, 88, 204, 265
Jerusalem Temple 210, 211
Jesuit(s) 112, 114–117, 119, 207, 269
Jesuit Anthony Lucas 113–115
Jesuit Francis Linus 113, 115
jesuit scholar 209
Jesus Christ 28, 32, 116, 145–147, 156, 159, 165, 176, 177, 186, 194, 200, 201, 238
Jewel, John 149
Jewish 148
Jewish influence 155
Jewish Revolt 205, 208
Jewish synagogue 155
Jewish temple 241, 249
Jewish-Christians 152
Jews 155, 190, 268
Job 89
Joel 265
Joel 2:30-31 272
Jonas, Hans 29
~ "Gnosticism and Modern Nihilism" 29
Jones, Richard Foster 6
~ *Ancients and Moderns: A Study of the Rise of the Scientific Movement in Seventeenth-Century England* 6
Jonson, Ben 243
Jonston, Joannes 49
~ *Historia naturalis* 49
Jordan, W.K. 156
~ *The Development of Religious Toleration in England* 156
Josephus 186, 239
Josten, C.H. 10
~ "A Translation of John Dee's 'Monas hieroglyphia' (Antwerp 1564), with an Introduction and Annotations" 11
Journal of the Warburg and Courtauld Institutes 8
Judaism 147, 155
Jude 205
Justel, Henri 84

Justin 33, 203

Kabbala denudata 33, 34
Kabbalah 28–36, 41, 42, 147, 163
~ Lurianic Kabbalah 29, 3
Kabbalistic texts 35
Kabbalists 33, 35, 42, 43, 163
Karaite Jews 90
Katz, David S. 85, 126
~ "Isaac Vossius and the English Biblical Critics" 126
Kelly, Edward 3, 13
Kepler 26, 239
Keynes, John Maynard 179
Keynes MS 2 102, 107
Keynes MS 3 146, 150, 153, 156, 157, 173, 176
~ *Irenicum* 146, 150, 157, 166, 167, 173–177, 253
Keynes MS 5 97, 105, 210
Keynes MS 121 118, 119
Keynes MS 130 157
Keynes Newton MS 5 171
King Charles I 124
King Charles II 260
King James I 149
King James II 101, 117–119
King, Richard 94
Kircher 186
Klauber, M.I. 79
~ "Between Protestant Orthodoxy and Rationalism: Fundamental Articles and the Early Career of Jean Le Clerc" 79
Kleinbrg, Aviad M. 19
~ "De Agone Christiano: The Preacher and his Audience" 19
Knoespel, Kenneth 73, 76, 121, 126, 127, 179, 183, 186, 197
~ "Milton and the Hermeneutics of Time: Seventeenth-Century Histories and the Science of History" 126
~ "Newton in the School of Time: The Chronology of Ancient Kingdoms Amended and the Crisis of Seventeenth-Century Historiography" 73, 121, 127, 183
~ *Narcissus and the Invention of Personal History* 186
~ "Reshaping the Earth: Olof Rudbeck and the Transformation of Sweden" 197
Knorr von Rosenroth 33
~ *Kabbala denudata* 33, 34
Kochavi, Matania Z. 259
~ "One Prophet Interprets Another: Sir Isaac Newton and Daniel" 259
Koyré, Alexander 34, 223, 224
~ *Newtonian Studies* 223
~ "Newton's Regulae Philosophandi" 223, 224
Koyré, Alexander and I.B. Cohen 34
~ "Newton and the Leibnitz-Clarke Correspondence" 34
Kroll, Richard 38
Kroll, R., R. Ashcraft and P. Zagorin 159
~ *Philosophy, Science and Religion in England, 1640–1700* 159
Kromayer 213
Kubrin, David 40

Lady Masham 168
Laertius 186, 192
Lake, P. 100
~ "Anti-popery: the Structure of a Prejudice" 100
Lamont, William 217, 218
~ *Godly Rule: Politics and Religion, 1603–60* 217, 218
Lamothe, Claude 79
~ *The Inspiration of the New Testament Asserted and Explained* 79
language of authenticity 89
language of infallibility 89
language of Nature 47
Laodicea 171
Laslett, P. 128
~ *see also* J. Harrison
Last Judgment (universal Judgment) 210, 287
Latitudinarian(s) 157, 159–162
Latitudinarianism 145, 160, 164
Lawenus 209
Lawrence, C. and S. Shapin 111

Leclerc, Jean 77–81, 83, 84, 91, 94, 96, 127, 168
~ *Compendium Historiae Universalis: A Compendium of Universal History from the Beginning of the World to the Reign of Charles the Great* 127
~ *Sentimens* 79, 80
~ *The Five Letters Concerning Inspiration* 79
Leers, Renier 85
Lefèvre détaples, Jacques 4
Legitimate use, Augustine's doctrine of 25
Lehmann, Helmut T. 206
Leibniz, Gottfried Wilhelm 30, 40, 41, 106, 142, 145, 162–164, 225, 244, 245
~ *Discourse on Metaphysics; Correspondence with Arnauld; and Monadology* 163
Leibnizians 162
Lemay, J.A.L. 159
Lestringant, Frank 126
~ *Mapping the Renaissance World: The Geographical Imagination in the Age of Discovery* 126
Levant 204
Levine, Joseph 239, 243
~ *The Battle of the Books. History and Literature in the Augustan Age* 238, 239, 242, 243
life sciences 47
light 113
light of nature 56, 57, 59, 232
Lightfoot, John 156, 239, 264, 281
Liljegren, S.B. 156
~ "Harrington and the Jews" 156
Lilly, Kate 13
Limborch 79, 85
Lindberg, David C. 46, 249
Lindberg, David C. and Robert S. Westman 249
~ *Reappraisals of the Scientific Revolution* 249
Lindroth, Sten 197
~ *Svensk lärdomshistoria: stormaktstiden* 197
linguarum varietas 67
linguistic diversity 67
linguistic variety 67, 68
Lobiensis 82
Locke, John 77–81, 83–87, 89, 92–99, 126–128, 134, 160, 168, 169, 171, 222, 234
~ *Correspondence* 78–80, 85, 94, 96, 169
~ *Essays Concerning Human Understanding* 232
~ "Locke, Law and the Laws of Nature" 234
~ *Paraphrases* 94
~ *Reasonableness of Christianity* 93
~ *Second Treatise of Government* 232
~ *Treatise of Government* 127, 128
~ *Two Treatises of Government* 128
Locke's Library 93
Logos 56
Lord (of God) 71, 102, 103, 136, 241, 242, 246, 248, 263, 272, 275
Lord George Digby 106
Lord God of Dominion 241
Lovelace collection of Locke MSS 169
Lowth, William 79
~ *A Vindication of the Divine Authority and Inspiration of the Wirtings of the Old and New Testament* 79
Lubac, Henri de 206
~ *Exégèse médiévale: Les quatre sens de l'écriture* 206
Lucretian cosmology 21
Lucretius 238
Lugli, Adalgisa 51
~ *Naturalia et mirablia. Il collezionsimo enciclopedico nelle Wunderkammern d'Europa* 51
Luke 21:25-26 272
lumen naturae 59
Luther, Martin 98, 205–207, 212, 218
~ "Vorhede auf die Offinbarung Sanct Johannis" 205
Lutheran attack 218
Lutheran Bibles 205
Lutheran Church 28
Lutheran reaction 217
Lutherans 205, 206, 218
Lutherian thought 24

Luxon, Thomas H. 260, 269
~ *Literal Figures: Puritan Allegory and the Reformation Crisis in Representation* 260
Lyra, Nicholas of 203, 206–208, 215
~ *Postille* 206
MacGregor, Arthur, *see* Impey, Oliver
Machiavelli 107
Maclean, G. 80
~ *Culture and Society in the Stuart Restoration: Literature, Drama, History* 80
Macrobius 186, 196
Madimi 12
Magdeburg Centuries 206
Magic 18, 21, 24, 108, 192, 200
Magical - Hermetic philosophy 22
Magical traditions of the Renaissance 27
Magnus, Albertus 19
Magus, Simon 32
Maier, Michael 187
~ Symbola auriae mensae duodecim nationaum 187
Maimonides 35, 42, 239
Makdishi, George 24
Maker 279
Malgaigne, J.F. 68
Malm, Matts 197
~ *Minervas äpple* 197
Mamiani, Maurizio 168, 177, 178, 203, 209
Man of Sin 146
Mandelbrote, S. 81, 150, 157
~ "'A Duty of the Greatest Moment': Isaac Newton and the Writing of Biblical Criticism" 81, 150, 157
Mandeville, Bernard 271
~ *The Fable of the Bees* 271
Mandlebrote, Scot 77
Manethus 186
Manichaeans 207
Manichaeism 27
manifest errors 88
Manilius, Marcus 7
~ *Astronomica* 7
Manuel, Frank 30, 31, 33, 36, 37, 40, 81, 93, 99, 102, 121, 150, 162, 163, 165, 176, 179, 203, 205, 210, 217, 233, 241, 243, 245, 246, 259, 263, 269, 274, 275, 279, 284, 285, 287
~ *A Portrait of Isaac Newton* 81, 99, 243
~ *Isaac Newton, Historian* 102, 121, 146, 165
~ *The Changing of the Gods* 241
~ *The Religion of Isaac Newton* 30–32, 40, 81, 93, 121, 150, 162, 165, 179, 203, 205, 210, 217, 233, 241, 243–246, 259, 269, 274, 275, 279, 284, 285, 287
Marcionite 91
Mares, Jeffrey Scott 262
Marian 107
Markley, Robert 75, 76, 121, 122, 141, 142
~ *Fallen Languages: Crises of Representation in Newtonian England, 1660–1740* 73, 76, 122, 142
~ "'So Inexhaustible a Treasure of Gold': Defoe, Capitalism, and the Romance of the South Seas" 141
Markley, Robert and Molly Rothenberg 142
~ "The Contestations of Nature: Aphra Behn's 'The Golden Age' and the Sexualizing of Politics" 142
Marlowe, Philip 24
~ *Dr. Faustus* 24
Marshall, John 77, 79, 83, 85, 98
~ *John Locke. Resistance, Religion and Responsibility* 78, 83, 98
Marsham, John 183, 239, 240
Martyr, Justin 36, 148, 216
martyrs 247, 277, 280
Mary 206
Masoretic system 88
Masoretic (Hebrew) texts 129
Material redemption of nature 8, 9
Mather, Cotton 260–262, 263, 264, 265, 275, 276, 278, 279, 281–289
~ *A Dissertation Concerning the Future Conversion of the Jewish Nation* 262
~ *Christianus per Ignem* 262
~ *Five Books Against Marcion* 282
~ *Magnalia Christi Americana* 262
~ *Memorable Providences* 283, 288
~ "Problema Theologicum" 262, 263,

278, 282, 283, 286
~ *The Present State of New England* 262
~ *Threefold Paradise* 260, 262, 264, 275, 276, 279, 281–283, 285–288
Matt 24:29 264, 279
Maximian 220
McGuire, J.E. 18, 19, 33, 41, 223, 248
~ "Atoms and the Analogy of Nature: Newton's Third Rule of Philosophizing" 223
~ "Force, Active Principles and Newton's Invisible Realm" 34
~ "Neoplatonism and Active Principles: Newton and the Corpus Hermeticum" 18, 33
~ *Tradition and Innovation. Newton's Metaphysics of Nature* 223
McGuire, J.E. and P.M. Rattansi 33, 36, 40, 41
~ "Newton and the Pipes of Pan" 33, 36, 40, 41
McGuire, J.E. and Robert S. Westman 18, 33
~ *Hermeticism and the Scientific Revolution* 18, 33
McKane, W. 89
~ *Selected Christian Hebraists* 89
McKenna, Antony 77
McLachlan, Herbert 150, 210, 214
~ *Sir Isaac Newton: Theological Manuscripts* 150, 210, 214
Mebane 26
~ *Renaissance Magic and the Return of the Golden Age: The Occult Tradition and Marlowe, Jonson, and Shakespeare* 26
Meddus, Samuel 261, 262
Mede, Joseph 86, 100, 119, 171–174, 178, 203, 207–216, 218, 219, 239, 260–264, 267–269, 273, 274, 276–279, 281, 283, 289
~ *An Explanation of the Grand Mystery of Godliness; or, A True and Faithful Representation of the Everlasting Gospel of our Lord and Saviour Jesus Christ, the Onely Begotten Son of God and Sovereign over Men and Angels* 219
~ *Paraleipomena. Remaines On Some Passages in the Revelation* 262, 277
~ *The Works* 209, 210, 213, 215, 216, 218, 260, 263, 274, 277
~ ~ *Clavis apocalyptica* 172, 203, 209, 210
~ ~ *Commentarius* 211, 213, 215, 216, 218
~ ~ *Fragmenta* 211
~ ~ *The Remains* 210, 211
Medicea 190
Medicean benefactor 190
medicine 47, 59
~ Galenic medicine 52
~ Renaissance medicine 53
Medieval 25, 207
Medieval attitudes 23
Medieval Church 24
Medieval commentators 215
Medieval culture 25
Medieval notion of "operativity" 26
Medieval period 23, 213, 216
medieval practice 269
Medieval thought 24
Medieval view 206
Medieval works 208
Medieval writings 22
Medicina dei 9, 10
Melchizedech 194
~ *The Death of Nature: Women, Ecology, and the Scientific Revolution* 129
Mercury van Helmont, Francis 33, 163
Merkel, Ingrid and Allen G. Debus 18
Merrens, Rebecca 129
~ "Exchanging Cultural Capital: Troping Women in Sidney and Bacon" 129
Mersenne 26, 71
Merton 26
Messiah 251, 281, 286
Mesue 59
metaphor 48
meteorite 272
Metzger, Bruce 204, 205
~ *The Canon of the New Testament: Its Origin, Development and Significance*

204, 205
Michael, the angel 5, 11
Middle Ages 23, 25, 53, 206, 212, 213
millenarian cosmologies 276
millenarian expectations 218
millenarian reading 216, 217
millenarian system 259, 287, 289
millenarian taxonomies 272
millenarian theory 216
millenarian thought 276
millenarianism 36, 216–218
millenarian(s) 40, 178, 203, 216, 217, 261, 264, 265, 267, 278, 282, 284, 285, 289
MIllenaries 281
Millenary 106
millennial reading 216
millennialism 260
millennialists 260, 265, 275, 288
Miller, Perry 282
Milton, John 122, 143, 184, 245, 286
Mirandola, Pico della 8, 29
Misinterpretation 101, 105
mistranscription 82
modern state of knowledge 243
Momiglliano, Arnaldo 126, 194
~ *On Pagans, Jews and Christians* 126
~ "Time in Ancient Historiography" 194
~ "The Origins of Universal History" 194
monarch 117
monster(s) 212, 215
Montanists 110
Moran, Bruce T. 51
~ *Patronage and Institutions: Science, Technology, and Medicine at the European Court, 1500–1750* 51
Morden, Robert 183
~ *Geography Rectified: or, a Description of the World* 183
More, Henry 31, 33–39, 100, 119, 168, 173, 174, 178, 203, 207, 208, 217, 219, 230, 287
~ *Annotations upon the two foregoing Treatises, Lux Orientalis ... and the Discourse of Truth* 35
~ *Antidote against Atheism* 289
~ *Apocalypsis Apocalypseos; or the Revelation of St. John the Divine Unveiled* 207
~ *Conjectura Cabbalistica* 35, 36
~ *Divine Catalogues* 174
~ *Exposition of Daniel* 219
~ "Fundamenta Philosophiae, sive Cabbala Aet-Paedo-Melissaea" 34
~ "Spirit of Nature" 34
~ *The Immortality of the Soul* 34, 288, 289
More, L.T. 252
~ *Whiggish biography* 252
More, Thomas 288
Moses 13, 102, 103, 116, 139, 156, 175, 187, 196, 232, 237, 240, 255, 272
Mott, Andrew 224
~ *Mathematical Principles of Natural Philosophy* 224, 232, 233
Moyne 86
MS Bodmer 30, 146, 147, 152–155, 161
MS Bodmer Newton 151
MS Locke 86, 171
Mulligan, Lotte 27
~ "Civil War Politics, Religion and the Royal Society" 27
Mumford, Lewis 23
~ *Technics and Civilization* 23
Munoz, Jeronimo 7
~ *Du nouveau comete de l'an 1572* 7
Murrin, Michael 203, 208–211, 216, 218, 219
~ "Revelation and Two Seventeenth Century Commentators" 208–211, 216, 218, 219
Muslims 158
myth 200
mythographer 184
mythographic commentaries 190
mythographic information 200
mythographic pantheons 183, 200
mythographic research 184, 200
mythographic traditions 50
mythographic work 186, 192, 194, 196
mythography 52, 181, 194
mythohistorical project 197
mythological learning 62

mythological narrative 186
mythological work 190
mythology 48, 186, 198
mythology, Mediterranean 183

NA 5
Nadav, Mordekai 180
Nag Hammadi 28
Nariel, the angel 5
narrative traditions 50
native Americans 50
natives 50
natural history 50
natural philosophy 41
natural world 2, 3, 11, 49, 51, 102, 103, 112, 119, 129, 135, 138, 139, 143, 203, 225, 232, 233, 245, 263, 274
Nature 2, 6–11, 27, 31, 37, 41, 46, 47, 50, 51, 54–65, 68–75, 101, 116, 129, 130, 132–137, 140, 141, 174, 186, 196, 200, 223, 224, 226, 227, 229, 231, 232, 234, 238–240, 244, 245, 248, 265, 266, 268, 269, 273, 275, 276
~ Demonized nature 127
nature's playfulness 65
Nazarenes 152, 155, 158
Neapolitan poet 24
need for critical restoration 87
Nelson, Jr., Lowry 24
Neoplatonic sympathies 33
Neoplatonism 18, 21, 28, 29
Neoplatonists 17, 20
Neuman, Kalman 156
Newman, William 23, 25
~ "Technology and Alchemical Debate in the late Middle Ages" 23, 25
(New) Jerusalem 11, 216, 217, 242, 246, 247, 260, 261, 276–287
New Science 134
New Testament 81, 82, 85, 87, 90–92, 99, 102, 156, 166, 183, 184, 190, 200, 205, 227, 270
Newton, Humphrey 108, 181, 240, 241
Newton, Isaac 1, 2, 11, 14, 15, 26, 27, 30–42, 45–47, 51, 71–76, 78–85, 87, 89, 91–93, 95–119, 121–123, 125, 126, 128, 129, 132, 134–143, 145–153, 154, 155, 157–159, 161–167, 168, 169, 171–181, 183, 184, 186–190, 192–220, 223–225, 229, 231–234, 237–256, 259, 261, 263, 265, 267, 269, 271, 273–275, 278, 279, 281, 283–285, 287–289
~ *A Treatise on the Apocalypse* 199
~ *Chronology of Ancient Kingdoms Amended* 73, 181, 183
~ *Correspondence* 78, 80, 81, 82, 84, 99, 103, 113–115, 169, 231
~ *Essays on Several Important Subjects in Philosophy and Religion* 229, 230
~ ~ Essay II: "Of Scepticism and Certainty" 229
~ ~ Essay 5: "The Agreement of Reason and Religion" 229
~ *Observations on the Book of Daniel and the Book of Revelation* 196, 205
~ *Observations upon the Prophecies of Daniel and the Apocalypse of St. John* 128, 147, 166, 171, 172, 176, 205, 209–220, 237, 259, 269, 271, 273, 274, 275
~ *Of the Church* 145, 146, 154, 157–159, 161, 177
~ ~ Chapter 1: "Of the Host of Heaven & The Prince of the Host" 147
~ ~ Chapter 2: "Of the Holy Covenant" 147
~ ~ Chapter 3: "Of the Faith once delivered to the Saints" 147
~ ~ Chapter 4: "Of the Theology of the Heathens, Cabbalists and ancient Hereticks" 147
~ ~ Chapter 5: "Of the Host of Heaven and the corruptions which crept into it" and "Of the disputable opinions which crept into the religion of the primitive Church" 147
~ ~ Chapter 6: "A Further Account of the Host of Heaven and the Corruption which crept into it" 148
~ ~ Chapter 7: "Of the Christian Religion and its corruption in Morals" 148
~ ~ "Of the corruption of the

Christian Religion in discipline & morality" 148
~ ~ Chapter 8: "Of the Corruption of the Church in Language and Opinions" 148
~ ~ Chapter 9 & 10: "Of the Sanctuary of Strength, or Of the Temple
~ ~ & Synagogues of the Jews & Churches of the Christians" 148
~ ~ Chapter 11: "Of the Roman Empire in relation to its Ecclesiastical Dominion and legislature in matters of Religion" 148
~ ~ Chapters 12 & 13: "Of the Rise and Dominion of the (Roman) Catholic Church" 148
~ ~ Chapter 14: "Of the Abomination of Desolation" 148
~ "Of the Day of Judgment and World to Come" 275
~ *Of the Prophetic Language* 274
~ *Opticks, or a Treatise of the Reflections, Refractions, Inflections and Colours of Light* 40, 101, 102, 138, 139, 225, 231–233, 239
~ *Original of Religions* 105, 122, 123, 139, 192, 194, 195
~ "Paradoxical Questions concerning ye morals & actions of Athanasius & his followers" 288
~ *Philosophical Transactions* 113
~ *Principia* 31, 36, 39, 103, 107, 117, 142, 157, 174, 181, 197, 200, 223, 239, 241, 244, 248, 256
~ *Regulae Philosophandi* 223, 232, 233
~ "Rules for Interpreting the Words and Language in Scripture" 259, 275
~ *Rules for Methodizing Scripture* 244
~ *Rules for Methodizing the Apocalypse* 243
~ "Rules of Reasoning" 244
~ *Scienza Nova* 189
~ *The Synchonisms of the Three Parts of the Prophetick Interpretation* 246
~ *Theologiae Gentilis Origines Philosophicae* 122, 134–136, 138, 140, 142, 154, 179–181, 183–186, 190, 192, 193, 195–200, 202, 240, 253
~ *Theological Manuscripts* 213–215, 219
~ *Trattato sull' Apocalisse* 166, 203, 207, 208, 212, 213–218
Newton MS 1 166, 203, 209
Newton MS 3 166
Newton MS 16 180, 181
Newton's antitrinitarian and anti-Catholic interpretation of revelation 100
Newton's library 158
Newtonian science 40
Newtonian style 140
Nice 108
Nicea 33
Nicene Creed 153, 154
Nicholas 215
Nicolson, Marjorie Hope 37
~ *The Conway Letters* 37
Nidditch, P.H. 222
~ *An Essay Concerning Human Understanding* 222
Ninus 195
Nisbet, Alexander 267
~ *An Exposition of 1 & 2 Peter* 267
No Protestant Plot 107
Noachian deluge 266
Noachian idolatry 185
Noachian monotheism 137, 139
Noachian religion 129, 135–137, 196
Noachian theology 140
Noachide commandments 151, 158
Noachide laws 155, 156
Noachide (natural) religion 145, 256
Noachid(e)s 183, 192, 194
Noah 41, 63, 73, 132–134, 136, 137, 139, 175, 181, 183, 185, 192, 195, 253–256, 272
Notarikon 10, 11
Nous 38
Nye, Stephen 162

Oakeshott, Michael 278
~ *Leviathan; or, The Matter, Forme and Power of a Commonwealth Ecclesiasticall and Evil* 278

Oakley, F. 174
~ *Omnipotence, Covenant and Order: An excursion in the History of Ideas from Abelard to Leibniz* 174
objective 210
Occult 8, 13, 18, 27, 45, 55, 57, 58, 60, 75, 187
Occult philosophy 17
occult properties 59
Occultism 18, 20
Ockham 234
Oecolampadius 205
Oecomenius 215
Officer, Charles and Jake Page 272
~ *The great Dinosaur Extinction* 272
Old Testament 83, 86–89, 91, 92, 100, 101, 126–129, 146, 167, 176, 183, 200, 217, 239, 269, 270, 276
Oldenburg 113–115
Olson, Richard 237, 244
~ *Science Deified & Science Defied: The Historical Significance of Science in Western Culture* 237, 244
Oratorian 92
Original sin 21
Origen 88, 263, 265
Originals of Religion 103
Orthodox church 42
orthodox biblical critics 82
Osiander 213
Osler, Margaret J. 234
~ *Divine Will and the Mechanical Philosophy* 234
Ovid 126, 184, 186, 286
~ *Metamorphosis* 286
Ovidian poetry 184
Ovitt, Jr., George 23
~ *The Restoration of Perfection: Labor and Technology in Medieval Culture* 23
Oxford Latin 193

Pagden, Anthony 50
~ *European Encounters with the New World* 50
~ *The Fall of Natural Man: The American Indian and the Origins of Comparative Ethnology* 50
Pagels, Elaine 28, 42
~ *The Gnostic Gospels* 28, 42
papacy 82, 172, 206, 208
papal authority 96
papal church 81
papal Rome 208, 212, 214
papal territories 262
papal tiara 205
Papias 203, 216
papists 98, 101, 118, 269
Paracelsian 54, 55, 58, 59
Paracelsianism 47
Paracelsus 10, 11, 53–60, 71, 74
~ *Astronomia Magna, oder die ganze Philosophia Sagax der grossen und kleinen Welt [1537–1538]* 57–59
~ *Die 9 Bucher De Natura Rerum* 53, 54
~ *Labyrinthus medicorum errantium [1537–1538]* 59
~ Paracelsus's "doctrine of signatures" 57
~ *The Hermetic and Alchemical Writings of Aureolus Philippus Theophrastus Bombast of Holenheim, Called Paracelsus the Great* 53
~ *Von den Natürlichen Dingen [1525–?]* 58
Paradice 62
Paradise 8, 9, 53, 56, 62, 63
Paré 68, 70
~ *Le Livre des animaux et de l'excellence de l'homme* 68
Pareus, David 207, 213, 216, 219
~ *A Commentary upon the Divine Revelation of the Apostle and Evangelist John* 207, 216, 219
Parliament 118, 159
Pascal 26
Patrides, C.A. and Joseph A. Wittreich 6, 208, 260
~ *The Apocalypse in English Renaissance Thought and Literature* 6, 208, 260
Patristic theology 19
Patterson, F. 245
~ *The Works* 245
Paul 24, 91, 96

Paulinus 204
Pausanius 186
Peachell, John 118
Pelagian ethic 27
Pelagianism 27–29
Pele 5
Pembroke, Thomas 252
Pentateuch 86, 145
Pentecost 53, 60, 65, 68, 70, 71
Pentecostal narrative of man's fall and redemption 71
Pentecôte 70
Pépin, Jean 199
~ *Mythe et Allégorie: Les origines grecques et les contestions judéo-chrétiennes* 199
Pergamus 171
Petau, Denys 239
Peterfreund, Stuart 121, 139
~ "Mosaic Self-Presentation in Newtonian Optics" 139
~ "Saving the Phenomenon or Saving the Hexameron? Mosaic Self-Presentation in Newtonian Optics" 121
Peters, E. 24
~ "Libertas Inquirendi an the Vitum Curiositatis in Medieval Thought" 24
Peters, Manfred 66
Petrarch 206
Petrine conflagration 263, 264, 267, 269, 272, 274, 276
Petrine prophecy 263, 265, 267, 276
Pfeiffer, Rudolf 190
~ *History of Classical Scholarship 1300–1850*
Pharaoh 102
Philadelphia 171
Philo 35, 186, 239
Philosopher's Stone 9, 12, 14, 137
Photius 186
physical world 228
Picatrix 26
Pico 20–22, 24
~ *Oration on the Dignity of Man* 20
Pictures & Images 97, 108
Piggot, Stuart 250, 251, 252, 256
~ *The Druids* 256
~ *William Stukeley. An Eighteenth-Century Antiquary* 251, 252
Piscator 262
Pitassi, M.C. 79, 94
~ *Entre Croire et Savoir. Le Probleme de la Methode Critique chez Jean Le Clerc* 79
~ "Le notion de communication dans l'exégèse biblique de la fin du XVIIe siècle" 94
~ "Le Philosophe et L'écriture. John Locke exégète de sainte Paul" 94
Pitman, John Rogers 264
~ *The Whole Works of the Rev. John Lightfoot, D.D.* 264
Plato 19, 30, 186
platonic 230
Platonism 30, 34, 147
Platonists 22, 30, 33, 162, 190
Plinie 132, 134
Pliny 186, 196
Plotinus 20, 28, 186
~ *Enneads* 28
Plutarch 186, 196
Pococke, Edward 239, 249
~ *The Ancient Constitution and the Feudal Law: A Study of English Historical*
~ *Thought in the Seventeenth Century* 249
Poiret, Pierre 286
~ *L'Oeconomie Divine* 286
Polybius 124, 126
Pomis De, David 286
~ *Zemah David* 286
Pompilius Azalus 4
~ *De omnibus rebus naturalibus* 4
Pope, Alexander 220
~ *An Essay on Criticism* 220
Pope Damasus 148
Pope Innocent III 206
Pope Pius II 107
Pope Valens 148
Pope Valentinian 148
Pope(s) 97, 107, 146, 148, 219

Popery 100, 107, 119
Popish 111, 288
Popish king 117
Popish plot 119
Popkin, Richard H. 31, 33, 35, 36, 73, 74, 77, 159, 165, 179, 186, 203, 205, 209, 221, 222, 241, 259
~ "Joseph Glanvill: A Precursor of David Hume" 222
~ "Newton as a Bible Scholar" 204, 205, 209, 259
~ "Newton's Biblical Theology and his Theological Physics" 35, 165, 179
~ "Polytheism, Deism and Newton" 186
~ "Spinoza, Neoplatonic Kabbalist?" 31, 33
~ *The High Road to Pyrrhonism* 222
~ "The 'Incurable Scepticism' of Henry More, Blaise Pascal and Søren Kierkegaard" 37
~ *The Third Force in Seventeenth-Century Thought* 36
~ "TheThird Force in 17th Century Philosophy: Skepticism, Science and Biblical Prophecy" 36
Popkin, Richard and Arjo Van der Jagt 126, 154
Popkin, Richard and Gordon M. Weiner 150, 241
~ *see also* Force, James
Porphyrius 186, 192
Postel 11
Postlapsarian 54, 56–58
post-millennialism 275
post-Newtonian 248
Praxis 141
Predestinarianism 161
premillennialist 245, 260
Presbyter, John the 204
Presbyterian Hebraist 156
Presbyterian orthodoxy 280
Presbytarianism 156
Presbyterians 159
Pre-Socratic 21
Preterists 264
Prigent, Pierre 213, 215
~ *Apocalypse 12: Histoire de l'exégèse* 213, 215
primitive Christianity 148–150
primitive Church 155
print technology 82
Principia Mathematica 101
Principles 151
Prisca Sapientia 35, 41
Prisca Theologia 35, 36, 41, 42
Pristine Christians 158
Post-biblical times 172
Proclus 186
Progress 27, 253, 256
progressive march of knowledge 243
Prophecy 11, 36–38, 96, 100, 128, 165, 67–169, 171, 173, 176, 177, 190, 194, 200, 205, 206, 212, 217, 219, 241, 242, 244, 248, 259, 264, 265, 268–271, 274, 280, 286
Prophet(s) 188, 200, 216, 269, 274, 276
Prophet Haggai 271
prophetic 27, 205, 219, 270
prophetic authority 139
prophetic books 43, 175
prophetic certainty 38
prophetic chronometry 275
prophetic function 208
prophetic history 244
prophetic intent 275
prophetic interpretation 36, 37, 264
prophetic language 208, 210, 243, 259, 271, 274, 281
prophetic literature 289
prophetic reading 219
prophetic scheme 171, 173
prophetic(k) scriptures 167, 169
prophetic sections 205
prophetic spirit 265, 275, 276
prophetic symbols 219
prophetic texts 198
prophetic visions 244
prophetic writings 239
prophets 116, 167, 171, 209, 210, 217, 245, 247, 270, 272
Protestant(s) 26, 84, 89, 90, 98–100, 106, 107, 110, 116, 118, 119, 135, 172, 205–

209, 212, 213, 215, 218–220
Protestant Christianity 252
Protestant humanism 88
Protestant interpretation of Revelation 203
Protestant polemic 209, 212, 214
Protestant speculation 218
Protestant textual certainties 89
Protestant theology 135
Protestant thinking 212
Protestant tradition 215
Protestant translations 88
Protestant version of the Old Testament 88, 89
Protestant view 117
Protestantism 26, 95, 100
Providential deity 1
Prytanea 104, 136–140, 194, 195–197, 253–255
prytanean fire 196, 200
prytaneum 193–196, 198, 255
prytaneum fire 194, 196
Purgatory 288
Pythagoras 36, 196, 256
Pythagorean 21
Pythagorian doctrine 36
Pythagorean philosophy 36

Queen Elizabeth 7, 124

Racin, John 126
~ *Sir Walter Raleigh as Historian: An Analysis of The History of The World* 126
Ralegh, Sir Walter 123, 125, 126, 132–134, 136,
138, 143
~ *The History of the World* 132–134, 143
Rangoni, Tomasso 4
rapture of the Saints 274
Rasis 59
Rattansi, P.M. 18, 33, 41, 239, 248
~ "Newton and 'The Pipes of Pan'" 248
~ "Newton and the Wisdom of the Ancients" 239, 248
~ *Notes and Records of the Royal Society* 248
~ "Reason in Sixteenth and Seventeenth-Century Natural Philosophy" 18
Ray, John 260–262, 264–267, 272, 276, 289
~ *Three Physico-Theological Discourses* 262, 264, 265, 276
Reedy, G. 85, 92
~ *The Bible and Reason: Anglicans and Scripture in Late Seventeenth-Century England* 85, 92
Reformation of the church 6, 24, 28, 100
~ first reformation 206
Reformers 206
Refrangibility 113
Relics 109, 148
religious corruption 141, 145
Renaissance 17, 25, 27–29, 50, 59
Renaissance cultural narratives 68
Renaissance Hermeticists 17
Renaissance language theory 70
Renaissance magicians 39
Renaissance natural history 49, 64
Renaissance natural medicine 52, 53
Renaissance natural philosophy 48
Renaissance naturalist(s) 48, 66, 69
Renaissance Neoplatonists 19, 25, 42
Renaissance Platonists 26
Renaissance writers 243
Restitution 2
~ Cosmic restitution
Restoration 143, 216, 251, 264, 267
Resurrection 247, 260, 261, 277, 278, 280
~ First Resurrection 262, 277, 278, 280
~ Saints and Martyrs of the First Resurrection 273, 276, 277, 280
~ Last Resurrection 262
~ Second or General Resurrection 261, 273, 277, 278
Reuchlin, Johannes 4, 5, 8
~ *De arte cabalistica* 5
~ *De verbo mirifico* 4, 5
Revelation 1 169, 171, 173
Revelation 5–7 171, 210
Revelation 8–16 171, 172, 210, 211, 214
Revelation 11–14 171, 209, 212, 214
Revelation 12-168, 215
Revelation 13-213, 215

Revelation 17-18 212–215, 268
Revelation 19-165, 168, 176, 268
Revelation 20-268, 286
Revelation 21-212, 279, 280, 281, 284
revelation 83, 87, 110, 119, 128, 135, 140, 167, 176, 178, 190, 203, 205–208, 219, 220, 239, 265, 267, 281, 284–286
~ Book of Revelation 100, 107, 119, 166, 167, 171– 173, 176–178
Revelations 35
revisionist Newtonian scholarship 45
Revolution 216
Rhodiensis 82
Rhodius, Apollonius 186
Ribera, Francisco de 203, 207–209, 213, 216
~ *In sacram beati Ioannis apostoli, et evangelistae Apocalypsin commentarii* 207, 209, 213
Rice, Eugene 24, 28
~ *The Renaissance Idea of Wisdom* 24, 28
Richards, Julian 249
Ridley 107
~ *De Cena Domini* 107
Righini Bonelli, L.M.and William R. Shea 18
~ *Reason, Experiment and Mysticism in the Scientific Revolution* 18
Roberts, Alexandra and James Donaldson 188
Roberts, Julian and Andrew G. Watson 3, 4, 7
~ *John Dee's Library Catalogue* 4, 7
Robinson, John M. 226, 227
~ *The New Organon* 226
~ *The Philosophical Works of Francis Bacon* 226
Rogers, G.A. 34, 221
~ "Locke, Newton and the Cambridge Platonists on Innate Ideas" 34
Roman Emperors 148
Romans 11.20 24
Rondelet, Guillaume 48, 69, 70
Rosenroth, Knorr von 33
~ *Kabbala denudata* 33
Rosicrucian Enlightenment 18, 36, 40
Rosicrucian Manifestos 17
Rosicrucian thinking 40
Ross, G. MacDonald 18
~ "Occultism and Philosophy in the Seventeenth Century" 18
~ "Reply to Simon Schaffer" 18
Rossi, Paolo 18, 39, 126
~ "Hermeticism, Rationality and the Scientific Revolution" 18, 39
~ The Dark Abyss of Time: The History of the Earth and the History of Nations from Hooke to Vico 126
Rosicrucians 30
Rothstein, Maria 61
~ "Etymology, Genealogy, and the Immutability of Origins" 61
Rotman, Brian 140
~ *Ad Infinitum: The Ghost in Turing's Machine* 140
Royal Society 14, 27, 112, 113, 160, 197, 238, 239, 242, 260, 269
Rudbeck, Olof 197, 198
~ *Atlantica* 197
Rudolph, Hartmut 57
~ "Einige Gesichtspunkte zum Thema 'Paracelsus und Luther'" 57
~ "Kosmosspekulation and Trinitaetslehre, Weltbild und Theologie bei Paracelsus" 57
~ *Paragranum [1529–1530]* 57
Rupp, G. 162
~ *Religion in England* 162
Russell, Bertrand 221
~ *Nightmares of Eminent Persons, and Other Stories* 221
Russell, C. 99

Sacred texts 129
Sacy 90
Saints 108, 109, 148, 262, 274, 278– 282, 284–288
~ Camp of the 281
~ Raised Saints 260, 262, 273, 277–279, 281– 286
~ of the First Resurrection 262

Salisbury, John of 243
~ *Metalogicon* 242
Salmatius (Claude de Saumaise, 1588–1653) 157
Salvation 20, 21, 62, 64
Samuel 255
Sanhedrin 158
Sardis 171
Sargent, Rose-Mary 73, 112, 260
~ *The Diffident Naturalist: Robert Boyle and the Philosophy of Experiment* 73, 112, 260
Sarton, George 180
Satan 150, 213, 247, 260, 261, 285, 286
Saul 255
Saul, Barnabas 3
Saved Nations 261–263, 274, 275, 277, 278, 283, 285
Saviour 268
Scaliger 157
Schaffer, Simon 1, 18, 112, 113
~ "Glass works: Newton's prisms and the uses of experiment" 113
~ "Occultism and Reason" 18
~ Schaffer's "Newton's Comets and the Transformation of Astrology" 1
~ *See also* S. Shapin 112
Scheurer, P.B. 35, 36
Scheurer, P.B. and G. DeBrock 179
~ *Newton's Scientific and Philosophical Legacy* 179
Schmitt, Charles 19
~ "Reappraisals in Renaissance Science" 19
Scientific progress 27, 237, 242, 243, 248, 256, 266
Scientific revolution 26, 27, 36, 46, 49, 60, 237
Scholasticism 23
scriptural inspiration 79
scriptural prophecies 245, 270
scriptural reports 245
Scripture(s) 30–32, 35, 37, 41, 43, 63, 73, 78–83, 87, 89, 90, 92–99, 105, 106, 116, 135, 150, 151, 153, 190, 204, 205, 227, 233, 242, 245, 266, 268, 269, 275, 279, 281–283
~ corruption of 80, 106
Scripture prophecy 267
seals of divinity 104
second coming of Christ 167, 176, 177, 200, 210, 246, 247, 260, 261, 263, 269, 274, 276, 285
Second and Third John 205
Second Peter 205
Selden, John 155–158, 239
~ *De jure naturali* 158
~ *De Synedriis* 156, 158
~ *see also* Ziskind, J.R.
self-censorship 79
sensible objects 104
Septuagint 88
Septuagint texts 129
Septuagint translation 95
Servius 186
Seznec, Jean 254
~ *The Survival of the Pagan Gods. The Mythological Tradition and Its Place in Renaissance Humanism and Art* 254
Shairani, H.M.K. 155
Shakelton, Francis 6
~ *A blazyng Starre* 6
Shapin, S. and S. Schaffer 112
~ *Leviathan and the Airpump: Hobbes, Boyle and the Experimental Life* 112
Shapiro, A. 114, 209
~ "The gradual acceptance of Newton's theory of Light and Color 1672–1727" 114
Shapiro, Barbara 27, 159
~ "Latitudinarianism and Science in Seventeenth-Century England" 27, 159
Shapiro, B. and M.C. Jacob 159
Sheltco à Geveren 6
~ *Of the Ende of this Worlde, and the second Commyng of Christ* 7
Sherlock, Thomas 264
~ *Discourses on the Use and Intent of Prophecy* 264
Sherlock, William 162
Sherman, William 4
~ *John Dee: the Politics of Reading and*

Writing in the English Renaissance 4
Sigillum Dei or seal of God 5
signatura 55
signatures 57–60
Silver age 126
Simeon 263
Simon, Richard 79, 83–96, 205, 239
~ *A Catalogue of the Chief Editions of the Bible* 95
~ *Critical History of the New Testament* 84, 90, 91, 93
~ *A Critical History of the Versions of the New Testament* 91, 95
~ *Critical History of the Old Testament* 86–89
~ ~ Book One 86–88
~ ~ Book Two 88, 89
~ ~ Book Three 89, 90
~ *Histoire Critique du Vieux Testament* 84–86, 95, 205
Singleton, Charles S. 17
~ *Art, Science, and History in the Renaissance* 17
~ CHNT 84
Sirin, Ahmad ibn 203, 214
~ *Oneirocriticon* 203
Sloane MS 3188-3189 1
Sloane MS 3191 1
~ Add. MS 36674 1
Sloane MS 3624-3628 12
Sloane, Sir Hans 12
Smith, Cyril Stanley
Smith, John 286
Smith, Pamela 137
~ *The Business of Alchemy: Science and Culture in the Holy Roman Empire* 137
Smolinski, Reiner 259, 260, 275, 283
~ "Israel Redivivus: The Eschatological Limtis of Puritan Typology in New England" 275, 282
~ *The Threefold Paradise of Cotton Mather* 260, 283
~ "When Shall These Things Be?" 260
Smyrna 171
Snider, Alvin 129
~ *Origin and Authority in Seventeenth-Century England: Bacon, Milton, Butler* 129
Socinianism 83, 161
Socinians 89, 159, 162
Socrates 238
Solomon 5, 13, 102, 210, 226, 227, 255
Solon's Temple 195
Sommerus, Joannes 7
~ *Refutatio scripti Petri Carolii editi Wittebergae* 7
Son 174, 176, 178
Son of Man 169, 171, 174, 177, 178
Sorcery 106–108
Sorelius, Gunnar 197
Sourdel, Dominique 24
Sourdel-Thomine, Janine 24
South, Robert 162
Spencer, Edmund 7, 86, 212
Spencer, Edmund and Gabriel Harvey 7
~ *Letters Touching the Earthquake* 7
Spencer, John 217, 239
Scripture(s), holy 1, 87, 89, 245
Spinoza, Baruch 31, 34, 79, 87, 91, 269, 274, 276
~ *Tractatus (Treatise)* 276
Sprat, Thomas 27, 72
Srigley, Michael 197
St John 81, 167, 176–178, 241, 281
~ St John's Temple 282
St Paul 196, 226, 271
St Peter 273
Staudenbauer, Craig A. 38
~ "Galileo, Ficino and Henry More's Psychathanasia" 38
Staynoe, Thomas 286
Steiner, Arpad 24
~ "The Faust Legend and the Christian Traditions" 24
Stillingfleet, Edward 85, 92, 119, 155–158
~ *Irenicum. A Weapon Sale for the Churches wounds. Or the Divine Right of Particular Forms of Church-Government; Discussed and examined according to the Principles of the Law of Nature, the positive Laws of God, the practice of the Apostles and the Primitive*

Church, and the judgment of Reformed Divines. Whereby a foundation is laid for the Churches peace, and the accommodation of our present differences. 157
Stillman, Robert 129
~ *Philosophy and Universal Languages in Seventeenth-Century England: Bacon, Hobbes, and Wilkins* 129
Stock, Brian 186
~ *Myth and Science in the Twelfth Century* 186
Stoic sources 21
Stoic view 19
Stoicism 20, 265
Stonehenge 242, 248–251, 253, 255, 256
Stow, John 249
Strabo 186
Stuart England 159
Stubbe, Henry 155–159
~ *An Account of the Rise and Progress of Mahometanism* 155, 156, 158
Stukeley, William 241, 249–253, 255, 256
~ *Memoirs of Sir Isaac Newton's Life* 250, 252
~ *Palaeographia sacra: or, Discourses on Monuments of Antiquity that Relate to Sacred History* 250, 256
~ "Romance in the Stones: William Stukeley's Curious Journeys" 249
~ *Stonehenge, A Temple Restored to the British Druids* 251
Sudhoff, Karl 53, 57
Sullivan, R.E. 162
~ *John Toland and the Deist Controversy* 162
superiority of modern knowledge 243
supreme being 135
Suspensions of the law 117
Swift, Jonathan 282
Sykes, Arthur Ahley 146
Sylvius, Aeneus (later Pope Pius II) 107
symbolic discourse 210
symbolic exegesis 64
symbolic literalism 53
symbolic meaning 60
symbolic order of nature 75
symbolic reading 75
symbol(s) 48, 59, 73–76, 106, 135, 210, 213–215
symbol system 212
symbolism 259

Tabernacle 195, 196
Tainter, Joseph A. 125, 128
~ *The Collapse of Complex Societies* 125, 128
Tallents, Francis 127
~ *A View of Universal History* 127
Talmudic scholars 239
Talmudic tradition 260
Tansillo 24
Teach Gnudi, Martha 9
Teich, Mikulas and Robert Young 18
~ *Changing Perspectives in the History of Science: Essays in Honour of Joseph Needham* 18
Temple, Sir William 239, 242
~ *Essay upon Ancient and Modern Learning* 239
Temple(s) 104, 135, 136, 140, 141, 148, 156, 157, 177, 195–198, 209–211, 248, 252, 281
Temple of God 137
Temple of Solomon 135, 195, 196, 252
Temple polities 155
Temurah 10, 11
Ten Commandments 153, 175
Tenison, Thomas 119
Tent 210
Tertullian 33, 110, 186, 262, 282
Test Act 117
Text 87, 88
textual accuracy of biblical books 205
textual authenticity 80
textual corruption 82, 135
textual criticism 81
textual perversions 99, 101, 106
The Faerie Queene 212
Theoderetus 33
Theodicy 31
Theodosius 171, 172

Theologia Platonica 20, 23
theological and hermeneutical opinion 79
theological interpretation 96
theological notebook 106
Thessalonians 267
Thirty-Nine Articles 159–161
Thomist 22
Thou, de 29
Thyatira 171
Thycydides 194
Tibullus 186
Tillotson, John 119
Tillinghast 218
Titus 267
Todd, Janet 142
~ *The Works of Aphra Behn* 142
Toinard, Nicholas 84, 85, 94
~ *Harmonia* 94
Toland, John 91, 155, 156, 158, 159
~ *Nazarenus: or, Jewish, Gentile and Mahometan Christianity* 155, 156, 158
Toleration Act 159, 161
Toon, Peter 278
Topsell, Edward 62–64
~ *Beastes* 63, 64
~ "Epistle Dedicatory" 63, 64
~ *The Historie of Foure-Footed Beastes (1607)* 62
Torah 210
Tornacensis 82
Tower of Bable 60, 62, 133, 134, 137
traditional creeds 161
Trajan's time 263
transcriptional errors 82
translations, bad 88
Transmigration of Souls 101, 105, 139
Transmutation of metaphysical emanation theories 148
Transubstantiation 106
Trevor-Roper 27, 28
~ *The Crisis of the Seventeenth Century: Religion, the Reformation and Social Change* 27, 28
Trinitarian apostasy 166
Trinitarian corruption of the biblical text 135
Trinitarian Creed 153, 162
Trinitarian doctrine 162
Trinitarian histories of the Church 21
Trinitarian theology 122
Trinitarianism 33, 108, 139, 143, 148, 154, 158, 173, 176
Trinitarian(s) 142, 149, 159
Trinity 82, 100, 153, 158, 160, 162, 173, 239
Trinkhaus, Charles 19, 20, 22, 27, 42
~ *In Our Image and Likeness: Humanity and Divinity in Italian Humanist Thought* 19, 22
Trismegistus, Hermes 26
Tritheim, Johan 4
~ *De Septem Secundeis* 4
Tritheism 162
Trithemius 4
~ *Liber Octo Quaestionum, Quas Illi Dissolvendas Proposuit Maximilianus Caesar* 4
~ *De Septem Secundeis* 4
Triune godhead 100
true church 213, 214
True names 11
true religion 248
Turnbull, H.W. 78, 169, 193, 224
~ *The Correspondence of Isaac Newton* 78, 98, 169, 193, 224, 225, 243
Tuveson, Ernest L. 6, 7
~ *Millennium and Utopia: A Study in the Background of the Idea of Progress* 6, 7
Twisse, William 261
Twyne, John 249
Tyconius 203, 204, 207, 210, 213

Ucko, Peter J., M. Hunter, A.J. Clark and A. David 250
~ *Avebury Reconsidered: From the 1660s to the 1990s* 250
ULC Add MS 3970 200
Ullmann-Margalit, Edna 149
Una 212
Unitarianism 162
Unitarians 159, 161, 162
universal language 70

universal historians 126, 131, 143
universal history 122–126, 128, 130, 132, 135, 138–143, 183, 184, 197
Uriel, the angel 5
Utzinger, Alexandrus 218

Vallarsius, Dominicus 204
~ *Opera* 204
Van den Berg, J. 171
~ "Continuity within a Changing Context: Henry More's Millenarianism Seen Against the Background of the Millenarian Concepts of Joseph Mede" 172
Van der Jagt, Arjo 241
~ *Scepticism for and against religion in the Seventeenth and Eighteenth Centuries* 241
Van Helmont, Francis Mercury 33
Van Kley, Edwin J. 133
~ "Europe's 'Discovery' of China and the Writing of World History" 133
Vaticanus 82
Vaughan 30
Vaux, Roland de 211
~ *Ancient Israel* 211
~ *Religious Instutions* 211
Veneration of reliques 109
Veneration of sacred animals and plants 105
verbum dei 55
Vernon, George 131
~ *The Life of the Learned and Reverend Dr. Peter Heylyn* 131
Verstegan, Richard 249
Vespasian 171
Vials of wrath 105
Vickers, Brian 18
~ "Analogy and Identity" 18
~ "Frances Yates and the Writing of History" 18
~ *Occult and Scientific Mentalities in the Renaissance* 18
~ "On the Function of Analogy in the Occult" 18
Vico 189
Victorinus 204, 210, 213
Vinci da, Leonardo 42
Virgil 186, 190
~ *Aeneis* 190
Vives, Juan Luis 243
Voegelin, Eric 29
~ *The New Science of Politics: An Introduction* 29
Voeltzel, R. 83
~ "Jean le Clerc (1657–1736) et la critique biblique" 84
Voltaire 239
~ "A Messieurs de l'Academie de Sciences" 239
~ "Ancients and Moderns" 239
Voluntarism 163, 173, 174, 234
Vossius, Isaac 127, 239
~ Dissertatio de vera aetate munid, qua ostenditur Natale munid tempus annis minimum 1400 vulgarem aeram anticipare 127
Vossius, Gerardus 127, 129, 134, 138, 183–186, 190, 195, 239, 240
~ *De Theoligia Gentili et Physiologia Christiana* 127, 184–186
~ *Qua est de cultu corporis caelestis* 184
vulgar edition 95
Vulgate 88, 89, 95

Wagar, W.W. 104, 121, 154, 241
~ *The Secular Mind: Essays Presented to Franklin L. Baumer* 241
Wainwright, A.W. 93
~ *John Locke: A Paraphrase and Notes on the Epistle of St Paul* 93
Waite, Arthur Edward 53, 54
Waldensians 206
Walker, Antony 91
Walker, D.P. 18
~ *The Ancient Theology* 18
Wallis, R.T. 20
~ *Neoplatonism* 20
Walton 90
Ward, Richard 38
~ Life of Henry More 38
Watson, Andrew G., *see* Roberts, Julian
Watts, M.R. 161

~ *The Dissenters, vol. I: From the Reformation to the French Revolution* 161
Waquet, F. 94
Webb, John 127, 133, 143
~ *An Historical Essay Endeavoring a Probability that the Language of the Empire of China is the Primitive Language* 127
~ *Language of the Empire of China* 133
Weber 26
Webster, Charles 18, 41, 159
~ *The Great Instauration: Science, Medicine and Reform, 1626–1660* 18, 41
Weemse, John 269
Weiner, G.M. 35
Westfall, R.S. 99, 104, 121, 122, 135, 142, 143, 150, 154, 165, 171, 173–177, 179, 181, 193, 203, 205, 209, 210, 217, 240–243, 252, 254, 273, 274
~ "Isaac Newton's *Theologiae Gentilis Origines Philosophicae*" 104, 121, 135, 154, 179, 193, 241, 254
~ *Never at Rest: A Biography of Isaac Newton* 121, 122, 142, 165, 171, 173–175, 177, 179, 203, 205, 209, 217, 240, 241, 243, 252, 274
~ "Newton: Theologian" 149
~ "Newton's Theological Manuscripts" 179, 241, 242
~ "Short-Writing and the State of Newton's Conscience" 99
Westman, Robert S.
~ *Chronologies and the Rhetoric of Relative Time* 127
~ *see also* Lindberg, David C.
Whiston, William 245, 259, 262, 267, 269, 270, 271–273, 275, 276, 282
~ *Astronomical Principles of Religion, Natural and Reveal'd* 245
~ "Introductory Discourse" 270, 271, 282
~ *New Theory of the Earth* 245, 262, 270, 272, 273
~ "Postulata" 259
~ *The Accomplishment of Scripture Prophecies* 269
Whitaker, William 269
Whitby, Daniel 275, 279
~ *Treatise of The True Millennium* 275, 279
whore 205, 208, 212, 214–216
whore Duessa 212
will of God 92
Williams, John 79, 92
Williams, Michael Allen 28
~ *Rethinking "Gnosticism": An Argument for Dismantling a Dubious Category* 28
Willson, David Harris and Stuart E. Prall 149
~ *A History of England* 149
Wilkins, John 40, 133, 238
~ *Essay Towards a Real Character and a Philosophical Language* 133
Witchcraft 110
Wittreich, Joseph A., *see* Patrides, C.A.
woman 208, 212–216
women 212, 216
Wood, Ellen Meiksins 124
~ *The Pristine Culture of Capitalism* 124
Word (of God) 45, 46, 52, 53, 55–63, 68–73, 75, 76, 89, 90, 92, 98, 167, 168, 175, 177
Works 38–40, 52
Worm, Ole 249
Wormwood, star 207
Worthington, John 216, 260
~ *The Works of the Pious and Profoundly Learned Joseph Mede* 260
Wotton, William 239, 248
~ *Reflections upon Ancient and Modern Learning* 239
~ *A Defense of the Reflections upon Ancient and Modern Learning* 239
Xenophon 186

Yahuda, A.S. 179, 180
Yahuda collection 146, 180, 181, 209
Yahuda MS 1 122, 166–168, 190, 199, 209, 243, 287
Yahuda MS 3 175
Yahuda MS 4 94, 205
Yahuda MS 6 263, 274, 284, 285, 287

Yahuda MS 7 200
Yahuda MS 8 171, 190
Yahuda MS 9 107, 263, 279, 284, 285
Yahuda MS 13 240
Yahuda MS 14 106, 174
Yahuda MS 15 146, 173, 174, 245
Yahuda MS 15.5 106, 152
Yahuda MS 15.7 30, 31, 100, 161
Yahuda MSS 17 and 16 104, 105, 181, 185, 187–189, 191–197, 240, 241, 255
Yahuda MS 21 103
Yahuda MS 41 104, 105, 135, 136, 138, 139, 142, 192–196, 254
Yahuda MSD 15.3 32, 100, 106, 116
Yahuda MSS (A Synopsis of the Synchronisms of the Apocalypse) 165, 169
Yahuda Newton MS 1 73, 167, 173, 176, 177
Yahuda Newton MS 15 177
Yahuda MS. Var. I, Newton MS 21 76, 168, 180, 181
Yahuda MS Var. 1, Newton MS 43 180
Yates, Frances 17–19, 21, 22, 25, 26, 40, 42
~ *Giordano Bruno and the Hermetic Tradition* 22
~ "The Hermetic Tradition in Renaissance Science" 17
Yates thesis 17, 18, 27, 30, 36, 42
Yolton, John 246
~ *The Blackwell Companion to the Enlightenment* 246

Zagorin, Perez 38
Zika, Charles 8
~ "Reuchlin's *De verbo mirifico* and the Magic Debate of the Late Fifteenth Century" 8
Ziskind, Jonathan R. 156
~ *John Selden on Jewish Marriage Law: the "Uxor Hebraica"* 156
Ziskind, Martha 156
Zohar 33
Zwingli 205

X Joh: 9.14 86
1 Thess 4:16-17 274
2 Kings 17 103
2 Peter 3:7 264
2 Peter 3:10 261–263, 267, 269
2 Reg IV 86
3 Reg XI 86
2 Thess 1:7 and 2:8 268
Jos. X 86
reg. ii. 16, Sanctius 86

ARCHIVES INTERNATIONALES D'HISTOIRE DES IDÉES
*
INTERNATIONAL ARCHIVES OF THE HISTORY OF IDEAS

---

1. E. Labrousse: *Pierre Bayle.* Tome I: *Du pays de foix à la cité d'Erasme.* 1963; 2nd printing 1984 ISBN 90-247-3136-4
   For Tome II *see below under Volume 6.*
2. P. Merlan: *Monopsychism, Mysticism, Metaconsciousness.* Problems of the Soul in the Neoaristotelian and Neoplatonic Tradition. 1963; 2nd printing 1969 ISBN 90-247-0178-3
3. H.G. van Leeuwen: *The Problem of Certainty in English Thought, 1630–1690.* With a Preface by R.H. Popkin. 1963; 2nd printing 1970 ISBN 90-247-0179-1
4. P.W. Janssen: *Les origines de la réforme des Carmes en France au 17ᵉ Siècle.* 1963; 2nd printing 1969 ISBN 90-247-0180-5
5. G. Sebba: *Bibliographia Cartesiana.* A Critical Guide to the Descartes Literature (1800–1960). 1964 ISBN 90-247-0181-3
6. E. Labrousse: *Pierre Bayle.* Tome II: *Heterodoxie et rigorisme.* 1964 ISBN 90-247-0182-1
7. K.W. Swart: *The Sense of Decadence in 19th-Century France.* 1964 ISBN 90-247-0183-X
8. W. Rex: *Essays on Pierre Bayle and Religious Controversy.* 1965 ISBN 90-247-0184-8
9. E. Heier: *L.H. Nicolay (1737–1820) and His Contemporaries.* Diderot, Rousseau, Voltaire, Gluck, Metastasio, Galiani, D'Escherny, Gessner, Bodmer, Lavater, Wieland, Frederick II, Falconet, W. Robertson, Paul I, Cagliostro, Gellert, Winckelmann, Poinsinet, Lloyd, Sanchez, Masson, and Others. 1965 ISBN 90-247-0185-6
10. H.M. Bracken: *The Early Reception of Berkeley's Immaterialism, 1710–1733.* [1958] Rev. ed. 1965 ISBN 90-247-0186-4
11. R.A. Watson: *The Downfall of Cartesianism, 1673–1712.* A Study of Epistemological Issues in Late 17th-Century Cartesianism. 1966 ISBN 90-247-0187-2
12. R. Descartes: *Regul ad Directionem Ingenii.* Texte critique établi par Giovanni Crapulli avec la version hollandaise du 17ᵉ siècle. 1966 ISBN 90-247-0188-0
13. J. Chapelain: *Soixante-dix-sept Lettres inédites à Nicolas Heinsius (1649–1658).* Publiées d'après le manuscrit de Leyde avec une introduction et des notes par B. Bray. 1966 ISBN 90-247-0189-9
14. C. B. Brush: *Montaigne and Bayle.* Variations on the Theme of Skepticism. 1966 ISBN 90-247-0190-2
15. B. Neveu: *Un historien à l'Ecole de Port-Royal.* Sébastien le Nain de Tillemont (1637–1698). 1966 ISBN 90-247-0191-0
16. A. Faivre: *Kirchberger et l'Illuminisme du 18ᵉ siècle.* 1966 ISBN 90-247-0192-9
17. J.A. Clarke: *Huguenot Warrior.* The Life and Times of Henri de Rohan (1579–1638). 1966 ISBN 90-247-0193-7
18. S. Kinser: *The Works of Jacques-Auguste de Thou.* 1966 ISBN 90-247-0194-5
19. E.F. Hirsch: *Damião de Gois.* The Life and Thought of a Portuguese Humanist (1502–1574). 1967 ISBN 90-247-0195-3
20. P.J.S. Whitemore: *The Order of Minims in 17th-Century France.* 1967 ISBN 90-247-0196-1
21. H. Hillenaar: *Fénelon et les Jésuites.* 1967 ISBN 90-247-0197-X
22. W.N. Hargreaves-Mawdsley: *The English Della Cruscans and Their Time, 1783–1828.* 1967 ISBN 90-247-0198-8
23. C.B. Schmitt: *Gianfrancesco Pico della Mirandola (1469–1533) and his Critique of Aristotle.* 1967 ISBN 90-247-0199-6
24. H.B. White: *Peace among the Willows.* The Political Philosophy of Francis Bacon. 1968 ISBN 90-247-0200-3

25. L. Apt: *Louis-Philippe de Ségur.* An Intellectual in a Revolutionary Age. 1969 ISBN 90-247-0201-1
26. E.H. Kadler: *Literary Figures in French Drama (1784–1834).* 1969 ISBN 90-247-0202-X
27. G. Postel: *Le Thrésor des prophéties de l'univers.* Manuscrit publié avec une introduction et des notes par F. Secret. 1969 ISBN 90-247-0203-8
28. E.G. Boscherini: *Lexicon Spinozanum.* 2 vols., 1970 Set ISBN 90-247-0205-4
29. C.A. Bolton: *Church Reform in 18th-Century Italy.* The Synod of Pistoia (1786). 1969 ISBN 90-247-0208-9
30. D. Janicaud: *Une généalogie du spiritualisme français.* Aux sources du bergsonisme: [Félix] Ravaisson [1813–1900] et la métaphysique. 1969 ISBN 90-247-0209-7
31. J.-E. d'Angers: *L'Humanisme chrétien au 17ᵉ siècle.* St. François de Sales et Yves de Paris. 1970 ISBN 90-247-0210-0
32. H.B. White: *Copp'd Hills towards Heaven.* Shakespeare and the Classical Polity. 1970 ISBN 90-247-0250-X
33. P.J. Olscamp: *The Moral Philosophy of George Berkeley.* 1970 ISBN 90-247-0303-4
34. C.G. Noreña: *Juan Luis Vives (1492–1540).* 1970 ISBN 90-247-5008-3
35. J. O'Higgens: *Anthony Collins (1676–1729), the Man and His World.* 1970 ISBN 90-247-5007-5
36. F.T. Brechka: *Gerard van Swieten and His World (1700–1772).* 1970 ISBN 90-247-5009-1
37. M.H. Waddicor: *Montesquieu and the Pilosophy of Natural Law.* 1970 ISBN 90-247-5039-3
38. O.R. Bloch: *La Philosophie de Gassendi (1592–1655).* Nominalisme, matérialisme et métaphysique. 1971 ISBN 90-247-5035-0
39. J. Hoyles: *The Waning of the Renaissance (1640–1740).* Studies in the Thought and Poetry of Henry More, John Norris and Isaac Watts. 1971 ISBN 90-247-5077-6
    For Henry More, *see also below under Volume 122 and 127.*
40. H. Bots: *Correspondance de Jacques Dupuy et de Nicolas Heinsius (1646–1656).* 1971 ISBN 90-247-5092-X
41. W.C. Lehmann: *Henry Home, Lord Kames, and the Scottish Enlightenment.* A Study in National Character and in the History of Ideas. 1971 ISBN 90-247-5018-0
42. C. Kramer: *Emmery de Lyere et Marnix de Sainte Aldegonde.* Un admirateur de Sébastien Franck et de Montaigne aux prises avec le champion des calvinistes néerlandais.[Avec le texte d'Emmery de Lyere:] *Antidote ou contrepoison contre les conseils sanguinaires et envinemez de Philippe de Marnix Sr. de Ste. Aldegonde.* 1971 ISBN 90-247-5136-5
43. P. Dibon: *Inventaire de la correspondance (1595–1650) d'André Rivet (1572–1651).* 1971 ISBN 90-247-5112-8
44. K.A. Kottman: *Law and Apocalypse.* The Moral Thought of Luis de Leon (1527?–1591). 1972 ISBN 90-247-1183-5
45. F.G. Nauen: *Revolution, Idealism and Human Freedom.* Schelling, Hölderlin and Hegel, and the Crisis of Early German Idealism. 1971 ISBN 90-247-5117-9
46. H. Jensen: *Motivation and the Moral Sense in Francis Hutcheson's* [1694–1746] *Ethical Theory.* 1971 ISBN 90-247-1187-8
47. A. Rosenberg: *[Simon] Tyssot de Patot and His Work (1655–1738).* 1972 ISBN 90-247-1199-1
48. C. Walton: *De la recherche du bien.* A study of [Nicolas de] Malebranche's [1638–1715] Science of Ethics. 1972 ISBN 90-247-1205-X

ARCHIVES INTERNATIONALES D'HISTOIRE DES IDÉES
*
INTERNATIONAL ARCHIVES OF THE HISTORY OF IDEAS

49. P.J.S. Whitmore (ed.): *A 17th-Century Exposure of Superstition.* Select Text of Claude Pithoys (1587–1676). 1972 ISBN 90-247-1298-X
50. A. Sauvy: *Livres saisis à Paris entre 1678 et 1701.* D'après une étude préliminaire de Motoko Ninomiya. 1972 ISBN 90-247-1347-1
51. W.R. Redmond: *Bibliography of the Philosophy in the Iberian Colonies of America.* 1972 ISBN 90-247-1190-8
52. C.B. Schmitt: *Cicero Scepticus.* A Study of the Influence of the *Academica* in the Renaissance. 1972 ISBN 90-247-1299-8
53. J. Hoyles: *The Edges of Augustanism.* The Aesthetics of Spirituality in Thomas Ken, John Byrom and William Law. 1972 ISBN 90-247-1317-X
54. J. Bruggeman and A.J. van de Ven (éds.): *Inventaire* des pièces d'Archives françaises se rapportant à l'Abbaye de Port-Royal des Champs et son cercle et à la Résistance contre la Bulle *Unigenitus* et à l'Appel. 1972 ISBN 90-247-5122-5
55. J.W. Montgomery: *Cross and Crucible.* Johann Valentin Andreae (1586–1654), Phoenix of the Theologians. Volume I: Andreae's Life, World-View, and Relations with Rosicrucianism and Alchemy; Volume II: The *Chymische Hochzeit* with Notes and Commentary. 1973 Set ISBN 90-247-5054-7
56. O. Lutaud: *Des révolutions d'Angleterre à la Révolution française.* Le tyrannicide & *Killing No Murder* (Cromwell, *Athalie*, Bonaparte). 1973 ISBN 90-247-1509-1
57. F. Duchesneau: *L'Empirisme de Locke.* 1973 ISBN 90-247-1349-8
58. R. Simon (éd.): *Henry de Boulainviller* – Œuvres Philosophiques, Tome I. 1973 ISBN 90-247-1332-3
    For Œuvres Philosophiques, Tome II *see below under Volume 70.*
59. E.E. Harris: *Salvation from Despair.* A Reappraisal of Spinoza's Philosophy. 1973 ISBN 90-247-5158-6
60. J.-F. Battail: *L'Avocat philosophe Géraud de Cordemoy (1626–1684).* 1973 ISBN 90-247-1542-3
61. T. Liu: *Discord in Zion.* The Puritan Divines and the Puritan Revolution (1640–1660). 1973 ISBN 90-247-5156-X
62. A. Strugnell: *Diderot's Politics.* A Study of the Evolution of Diderot's Political Thought after the *Encyclopédie.* 1973 ISBN 90-247-1540-7
63. G. Defaux: *Pantagruel et les Sophistes.* Contribution à l'histoire de l'humanisme chrétien au 16ᵉ siècle. 1973 ISBN 90-247-1566-0
64. G. Planty-Bonjour: *Hegel et la pensée philosophique en Russie (1830–1917).* 1974 ISBN 90-247-1576-8
65. R.J. Brook: [*George*] *Berkeley's Philosophy of Science.* 1973 ISBN 90-247-1555-5
66. T.E. Jessop: *A Bibliography of George Berkeley.* With: *Inventory of Berkeley's Manuscript Remains* by A.A. Luce. 2nd revised and enlarged ed. 1973 ISBN 90-247-1577-6
67. E.I. Perry: *From Theology to History.* French Religious Controversy and the Revocation of the Edict of Nantes. 1973 ISBN 90-247-1578-4
68. P. Dibbon, H. Bots et E. Bots-Estourgie: *Inventaire de la correspondance (1631–1671) de Johannes Fredericus Gronovius* [1611–1671]. 1974 ISBN 90-247-1600-4
69. A.B. Collins: *The Secular is Sacred.* Platonism and Thomism in Marsilio Ficino's *Platonic Theology.* 1974 ISBN 90-247-1588-1

ARCHIVES INTERNATIONALES D'HISTOIRE DES IDÉES
*
INTERNATIONAL ARCHIVES OF THE HISTORY OF IDEAS

70. R. Simon (éd.): *Henry de Boulainviller.* Œuvres Philosophiques, Tome II. 1975 ISBN 90-247-1633-0
    For Œuvres Philosophiques, Tome I *see under Volume 58.*
71. J.A.G. Tans et H. Schmitz du Moulin: *Pasquier Quesnel devant la Congrégation de l'Index.* Correspondance avec Francesco Barberini et mémoires sur la mise à l'Index de son édition des Œuvres de Saint Léon, publiés avec introduction et annotations. 1974 ISBN 90-247-1661-6
72. J.W. Carven: *Napoleon and the Lazarists (1804–1809).* 1974 ISBN 90-247-1667-5
73. G. Symcox: *The Crisis of French Sea Power (1688–1697).* From the *Guerre d'Escadre* to the *Guerre de Course.* 1974 ISBN 90-247-1645-4
74. R. MacGillivray: *Restoration Historians and the English Civil War.* 1974 ISBN 90-247-1678-0
75. A. Soman (ed.): *The Massacre of St. Bartholomew.* Reappraisals and Documents. 1974 ISBN 90-247-1652-7
76. R.E. Wanner: *Claude Fleury (1640–1723) as an Educational Historiographer and Thinker.* With an Introduction by W.W. Brickman. 1975 ISBN 90-247-1684-5
77. R.T. Carroll: *The Common-Sense Philosophy of Religion of Bishop Edward Stillingfleet (1635–1699).* 1975 ISBN 90-247-1647-0
78. J. Macary: *Masque et lumières au 18ᵉ [siècle].* André-François Deslandes, Citoyen et philosophe (1689–1757). 1975 ISBN 90-247-1698-5
79. S.M. Mason: *Montesquieu's Idea of Justice.* 1975 ISBN 90-247-1670-5
80. D.J.H. van Elden: *Esprits fins et esprits géométriques dans les portraits de Saint-Simon.* Contributions à l'étude du vocabulaire et du style. 1975 ISBN 90-247-1726-4
81. I. Primer (ed.): *Mandeville Studies.* New Explorations in the Art and Thought of Dr Bernard Mandeville (1670–1733). 1975 ISBN 90-247-1686-1
82. C.G. Noreña: *Studies in Spanish Renaissance Thought.* 1975 ISBN 90-247-1727-2
83. G. Wilson: *A Medievalist in the 18th Century.* Le Grand d'Aussy and the Fabliaux ou Contes. 1975 ISBN 90-247-1782-5
84. J.-R. Armogathe: *Theologia Cartesiana.* L'explication physique de l'Eucharistie chez Descartes et Dom Robert Desgabets. 1977 ISBN 90-247-1869-4
85. Bérault Stuart, Seigneur d'Aubigny: *Traité sur l'art de la guerre.* Introduction et édition par Élie de Comminges. 1976 ISBN 90-247-1871-6
86. S.L. Kaplan: *Bread, Politics and Political Economy in the Reign of Louis XV.* 2 vols., 1976 Set ISBN 90-247-1873-2
87. M. Lienhard (ed.): *The Origins and Characteristics of Anabaptism / Les débuts et les caractéristiques de l'Anabaptisme.* With an Extensive Bibliography / Avec une bibliographie détaillée. 1977 ISBN 90-247-1896-1
88. R. Descartes: *Règles utiles et claires pour la direction de l'esprit en la recherche de la vérité.* Traduction selon le lexique cartésien, et annotation conceptuelle par J.-L. Marion. Avec des notes mathématiques de P. Costabel. 1977 ISBN 90-247-1907-0
89. K. Hardesty: *The 'Supplément' to the 'Encyclopédie'.* [Diderot et d'Alembert]. 1977 ISBN 90-247-1965-8
90. H.B. White: *Antiquity Forgot.* Essays on Shakespeare, [Francis] Bacon, and Rembrandt. 1978 ISBN 90-247-1971-2
91. P.B.M. Blaas: *Continuity and Anachronism.* Parliamentary and Constitutional Development in Whig Historiography and in the Anti-Whig Reaction between 1890 and 1930. 1978 ISBN 90-247-2063-X

ARCHIVES INTERNATIONALES D'HISTOIRE DES IDÉES
*
INTERNATIONAL ARCHIVES OF THE HISTORY OF IDEAS

92. S.L. Kaplan (ed.): *La Bagarre.* Ferdinando Galiani's (1728–1787) 'Lost' Parody. With an Introduction by the Editor. 1979 ISBN 90-247-2125-3
93. E. McNiven Hine: *A Critical Study of [Étienne Bonnot de] Condillac's* [1714–1780]*'Traité des Systèmes'*. 1979 ISBN 90-247-2120-2
94. M.R.G. Spiller: *Concerning Natural Experimental Philosphy.* Meric Casaubon [1599–1671] and the Royal Society. 1980 ISBN 90-247-2414-7
95. F. Duchesneau: *La physiologie des Lumières.* Empirisme, modèles et théories. 1982 ISBN 90-247-2500-3
96. M. Heyd: *Between Orthodoxy and the Enlightenment.* Jean-Robert Chouet [1642–1731] and the Introduction of Cartesian Science in the Academy of Geneva. 1982 ISBN 90-247-2508-9
97. James O'Higgins: *Yves de Vallone* [1666/7–1705]: *The Making of an Esprit Fort.* 1982 ISBN 90-247-2520-8
98. M.L. Kuntz: *Guillaume Postel* [1510–1581]. Prophet of the Restitution of All Things. His Life and Thought. 1981 ISBN 90-247-2523-2
99. A. Rosenberg: *Nicolas Gueudeville and His Work (1652–172?).* 1982 ISBN 90-247-2533-X
100. S.L. Jaki: *Uneasy Genius: The Life and Work of Pierre Duhem* [1861-1916]. 1984 ISBN 90-247-2897-5; Pb (1987) 90-247-3532-7
101. Anne Conway [1631–1679]: *The Principles of the Most Ancient Modern Philosophy.* Edited and with an Introduction by P. Loptson. 1982 ISBN 90-247-2671-9
102. E.C. Patterson: *[Mrs.] Mary [Fairfax Greig] Sommerville* [1780–1872] *and the Cultivation of Science (1815–1840).* 1983 ISBN 90-247-2823-1
103. C.J. Berry: *Hume, Hegel and Human Nature.* 1982 ISBN 90-247-2682-4
104. C.J. Betts: *Early Deism in France.* From the so-called 'déistes' of Lyon (1564) to Voltaire's 'Lettres philosophiques' (1734). 1984 ISBN 90-247-2923-8
105. R. Gascoigne: *Religion, Rationality and Community.* Sacred and Secular in the Thought of Hegel and His Critics. 1985 ISBN 90-247-2992-0
106. S. Tweyman: *Scepticism and Belief in Hume's 'Dialogues Concerning Natural Religion'.* 1986 ISBN 90-247-3090-2
107. G. Cerny: *Theology, Politics and Letters at the Crossroads of European Civilization.* Jacques Basnage [1653–1723] and the Baylean Huguenot Refugees in the Dutch Republic. 1987 ISBN 90-247-3150-X
108. Spinoza's *Algebraic Calculation of the Rainbow & Calculation of Changes.* Edited and Translated from Dutch, with an Introduction, Explanatory Notes and an Appendix by M.J. Petry. 1985 ISBN 90-247-3149-6
109. R.G. McRae: *Philosophy and the Absolute.* The Modes of Hegel's Speculation. 1985 ISBN 90-247-3151-8
110. J.D. North and J.J. Roche (eds.): *The Light of Nature.* Essays in the History and Philosophy of Science presented to A.C. Crombie. 1985 ISBN 90-247-3165-8
111. C. Walton and P.J. Johnson (eds.): *[Thomas] Hobbes's 'Science of Natural Justice'.* 1987 ISBN 90-247-3226-3
112. B.W. Head: *Ideology and Social Science.* Destutt de Tracy and French Liberalism. 1985 ISBN 90-247-3228-X
113. A.Th. Peperzak: *Philosophy and Politics.* A Commentary on the Preface to Hegel's *Philosophy of Right.* 1987 ISBN Hb 90-247-3337-5; Pb ISBN 90-247-3338-3

ARCHIVES INTERNATIONALES D'HISTOIRE DES IDÉES
*
INTERNATIONAL ARCHIVES OF THE HISTORY OF IDEAS

114. S. Pines and Y. Yovel (eds.): *Maimonides* [1135-1204] *and Philosophy.* Papers Presented at the 6th Jerusalem Philosophical Encounter (May 1985). 1986 ISBN 90-247-3439-8
115. T.J. Saxby: *The Quest for the New Jerusalem, Jean de Labadie* [1610–1674] *and the Labadists (1610–1744).* 1987 ISBN 90-247-3485-1
116. C.E. Harline: *Pamphlets, Printing, and Political Culture in the Early Dutch Republic.* 1987 ISBN 90-247-3511-4
117. R.A. Watson and J.E. Force (eds.): *The Sceptical Mode in Modern Philosophy.* Essays in Honor of Richard H. Popkin. 1988 ISBN 90-247-3584-X
118. R.T. Bienvenu and M. Feingold (eds.): *In the Presence of the Past.* Essays in Honor of Frank Manuel. 1991 ISBN 0-7923-1008-X
119. J. van den Berg and E.G.E. van der Wall (eds.): *Jewish-Christian Relations in the 17th Century.* Studies and Documents. 1988 ISBN 90-247-3617-X
120. N. Waszek: *The Scottish Enlightenment and Hegel's Account of 'Civil Society'.* 1988 ISBN 90-247-3596-3
121. J. Walker (ed.): *Thought and Faith in the Philosophy of Hegel.* 1991 ISBN 0-7923-1234-1
122. Henry More [1614–1687]: *The Immortality of the Soul.* Edited with Introduction and Notes by A. Jacob. 1987 ISBN 90-247-3512-2
123. P.B. Scheurer and G. Debrock (eds.): *Newton's Scientific and Philosophical Legacy.* 1988 ISBN 90-247-3723-0
124. D.R. Kelley and R.H. Popkin (eds.): *The Shapes of Knowledge from the Renaissance to the Enlightenment.* 1991 ISBN 0-7923-1259-7
125. R.M. Golden (ed.): *The Huguenot Connection.* The Edict of Nantes, Its Revocation, and Early French Migration to South Carolina. 1988 ISBN 90-247-3645-5
126. S. Lindroth: *Les chemins du savoir en Suède.* De la fondation de l'Université d'Upsal à Jacob Berzelius. Études et Portraits. Traduit du suédois, présenté et annoté par J.-F. Battail. Avec une introduction sur Sten Lindroth par G. Eriksson. 1988 ISBN 90-247-3579-3
127. S. Hutton (ed.): *Henry More (1614–1687). Tercentenary Studies.* With a Biography and Bibliography by R. Crocker. 1989 ISBN 0-7923-0095-5
128. Y. Yovel (ed.): *Kant's Practical Philosophy Reconsidered.* Papers Presented at the 7th Jerusalem Philosophical Encounter (December 1986). 1989 ISBN 0-7923-0405-5
129. J.E. Force and R.H. Popkin: *Essays on the Context, Nature, and Influence of Isaac Newton's Theology.* 1990 ISBN 0-7923-0583-3
130. N. Capaldi and D.W. Livingston (eds.): *Liberty in Hume's 'History of England'.* 1990 ISBN 0-7923-0650-3
131. W. Brand: *Hume's Theory of Moral Judgment.* A Study in the Unity of *A Treatise of Human Nature.* 1992 ISBN 0-7923-1415-8
132. C.E. Harline (ed.): *The Rhyme and Reason of Politics in Early Modern Europe.* Collected Essays of Herbert H. Rowen. 1992 ISBN 0-7923-1527-8
133. N. Malebranche: *Treatise on Ethics* (1684). Translated and edited by C. Walton. 1993 ISBN 0-7923-1763-7
134. B.C. Southgate: *'Covetous of Truth'.* The Life and Work of Thomas White (1593–1676). 1993 ISBN 0-7923-1926-5
135. G. Santinello, C.W.T. Blackwell and Ph. Weller (eds.): *Models of the History of Philosophy.* Vol. 1: From its Origins in the Renaissance to the 'Historia Philosophica'. 1993 ISBN 0-7923-2200-2
136. M.J. Petry (ed.): *Hegel and Newtonianism.* 1993 ISBN 0-7923-2202-9

ARCHIVES INTERNATIONALES D'HISTOIRE DES IDÉES
*
INTERNATIONAL ARCHIVES OF THE HISTORY OF IDEAS

137. Otto von Guericke: *The New (so-called Magdeburg) Experiments* [Experimenta Nova, Amsterdam 1672]. Translated and edited by M.G. Foley Ames. 1994 ISBN 0-7923-2399-8
138. R.H. Popkin and G.M. Weiner (eds.): *Jewish Christians and Cristian Jews.* From the Renaissance to the Enlightenment. 1994 ISBN 0-7923-2452-8
139. J.E. Force and R.H. Popkin (eds.): *The Books of Nature and Scripture.* Recent Essays on Natural Philosophy, Theology, and Biblical Criticism in the Netherlands of Spinoza's Time and the British Isles of Newton's Time. 1994 ISBN 0-7923-2467-6
140. P. Rattansi and A. Clericuzio (eds.): *Alchemy and Chemistry in the 16th and 17th Centuries.* 1994 ISBN 0-7923-2573-7
141. S. Jayne: *Plato in Renaissance England.* 1995 ISBN 0-7923-3060-9
142. A.P. Coudert: *Leibniz and the Kabbalah.* 1995 ISBN 0-7923-3114-1
143. M.H. Hoffheimer: *Eduard Gans and the Hegelian Philosophy of Law.* 1995 ISBN 0-7923-3114-1
144. J.R.M. Neto: *The Christianization of Pyrrhonism.* Scepticism and Faith in Pascal, Kierkegaard, and Shestov. 1995 ISBN 0-7923-3381-0
145. R.H. Popkin (ed.): *Scepticism in the History of Philosophy.* A Pan-American Dialogue. 1996 ISBN 0-7923-3769-7
146. M. de Baar, M. Löwensteyn, M. Monteiro and A.A. Sneller (eds.): *Choosing the Better Part.* Anna Maria van Schurman (1607–1678). 1995 ISBN 0-7923-3799-9
147. M. Degenaar: *Molyneux's Problem.* Three Centuries of Discussion on the Perception of Forms. 1996 ISBN 0-7923-3934-7
148. S. Berti, F. Charles-Daubert and R.H. Popkin (eds.): *Heterodoxy, Spinozism, and Free Thought in Early-Eighteenth-Century Europe.* Studies on the *Traité des trois imposteurs.* 1996 ISBN 0-7923-4192-9
149. G.K. Browning (ed.): *Hegel's* Phenomenology of Spirit: *A Reappraisal.* 1997 ISBN 0-7923-4480-4
150. G.A.J. Rogers, J.M. Vienne and Y.C. Zarka (eds.): *The Cambridge Platonists in Philosophical Context.* Politics, Metaphysics and Religion. 1997 ISBN 0-7923-4530-4
151. R.L. Williams: *The Letters of Dominique Chaix, Botanist-Curé.* 1997 ISBN 0-7923-4615-7
152. R.H. Popkin, E. de Olaso and G. Tonelli (eds.): *Scepticism in the Enlightenment.* 1997 ISBN 0-7923-4643-2
153. L. de la Forge. Translated and edited by D.M. Clarke: *Treatise on the Human Mind (1664).* 1997 ISBN 0-7923-4778-1
154. S.P. Foster: *Melancholy Duty.* The Hume-Gibbon Attack on Christianity. 1997 ISBN 0-7923-4785-4
155. J. van der Zande and R.H. Popkin (eds.): *The Skeptical Tradition Around 1800.* Skepticism in Philosophy, Science, and Society. 1997 ISBN 0-7923-4846-X
156. P. Ferretti: *A Russian Advocate of Peace: Vasilii Malinovskii (1765–1814).* 1997 ISBN 0-7923-4846-6
157. M. Goldish: *Judaism in the Theology of Sir Isaac Newton.* 1998 ISBN 0-7923-4996-2
158. A.P. Coudert, R.H. Popkin and G.M. Weiner (eds.): *Leibniz, Mysticism and Religion.* 1998 ISBN 0-7923-5223-8
159. B. Fridén: *Rousseau's Economic Philosophy.* Beyond the Market of Innocents. 1998 ISBN 0-7923-5270-X
160. C.F. Fowler O.P.: *Descartes on the Human Soul.* Philosophy and the Demands of Christian Doctrine. 1999 ISBN 0-7923-5473-7

GPSR Compliance
The European Union's (EU) General Product Safety Regulation (GPSR) is a set of rules that requires consumer products to be safe and our obligations to ensure this.

If you have any concerns about our products, you can contact us on

ProductSafety@springernature.com

In case Publisher is established outside the EU, the EU authorized representative is:

Springer Nature Customer Service Center GmbH
Europaplatz 3
69115 Heidelberg, Germany

www.ingramcontent.com/pod-product-compliance
Ingram Content Group UK Ltd.
Pitfield, Milton Keynes, MK11 3LW, UK
UKHW020135160926
475607UK00003B/48
9780792357445